Sustainability Sciences in Asia and Africa

Sustainable Agriculture and Food Security

Series Editor

Rajeev K. Varshney, Semi-Arid Tropics, International Crops Research Institute, Patancheru, Telangana, India

This book series support the global efforts towards sustainability by providing timely coverage of the progress, opportunities, and challenges of sustainable food production and consumption in Asia and Africa. The series narrates the success stories and research endeavors from the regions of Africa and Asia on issues relating to SDG 2: Zero hunger. It fosters the research in transdisciplinary academic fields spanning across sustainable agriculture systems and practices, post- harvest and food supply chains. It will also focus on breeding programs for resilient crops, efficiency in crop cycle, various factors of food security, as well as improving nutrition and curbing hunger and malnutrition. The focus of the series is to provide a comprehensive publication platform and act as a knowledge engine in the growth of sustainability sciences with a special focus on developing nations. The series will publish mainly edited volumes but some authored volumes. These volumes will have chapters from eminent personalities in their area of research from different parts of the world.

Hupenyu Allan Mupambwa •
Lydia Ndinelao Horn •
Pearson Nyari Stephano Mnkeni
Editors

Vermicomposting for Sustainable Food Systems in Africa

Editors
Hupenyu Allan Mupambwa
Sam Nujoma Marine and Coastal
Resources Research Center
University of Namibia
Henties Bay, Namibia

Lydia Ndinelao Horn
Centre for Research Services
University of Namibia
Windhoek, Namibia

Pearson Nyari Stephano Mnkeni
Faculty of Science and Technology
University of Arusha
Arusha, Tanzania

ISSN 2730-6771 ISSN 2730-678X (electronic)
Sustainability Sciences in Asia and Africa
ISSN 2730-6798 ISSN 2730-6801 (electronic)
Sustainable Agriculture and Food Security
ISBN 978-981-19-8079-4 ISBN 978-981-19-8080-0 (eBook)
https://doi.org/10.1007/978-981-19-8080-0

© The Editor(s) (if applicable) and The Author(s), under exclusive license to Springer Nature Singapore
Pte Ltd. 2023
This work is subject to copyright. All rights are solely and exclusively licensed by the Publisher, whether
the whole or part of the material is concerned, specifically the rights of translation, reprinting, reuse of
illustrations, recitation, broadcasting, reproduction on microfilms or in any other physical way, and
transmission or information storage and retrieval, electronic adaptation, computer software, or by
similar or dissimilar methodology now known or hereafter developed.
The use of general descriptive names, registered names, trademarks, service marks, etc. in this publication
does not imply, even in the absence of a specific statement, that such names are exempt from the relevant
protective laws and regulations and therefore free for general use.
The publisher, the authors, and the editors are safe to assume that the advice and information in this
book are believed to be true and accurate at the date of publication. Neither the publisher nor the authors or
the editors give a warranty, expressed or implied, with respect to the material contained herein or for any
errors or omissions that may have been made. The publisher remains neutral with regard to jurisdictional
claims in published maps and institutional affiliations.

This Springer imprint is published by the registered company Springer Nature Singapore Pte Ltd.
The registered company address is: 152 Beach Road, #21-01/04 Gateway East, Singapore 189721,
Singapore

Contents

Part I   Science of Vermicomposting

1   A Decade of Vermicomposting Research at the University of Fort Hare: Selected Insights 3
Pearson Nyari Stephano Mnkeni and Hupenyu A. Mupambwa

2   State-of-the-Art and New Perspectives on Vermicomposting Research: 18 Years of Progress 27
Jorge Domínguez

3   Experiences on Methods of Vermicompost Analysis for Plant and Soil Nutrition 45
Hupenyu A. Mupambwa and Pearson Nyari Stephano Mnkeni

4   An Outstanding Perspective on Biological Dynamics in Vermicomposting Matrices 59
Jerikias Marumure, Zakio Makuvara, Claudious Gufe, Richwell Alufasi, Ngavaite Chigede, and Rangarirayi Karidzagundi

5   Insights into Earthworm Biology for Vermicomposting 89
Rutendo Nyamusamba, Reagan Mudziwapasi, Fortune Jomane, Unity Mugande, Abigarl Ndudzo, Sicelo Sebata, and Morleen Muteveri

Part II   Vermicompost Production

6   Vermicomposting as an Eco-Friendly Approach for Recycling and Valorization Grape Waste 111
María Gómez-Brandón, Manuel Aira, and Jorge Domínguez

7   Vermitechnology: An Underutilised Agro-tool in Africa 127
Ebenezer Olasunkanmi Dada and Yusuf Olamilekan Balogun

8 Prospects of Vermicompost and Biochar in Climate Smart Agriculture 145
P. Nyambo, L. Zhou, T. Chuma, A. Sokombela, M. E. Malobane, and M. Musokwa

9 Wild Birds Animal Manure Vermicomposting: Experiences from Namibia 161
Brendan M. Matomola, Simon T. Angombe, Rhoda Birech, and Hupenyu A. Mupambwa

10 Rock Phosphate Vermicompost with Microbial Inoculation Potential in Organic Soil Fertility 171
Sinazo Ajibade, Hupenyu Allan Mupambwa, Barbara Simon, and Miklos Gulyas

Part III Vermicomposts on Soil Quality and Crop Growth

11 A Farmers' Synthesis on the Effects of Vermicomposts on Soil Properties 189
Adornis D. Nciizah, Hupenyu A. Mupambwa, Patrick Nyambo, and Binganidzo Muchara

12 Vermicompost as a Possible Solution to Soil Fertility Problems and Enrichment in the Semiarid Zones of Namibia 203
S. N. Nghituwamhata, L. N. Horn, and S. N. Ashipala

13 Role of Vermicompost in Organic Vegetable Production Under Resource-Constrained Famers in Zimbabwe 213
Cosmas Parwada and Justin Chipomho

14 Co-application of Vermicompost with Other Amendments for the Improvement of Infertile/Degraded Soils 227
Akinson Tumbure, Charity Pisa, and Pardon Muchaonyerwa

15 Sustainable Enhancement of Soil Fertility Using Bioinoculants 249
Mukelabai Florence and Chimwamurombe Percy

16 The Potential of Vermicomposts in Sustainable Crop Production Systems 261
M. T. Mubvuma, I. Nyambiya, K. Makaza, P. Chidoko, C. P. Mudzengi, E. Dahwa, X. Poshiwa, R. Nyamusamba, A. M. Manyanga, M. Muteveri, and H. A. Mupambwa

17 Vermicompost and Vermi-leachate in Pest and Disease Management 279
K. Sivasabari, S. Parthasarathy, Deepak Chandran, S. Sankaralingam, and R. Ajaykumar

Contents vii

Part IV Vermicomposting and Wastes

18 Vermicompost: A Potential Reservoir of Antimicrobial Resistant Microbes (ARMs) and Genes (ARGs) 307
Zakio Makuvara, Jerikias Marumure, Rangarirayi Karidzagundi, Claudious Gufe, and Richwell Alufasi

19 Potential Transformation of Organic Waste in African Countries by Using Vermicomposting Technology 335
Parveen Fatemeh Rupani, Asha Embrandiri, Hupenyu Allan Mupambwa, and Jorge Domínguez

20 Earthworms in Bioremediation of Soils Contaminated with Petroleum Hydrocarbons 349
Abdullah Ansari, Jonathan Wrights, and Sirpaul Jaikishun

About the Editors

Hupenyu Allan Mupambwa holds a PhD in Soil Science, where his research focused on vermicomposting as a waste beneficiation technology important in driving organic soil fertility. Currently, he leads the Desert and Coastal Agriculture research program at the University of Namibia. He has more than 14 years' experience in agriculture research and training as well as university undergraduate and postgraduate teaching and research. He has published more than 25 peer-reviewed international publications, 11 book chapters, and edited 1 book. His research focuses on waste beneficiation, vermi-technology, and organic soil fertility management with a drive towards improving the soil quality among resource-poor farmers.

Lydia Ndinelao Horn is a Research Coordinator at the Multidisciplinary Research Services (MRS) Department, University of Namibia (UNAM). She holds a BSc in Agriculture (Crop Science) and an MSc in Biodiversity Management and Research from UNAM. She obtained a PhD in Plant Breeding from the University of Kwa-Zulu-Natal under the TC Africa sponsorship in promoting women in nuclear science and technology. As a plant breeder, her role in crop improvement using nuclear science and technology led to the development of Namibia's first-ever mutant the development of the first-ever germplasms of cowpea, Bambara groundnut pearl millet, and sorghum.

Pearson Nyari Stephano Mnkeni worked as a professor of soil science at the University of Fort Hare (UFH), South Africa, until his retirement in 2018. Before 1997, he served as an Associate Professor in Soil Science at the Sokoine University of Agriculture in Morogoro, Tanzania. He is presently Dean of Science and Technology and Acting Deputy Vice Chancellor for Planning, Administration and Finance at the University of Arusha, Tanzania. His research focus has been on ways of improving soil fertility using ecological approaches that minimize the carbon, energy, and water footprints of crop production. He initiated and led climate-smart agriculture research at UFH. He led the development of climate-smart agriculture guidelines for South Africa as a consultant for the United Nations Environmental Programme (UNEP) in 2018. He has published over 90 articles and is ranked as an established researcher by the National Research Foundation of South Africa.

Part I
Science of Vermicomposting

Chapter 1
A Decade of Vermicomposting Research at the University of Fort Hare: Selected Insights

Pearson Nyari Stephano Mnkeni and Hupenyu A. Mupambwa

Abstract The work summarized in this chapter highlights work done at the University of Fort Hare (UFH) in South Africa over approximately 10 years focussing on the optimization of the vermicomposting process to produce stabilized and nutrient rich organic fertilizer free of pathogens. The potential of the resultant optimized vermicomposts as soil amendments in agriculture is also highlighted. The highlighted findings showed that: (1) optimizing the C: N ratio of feedstock materials and earthworm stocking density is critically important for effective vermicomposting, (2) the optimum feedstock C: N ratio for vermicomposting varied with the type of waste mixture, the C: N ratio for cow dung-waste paper mixtures was found to be 30 whilst that for chicken manure-waste paper mixtures was 40, (3) the optimum earthworm stocking also varied with type of feedstock material but generally an earthworm stocking density range of 22.5–37.5 g-worm kg^{-1} feedstock mixture was found to be ideal for cow dung waste paper mixtures as well as fly ash-cow dung-waste paper mixtures, (4) a precomposting period of 1 week was ideal for the vermicomposting and sanitization of cow dung—waste paper mixtures, (5) a combination of thermophilic composting and vermicomposting using *Eisenia fetida* was effective in substantially reducing the concentrations of oxytetracycline and its metabolites 4-epi-oxytetracycline, a-apo-oxytetracycline and b-apooxytetracycline in chicken manures mixed with shredded waste paper optimized at a C: N ratio of 40, (6) incorporation of phosphate rock and coal fly ash in cattle manure-waste paper feedstock mixtures improved the biodegradation of the mixtures and nutrient value of the resulting vermicomposts, (7) the combination of *Eisenia fetida* plus *Pseudomonas fluorescens* accelerated the biodegradation process in vermicomposting, (8) optimized vermicomposts produced through these studies proved to be good components of vegetable and ornamental plants growing media,

---

P. N. S. Mnkeni (✉)
Faculty of Science and Technology, University of Arusha, Usa River, Arusha, Tanzania

H. A. Mupambwa
Sam Nujoma Marine and Coastal Resources Research Center, University of Namibia, Henties Bay, Namibia

© The Author(s), under exclusive license to Springer Nature Singapore Pte Ltd. 2023
H. A. Mupambwa et al. (eds.), *Vermicomposting for Sustainable Food Systems in Africa*, Sustainable Agriculture and Food Security,
https://doi.org/10.1007/978-981-19-8080-0_1

(9) Fly ash enriched vermicompost applied at 40 mg-P/kg FV proved to be an effective amendment for the growth of *Chloris gayana* grown on a gold mine waste affected soil thereby indicating its potential for the phyto-stabilization of soils degraded by mine wastes. It is hoped that these findings will prove to be a stepping stone for further research work on vermicomposting in South Africa and other parts of the continent.

Keywords Vermicomposting · C: N ratio · *Eisenia fetida* · Earthworm stocking density · Manure · Fly ash · Sanitization · Antibiotics · Growing media

1.1 Introduction

The rapidly growing world population has prompted an intensification of agricultural and associated industrial activities to meet the growing demand for food (Mupambwa and Mnkeni 2018). The intensification of crop production is being done mostly through extensive use of synthetic inorganic fertilizers which are known to have negative consequences on soil health and its function. These negative impacts of inorganic fertilizers on soil health have stimulated interest in the use of soil health compatible organic nutrient sources like vermicomposts (Arancon et al. 2004, 2008).

Vermicomposts are a product of the action of mainly epigeic earthworm species and microorganisms which drive the mineralization of organic nutrients in organic materials (Mupambwa and Mnkeni 2018). Unlike traditional thermophilic composting, vermicomposting only occurs under mesophilic conditions. Vermicomposts provide not only nutrients for plant growth but are also humic acids and plant growth hormones (Singh et al. 2008). The greatest limitation to the adoption of vermicompost in commercial agriculture is the wide variation in their quality which is dependent on the quality of organic materials used, method of preparation, earthworm species and stocking density, among other factors. Research on optimization and standardization of the production procedures of vermicomposts is thus critical so as to enable the production of vermicomposts with predictable biological parameters and nutrient composition.

Animal manures which are the main organic wastes used in vermicomposting contain pathogenic faecal bacteria such as *Escherichia coli* 0157 (O'Connor 2002). As a result, raw animal manures and manure-based soil amendments have been implicated in outbreaks of *E. coli* 0157:H7 infections from eating spinach, lettuce and other crop produce contaminated with these organisms (Ackers et al. 1998; Rangel et al. 2005; CDCP 2006). Parasitic protozoa can also be transmitted to food crops when manure containing these organisms is used as a fertilizer or soil amendment or through other inadvertent contact (Cieslak et al. 1993; Natvig et al. 2002). Management strategies that can eliminate *E. coli* and parasitic protozoa in vermicomposts are therefore necessary.

The intensive use of veterinary antibiotics (VAs) is a common practice for both prophylactic and therapeutic purposes in the livestock industry (Selvam et al. 2012;

Tzeng et al. 2016). However, as much as 30–90% of veterinary antibiotics may be excreted as parent compounds or bioactive metabolites through animal faeces or urine due to poor absorption by animals (Nelson et al. 2011). It is, therefore, important that the biodegradability of antibiotics and their metabolites during vermicomposting be established.

The vermicomposting process involves synergy between earthworms and microbes and requires a balanced supply of mainly carbon, nitrogen and phosphorus, which are the main elements required in protein formation, energy generation and reproduction in every living organism (Bernal et al. 2009; Ndegwa and Thompson 2000). Due to the link between C, N and P to the vermicomposting process, it is likely that processes like nutrient mineralization, bio-stabilization and earthworm development will be compromised if these are not balanced during the vermicomposting process (Chang and Cheng 2010).

Several researchers have used different organic materials for vermicomposting such as cow dung (Lazcano et al. 2008), pig manure (Aira et al. 2006), chicken manure (Ravindran and Mnkeni 2016), household wastes (Frederickson et al. 2007), sewage sludge (Maboeta and Van Rensburg 2003), waste paper (Gupta and Garg 2009) among others. Few studies have, however, tried to optimize these different organic materials before vermicomposting to allow for effective vermi-degradation. The stocking density and type of earthworm species during vermicomposting is also important as it affects the growth and reproduction of earthworms, thus directly influencing the bio-chemical processes taking place within the vermicompost (Mupambwa and Mnkeni 2016).

The low macro nutrient concentrations of vermicomposts relative to inorganic fertilizers is among the major limitations to the sustained use of vermicomposts in agriculture. Among the macro nutrients, phosphorus (P) is the least mobile and least available, with its soil bioavailability being a major bottleneck in crop production (Gichangi et al. 2008). In organic based soil fertility management, improving the availability of soil phosphorus has been reported to be critical (Edwards et al. 2010) hence it needs to be factored in strategies for the nutrient enrichment of vermicomposts. The inclusion of inorganic materials with very high total concentrations of P is one such strategy for improving the P nutrition of vermicomposts (Mupambwa et al. 2016; Busato et al. 2012; Unuofin and Mnkeni 2014). It is believed that the action of earthworms and microbes during vermicomposting can be used to enhance mineralization of the P bound in the inorganic materials, thus generating a phosphorus rich vermicompost.

This book chapter summarizes highlights of research work done at the University of Fort Hare (UFH) in South Africa over a 10 years' period addressing some of the gaps in knowledge pointed above. The main focus of the work was optimization of the vermicomposting process and sought to address the following questions:

1. What are the most optimum conditions for the vermicomposting that will result in stabilized and nutrient rich organic fertilizer free of pathogens over the shortest period?

2. What opportunities are available for improving the vermicomposting process in terms of nutrient composition, biodegradation and the biological safety of the resulting compost?
3. What is the potential of the vermicomposts produced as soil amendments in agriculture?

1.2   Optimization of the Vermi-degradation Process

The vermicomposting research at UFH was done using locally available materials in the Eastern Cape as well as other parts of South Africa. Among these materials are various types of animal manures such as cattle, goat, sheep, chicken and pig manures. The most common type of animal manure is cattle manure which Malherbe (1964) differentiated as:

(i) Stable manure which has large amounts of bedding and is not older than a year. It is found mainly in the winter-rainfall regions of the Western Cape, South Africa.
(ii) Kraal manure which contains little, if any, bedding and has accumulated over a period of a year or longer. It is found in the summer rainfall regions of South Africa for example, Free State, Transvaal and the Eastern Cape. This is the main organic material used in studies reported herein.
(iii) Karoo manure or dung which is normally found in kraals in the arid sheep (or goat) areas. It is usually fairly old and contains no bedding material at all.

The UFH like other academic institutions generates large volumes of waste paper which is often incinerated, contributing to greenhouse gases in the atmosphere. This is another material that was extensively used in studies reported herein with a view to encourage its recycling and valorization.

The inorganic materials used as amendments in some studies summarized in this chapter are rock phosphate and coal fly ash. The rock phosphate is of igneous origin mined by Foskor Ltd in Phalaborwa, Limpopo Province, South Africa (Mupondi et al. 2018). Coal fly ash was obtained from coal-powered electricity stations in Mpumalanga, South Africa (Mupambwa and Mnkeni 2015).

The critical parameters that drive the earthworm-microbe driven process whose end product is vermicompost are C: N and C: P ratios of the organic materials to be vermicomposted, substrate biodegradation fraction, earthworm species and stocking density. Some of these parameters had not been adequately optimized for the materials used for composting in South Africa so the work done at UFH and summarized in this chapter sought to address this need.

1.2.1 Optimizing the C: N Ratio

The vermicomposting process is driven by microorganisms and these need carbon as a source of energy and for cell development and nitrogen for building cell materials. However, if there is too much carbon relative to nitrogen during vermicomposting, decomposition is slowed down as there will be less N for protein synthesis by the microbes leading to the gradual death of the microorganisms (Tuomela et al. 2000). For proper nutrition of both earthworms and microorganisms, it is of paramount importance that carbon and nitrogen be present at an appropriate ratio (Ndegwa and Thompson 2000).

Two studies have addressed the optimization of the C: N ratio of vermicomposting mixtures at UFH. A study by Mupondi (2010) investigated the effect of C: N ratio and composting method on the biodegradation of mixtures of dairy cow manure and waste paper. The study compared vermicomposting and combined thermophilic composting and vermicomposting on the vermi-degradation of mixtures of dairy manure and waste paper mixed to have C: N ratios of 30 and 45 as feedstock. Results of selected maturity parameters obtained after 8 weeks of composting (Table 1.1) showed that greater vermi-degradation by both combined composting and vermicomposting was observed with feedstock dairy manure-paper waste mixtures with C: N ratio of 30 than those with a C: N ratio of 45. The C: N ratios of feedstock mixtures decreased to 29, 22 and 23 after 4 weeks and further decreased to 25, 14 and 15 in final (week 8) composts in the control, vermicompost and combined thermophilic and vermicomposting treatments, respectively (Table 1.1). By contrast, the C: N ratios of dairy manure-waste paper mixtures with a C: N ratio of 45 also decreased with time reaching values of 38, 36 and 33 in final (week 8) for the control, vermicompost and combined compost vermicompost treatments, respectively (Table 1.1). According to Bernal et al. (2009), composts with a final C: N ratios of 14 and 15 could be added to soil without altering the microbiological equilibrium of the soil so vermicomposts from dairy manure-waste paper mixtures mixed to provide an initial C: N ratio of 30 could safely be applied to the soil. The humification parameters (HI) and (HR) in the final composts were also highest in the dairy manure-waste paper mixtures whose initial C: N ratio was 30. Therefore, the feedstock mixtures of dairy manure and paper waste with C: N ratio of 30 proved to be more suitable for composting/ vermicomposting as these produced more mature and humified compost than wastes with a C: N ratio of 45. Both vermicomposting and combined composting and vermicomposting were effective methods for the biodegradation of mixtures of dairy manure and paper waste with C: N ratio of 30.

The second study sought to determine the optimum C: N ratio for converting waste paper and chicken manure to a nutrient rich manure with minimum toxicity (Ravindran and Mnkeni 2016). Six treatments of feedstock C: N ratio 20 (T1), 30 (T2), 40 (T3), 50 (T4), 60 (T5) and 70 (T6) achieved by mixing chicken manure with shredded paper were used. The study involved a thermophilic precomposting phase for 20 days followed by vermicomposting with *Eisenia fetida* for 7 weeks. The

Table 1.1 Effect of composting method and C: N ratio of dairy manure and paper waste mixtures on selected compost maturity parameters

Composting method	Composting stage (Weeks)	Selected maturity parameters				
		C: N	C_{HA}: C_{FA}	HI (%)	HR (%)	NH_4^+:NO_3^- ratio
Feedstock C: N ratio = 30						
C	0	31	0.2	1.4	13	2.5
	4	29	0.4	9.4	13	5.8
	8	25	1.0	10	19	0.2
V	0	30	0.1	1.5	14	2.5
	4	22	1.0	11.2	22	4.5
	8	14	2.8	39	53	0.1
CV	0	30	0.1	1.4	13	2.5
	4	23	1.4	9.7	24	4.1
	8	15	3.6	40	52	0.1
Feedstock C: N ratio = 45						
C	0	46	0.1	0.6	6	1.3
	4	43	0.4	5.3	7	6.5
	8	38	0.8	4.4	9	0.3
V	0	46	0.1	0.4	6	1.2
	4	41	0.5	5.4	8	3.5
	8	36	1.3	6.6	12	0.1
CV	0	45	0.1	0.5	6	1.3
	4	37	0.7	5.1	9	3.4
	8	33	1.6	8.1	13	0.1

C Control (Dairy manure-paper waste mixtures allowed to decompose on their own), *V* Vermicomposting, *CV* Combined composting and Vermicomposting, C_{HA} Extractable humic acid carbon, C_{FA} Extractable fulvic acid carbon, *HI* Humification index, *HR* Humification ratio
Adapted from Mupondi (2010)

results revealed that 20 days of precomposting considerably degraded the organic waste mixtures from all treatments and a further 7 weeks of vermicomposting significantly improved the bioconversion and nutrient value of all treatments (Fig. 1.1). The feedstock C: N ratio of 40 (T3) resulted in the best quality vermicompost compared to the other treatments. Earthworm biomass (Fig. 1.2) was highest in treatments T3 and T4 possibly due to a greater reduction of toxic substances in these feedstock waste mixtures.

The total N, total P and total K concentrations increased with time whilst total carbon, C: N ratio, EC and heavy metal content gradually decreased with time during the vermicomposting process (Ravindran and Mnkeni 2016). Scanning electron microscopy (SEM) not shown in this chapter but reported in Ravindran and Mnkeni (2016) revealed the intrastructural degradation of the chicken manure and shredded paper matrix confirming the extent of biodegradation of treatment mixtures as result of the composting and vermicomposting processes. Phytotoxicity evaluation of final vermicomposts using tomato (*Lycopersicon esculentum*), radish (*Raphanus sativus*),

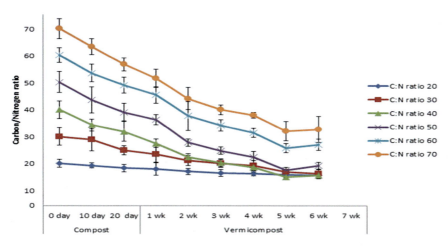

Fig. 1.1 Changes of C: N ratio during different stages of compost and vermicompost. Error bars indicate standard deviation. [Source: Ravindran and Mnkeni (2016)]

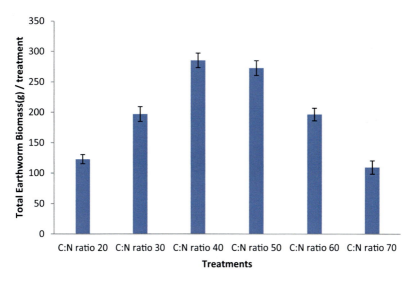

Fig. 1.2 Treatments effects on earthworm biomass (g) at the end of the vermicomposting phase. Error bars indicate standard deviation.

carrot (*Daucus carota*) and onion (*Allium cepa*) as test crops showed the non-phytotoxicity of the vermicomposts in the order: T3 > T4 > T2 > T1 > T5 > T6 (Ravindran and Mnkeni 2016). These results indicated that the combination of composting and vermicomposting processes is a good strategy for the management of chicken manure/paper waste mixtures and that the ideal feedstock C: N ratio of the waste mixture is 40 (T3). The fact that dairy manure and chicken manure had different optimal feedstock C: N ratios for vermicomposting and/or combined thermophilic composting and vermicomposting underscored the need for optimizing these parameters for different materials.

1.2.2 Optimizing the Sanitization of Vermicomposts

Animal manures and manure-based soil amendments have been implicated in outbreaks of *E. coli* 0157:H7 infections from eating spinach, lettuce and other crop produce contaminated with these organisms (Ackers et al. 1998; Rangel et al. 2005; CDCP 2006). This is attributed to the fact that animal manures contain pathogenic faecal bacteria such as *Escherichia coli* 0157 (O'Connor 2002). Because animal manures are the main organic wastes used in vermicomposting, it is important that measures are taken to ensure that vermicomposts prepared from these manures are free of these pathogens. Thermophilic composting is being promoted as a means of sanitizing wastes prior to vermicomposting (Gunadi et al. 2002). Two studies explored the effectiveness of introducing a precomposting step prior to vermicomposting in efforts to sanitizing the resultant vermicomposts. In one study Mupondi et al. (2010) compared vermicomposting and combined thermophilic composting and vermicomposting using dairy manure-waste paper mixtures with C: N ratios of 30 and 45. The pre-vermicomposting thermophilic phase lasted 28 days. The results obtained (Table 1.2) showed that vermicomposting preceded by a thermophilic composting phase (CV) eliminated the indicator pathogen *E. coli* 0157 from the final composts, whereas vermicomposting alone (V) only managed to reduce the pathogen population. By contrast, in the control treatment (C) where the dairy manure-paper waste mixtures were allowed to decompose on their own the pathogen population increased steadily. The increase in *E. coli* 0157 in the control with composting time could be attributed to creation of a good environment for multiplication of this pathogen through rehydration and subsequent availability of easily degradable substrates by dissolution following rehydration. These results indicate that dairy manure-waste paper mixtures allowed to simply compost in place could pose a health hazard to users and that alternative ways of handling the wastes that will result in a safer product are necessary.

A follow-up study with a wider range of precomposting periods was done so as to come up with a suitable precomposting period to use when composting and vermicomposting dairy manure-waste paper mixtures (Mupondi et al. 2011). Results of this study (Table 1.3) showed that over 95% of faecal coliforms, *E. coli* and of *E. coli* 0157 were eliminated from the wastes within 1 week of precomposting and total elimination of these and protozoan (oo)cysts achieved after 3 weeks of precomposting. However, vermicomposts from wastes precomposted for over

1 A Decade of Vermicomposting Research at the University of Fort Hare:...

Table 1.2 Effects of composting method on changes of *E. coli* 0157H: 7 numbers with composting time

Composting period (weeks)	Composting method		
	CV	V	C
0	2330 a	2340 a	2330 a
4	0 b	2120 a	2890 a
8	0 c	1120 b	4820 a

C Control (Dairy manure-paper waste mixtures allowed to decompose on their own), *V* Vermicomposting, *CV* Combined composting and Vermicomposting

Means in the same row followed by the same letter are not significantly different according to LSD at $P \leq 0.05$

Adapted from Mupondi (2010)

2 weeks were less stabilized, less humified and had less nutrient contents compared to vermicomposts from wastes that were precomposted for 1 week or less (Mupondi et al. 2011). These findings suggested that a precomposting period of 1 week is ideal for the effective vermicomposting of dairy manure-waste paper mixtures.

1.3 Optimizing Earthworm Stocking Density in Vermicomposting

The successful implementation of vermiculture requires that key operational parameters like earthworm stocking density be optimized for each target waste/waste mixture. One target waste mixture in South Africa is waste paper mixed with cow dung and rock phosphate (RP) for P enrichment. A study by Unuofin and Mnkeni (2014) sought to establish the optimal *Eisenia fetida* stocking density for maximum P release and rapid bioconversion of RP enriched cow dung- paper waste mixtures. *E. fetida* stocking densities of 0, 7.5, 12.5, 17.5 and 22.5 g-worms kg^{-1} dry weight of cow dung-waste paper mixtures were evaluated. The results obtained (Fig. 1.3) showed that the vermi-degradation of cow dung waste paper mixtures enriched with RP as measured by the C: N ratio and humification parameters, was highly dependent on *E. fetida* stocking density and that this effect varied with time. The stocking density of 12.5 g-worms kg^{-1} resulted in the highest earthworm growth rate and humification of the RP enriched waste mixture as reflected by a C: N ratio of <12 (Fig. 1.3a) and a humic acid/fulvic acid ratio of >1.9 in final vermicomposts (Fig. 1.3b). A germination test revealed that the resultant vermicompost had no inhibitory effect on the germination of tomato, carrot and radish (Unuofin and Mnkeni 2014). Extractable P increased with stocking density up to 22.5 g-worm kg^{-1} feedstock suggesting that for maximum P release from RP enriched wastes a high stocking density should be considered (Unuofin and Mnkeni 2014).

Another study sought to optimize *Eisenia fetida* stocking density for effective biodegradation and nutrient release in fly ash-cow dung-waste paper mixtures (Mupambwa and Mnkeni 2016). It evaluated four stocking densities of 0 g, 12.5 g, 25 g and 37.5 g-worm kg^{-1}. Though the treatments 12.5 g, 25 g and

Table 1.3 Effects of precomposting period on numbers of Faecal Coliforms (FC), *E. coli*, *E. coli* 157, microbial biomass (MBC) (shown for PW only), and presence (P) or absence (A) of *Cryptosporidium* cysts and *Giardia* oocysts in precomposted wastes (PW) and in final vermicomposts (V)

Precomposting period (weeks)	Faecal Coliforms (MPN gdw^{-1})		*E. coli* (MPN gdw^{-1})		*E. coli* 0157 (CFU g^{-1})		*Cryptosporidium* cysts (P or A 10 g dw^{-1})		*Giardia* oocysts (P or A 10g dw^{-1})		MBC (mg g^{-1})
	PW	V	PW	V	PW	V	PW	V	PW	V	PW
0	9125 a	4375 b	7608 a	3625 b	2516 a	820 b	P	P	P	P	10.9 a
1	525 c	417 c	400 c	292 c	76 c	32 c	P	P	P	P	8.6 b
2	167 c	108 c	67 c	33 c	31 c	12 c	P	P	P	P	7.7 c
3	0 c	0 c	0 c	0 c	0 c	0 c	A	A	A	A	6.3 d
4	0 c	0 c	0 c	0 c	0 c	0 c	A	A	A	A	6.1 c

Means in each column followed by different letters are significantly different according to LSD at $P < 0.05$
Adapted from Mupondi (2010)

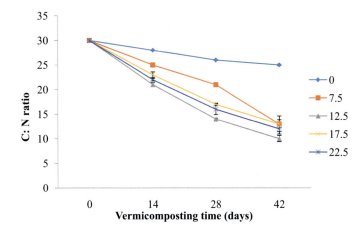

(a)

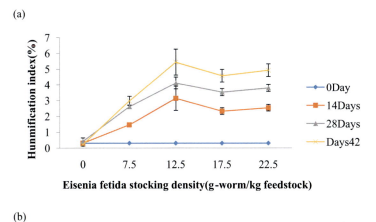

(b)

Fig. 1.3 *Eisenia fetida* stocking density effects on the C: N ratio (**a**) and humification index (**b**) of cow dung-waste paper mixtures during vermicomposting of cow dung waste paper mixtures. Error bars indicate standard deviations. [Source: Unuofin and Mnkeni (2014)]

37.5 g-worm kg^{-1} all resulted in a mature vermicompost, stocking densities of 25 and 37.5 g-worm kg^{-1} resulted in faster maturity, higher humification parameters and a significantly lower final C: N ratio range (11.1–10.4) (Fig. 1.4). The humification parameters (Fig. 1.5) showed that the stocking densities of 25 and 37.5 g-worm kg^{-1} resulted in highly humified vermicomposts but for practical purposes the lower stocking density of 25 g-worm kg^{-1} seemed most appropriate for the fly ash-cow dung-waste paper vermicompost.

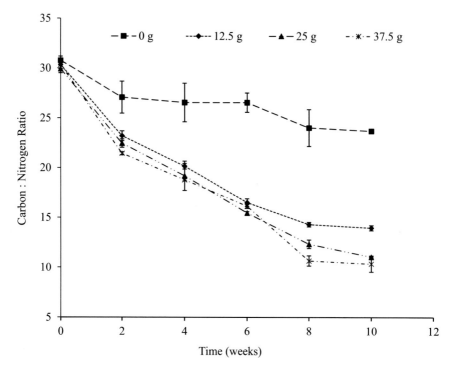

Fig. 1.4 Changes in Carbon: Nitrogen ratio during vermicomposting of fly ash-cow dung-waste paper mixture at different *E. fetida* stocking densities. Error bars indicate standard deviation. [Source Mupambwa and Mnkeni (2016)]

1.3.1 Optimizing Vermicomposting Through Incorporation of Inorganic Amendments

1.3.1.1 Rock Phosphate

The dairy manure-waste paper vermicomposts reported by Mupondi et al. (2010) were low in total P. In order to increase acceptability of dairy manure-waste paper vermicomposts as sources of nutrients or growing medium, it is necessary to increase their P content and availability. Biswas and Narayanasamy (2006) reported that enrichment of thermophilic composts with PR significantly increased their content of total P to 2.20% compared to 0.3% in straw composts where no PR was added. The PR-compost also had higher citric acid soluble P (0.72% P) compared to straw compost (0.1% P).

As there was little or no information on the effectiveness of PR incorporation to improve the P content and availability in vermicomposts, Mupondi et al. (2018) evaluated the effectiveness of a low grade South African phosphate rock (PR) in enriching dairy manure-waste paper vermicomposts with P and its possible effects

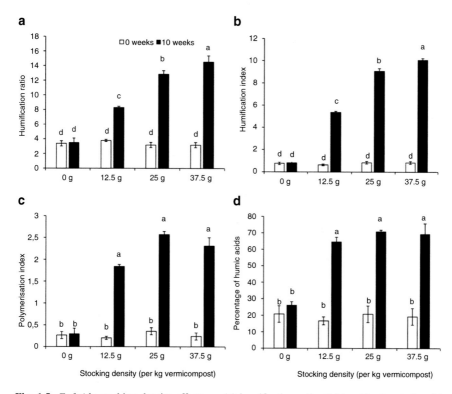

Fig. 1.5 *E. fetida* stocking density effects on (**a**) humification ratio, (**b**) humification index, (**c**) polymerization index, (**d**) percentage humic acids, following vermicomposting of fly ash-cow dung-waste paper mixture. Error bars indicate standard deviation. Different lower case letters indicate significant difference at $P \leq 0.05$. [Source: Mupambwa and Mnkeni (2016)]

on stabilization, nutrient content and physical properties of the vermicomposts. The dairy manure was optimized to a C: N ratio of 30 using waste paper and amended with RP to provide 0, 2, 4 and 8% of elemental phosphorus on a dry w/w basis. The mixtures were precomposted for 1 week before vermicomposting as recommended by Mupondi et al. (2011).

Incorporation of RP at 2 and 8% P enhanced compost biodegradation resulting in a significantly ($P < 0.001$) lower final C: N ratios of 12 and 10, respectively (Fig. 1.6), together with high humification parameters (Fig. 1.7). Amending the dairy manure—waste paper mixture with 2, 4 and 8% P as rock phosphate, resulted in increases of the bioavailable P fraction measured as bicarbonate extractable P of 19, 28 and 33%, respectively relative to the control (Mupondi et al. 2018). Additionally, vermicomposting with *E. fetida* significantly reduced heavy metals by up to 45% to levels below the maximum permissible concentration of potentially toxic elements in soils after 8 weeks. This study demonstrated the potential of optimized vermicomposting with igneous RP for generating nutrient rich organic fertilizers.

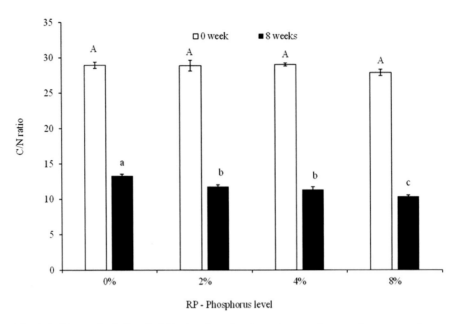

Fig. 1.6 Changes in C: N ratio following 8 weeks of vermicomposting cow dung-paper mixture amended with increasing levels of phosphorus from igneous rock phosphate. *Error bars represent standard deviation. *Different uppercase letters indicate significant differences ($P < 0.05$) between treatments at week 0; whilst different lowercase letters indicate significant differences ($P < 0.05$) between treatments at week 8. [Source: Mupondi (2010)]

1.3.1.2 Fly Ash

South Africa is highly dependent upon coal-fired power stations for electricity generation. Fly ash, a byproduct of coal combustion, contains a high total content of essential plant nutrients such as phosphorus but with low bioavailability. If the plant nutrient bioavailability in fly ash could be improved, and the toxic element content reduced, fly ash could contribute significantly as a fertilizer source in South African agriculture. One possible way of doing this is through co-vermicomposting fly ash with organic wastes such as manure and waste papers. This entailed establishing an appropriate mix of coal fly ash and the organic wastes that can be vermicomposted. To this end a study was conducted at UFH to establish an appropriate mixture ratio of fly ash (F) to optimized cow dung-waste paper mixtures (CP) in order to develop a high quality vermicompost using earthworms (*Eisenia fetida*) (Mupambwa and Mnkeni 2015). Fly ash was mixed with cow dung-waste paper mixtures at ratios of (F: CP) 1:1, 1:2, 1:3, 2:1, 3:1 and CP alone and composted for 14 weeks. Among the parameters measured were C: N ratio, humification parameters and scanning electron micrograph (SEM) images. Based on C: N ratio, the extent of vermi-degradation of the waste mixtures followed the decreasing order (F: CP) of 1:3 > 1:2 > 1:1 > CP alone > 2:1 > 3:1 (Fig. 1.7). Olsen P was significantly higher ($P < 0.05$) where earthworms were added. SEM images reported

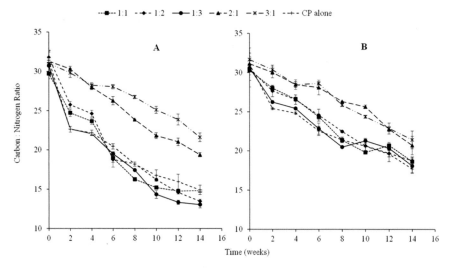

Fig. 1.7 Changes in C: N ratio over time during composting of fly ash incorporated into cow dung-waste paper mixtures (**a**) with earthworm presence; (**b**) without earthworms. Error bars represent standard deviation. [Source: Mupambwa and Mnkeni (2015)]

in Mupambwa and Mnkeni (2015) confirmed the extent of vermi-degradation reflected by the various humification parameters determined. Fly ash incorporation at the 1:2 ratio proved to be the most appropriate as it would allow processing of more fly ash whilst giving a vermicompost with desirable maturity and nutritional properties. This study demonstrated the nutritional potential of fly ash through vermicomposting.

1.3.2 Optimizing Vermicomposting Through Incorporation of Microbial Inoculants

1.3.2.1 Phosphate Solubilizing Bacteria

Due to the crucial role played by microbes during vermicomposting, deliberate inoculation of composts with specialized microbes as a way of further optimizing the vermicomposting process has been suggested. Lukashe et al. (2019) evaluated the potential of inoculating fly ash-cow dung-waste paper vermicompost with phosphate solubilizing bacteria (*P. fluorescens*) in improving vermi-degradation, nutrient mineralization and biological activity. Incorporation of *E. fetida* plus *P. fluorescens* accelerated the biodegradation process as indicated by the significant decrease in C: N ratio ($P = 0.0012$) resulting in a final C: N ratio of 11 compared to the control which had C: N ratio of 18 (Fig. 1.8). Inoculation with *P. fluorescens* resulted in improved availability of Olsen extractable P which amounted to 48.3%

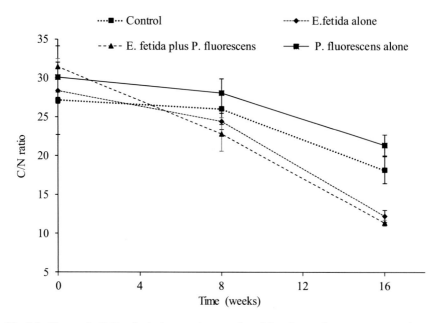

Fig. 1.8 Changes in C: N ratio during vermicomposting of fly ash-cow dung-waste paper mixture using *E. fetida* earthworms and *P. fluorescens* inoculation. [Source: Lukashe et al. (2019)]

more Olsen P relative to the control (Lukashe et al. 2019). The inoculation also caused a big decrease in alkaline phosphatase activity but yielded the highest FDA activity. The study concluded that the interaction of *E. fetida* earthworms with *P. fluorescens* can optimize vermi-degradation, nutrient release and biological activity during vermicomposting of fly ash-cow dung-waste paper substrate.

1.3.3 Degradation of Antibiotic Residues

The intensive use of veterinary antibiotics (VAs) is a common practice for both prophylactic and therapeutic purposes in the livestock industry (Selvam et al. 2012; Tzeng et al. 2016). However, according to Nelson et al. (2011) as much as 30–90% of veterinary antibiotics may be excreted as parent compounds or bioactive metabolites through animal faeces or urine due to poor absorption by animals. There are concerns that when such antibiotic laden manures are applied to agricultural lands they may have negative consequences for the environment as well as human health.

Ravindran and Mnkeni (2017) investigated an integrated system involving a combination of thermophilic composting and vermicomposting using *Eisenia fetida* as a strategy for reducing the concentrations of oxytetracycline (OTC) and its metabolites (4-epi-oxytetracycline [EOTC], a-apo-oxytetracycline [a-Apo-OTC] and b-apooxytetracycline [b-Apo-OTC]) (Fig. 1.9) in chicken manure. Treatments

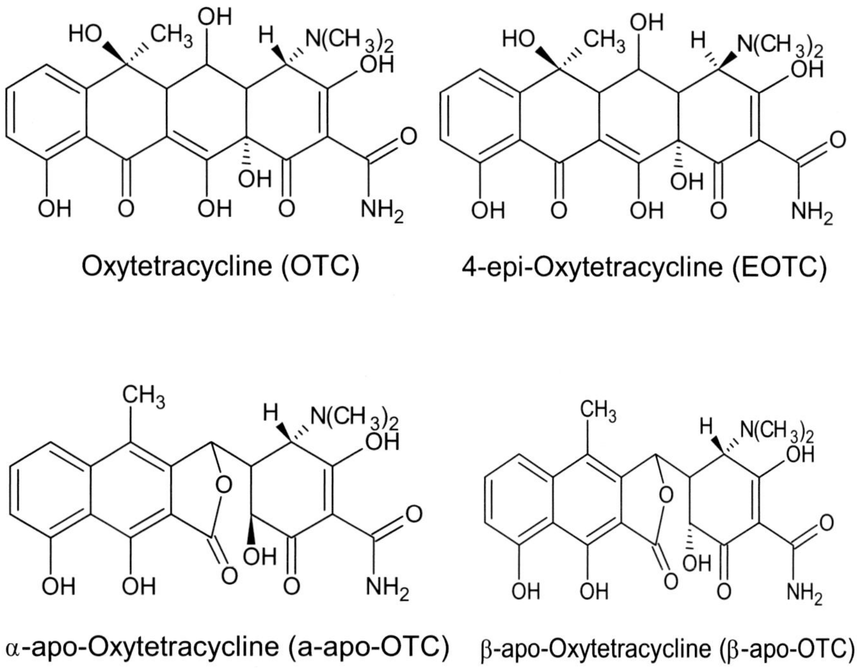

Oxytetracycline (OTC)   **4-epi-Oxytetracycline (EOTC)**

α-apo-Oxytetracycline (a-apo-OTC)   **β-apo-Oxytetracycline (β-apo-OTC)**

Fig. 1.9 Chemical structure of oxytetracycline and three degradation/transformation products of 4-epi-Oxytetracycline (EOTC), α-apo-Oxytetracycline (α-apo-OTC) and β-apo-Oxytetracycline (β-apo-OTC). [Source: Ravindran and Mnkeni (2017)]

consisted of combinations of chicken manure and waste paper mixed to produce feedstock waste mixtures with C: N ratios of 20 (T1), 30 (T2), 40 (T3), 50 (T4), 60 (T5) and 70 (T6). These were subjected to thermophilic composting for 20 days followed by vermicomposting with *E. fetida* for 7 weeks. The oxytetracycline concentration at the start of the experiment was in the range of 123.3 mg/kg to 35.2 mg/kg indicating that the chicken manure samples were contaminated with oxytetracycline and its metabolites. However, after the pre-vermicomposting thermophilic phase the initial levels were reduced to a range of 44 mg/kg to 25.3 mg/kg and thereafter to a range of 35.4 mg/kg to 20.7 mg/kg at the end of the vermicomposting stage (Fig. 1.10).

These results indicated that the combination of thermophilic composting and vermicomposting is ideal for the management of chicken manure mixed with shredded waste paper at a C: N ratio of 40. Generally, these findings suggest that combined thermophilic composting and vermicomposting not only enhances the bioconversion of organic wastes but also helps to reduce the veterinary antibiotic footprint of livestock manures.

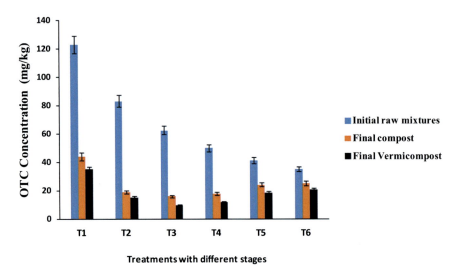

Fig. 1.10 Changes of oxytetracycline (mg/kg) through composting and vermicomposting processes (Error bars indicate standard deviations (n = 3). [Source: Ravindran and Mnkeni (2017)]

1.4 Vermicomposts and Plant Nutrition

Vermicomposts can be used as components of plant growing media or as soil amendments. According to Peyvast et al. (2007), vermicomposts have the ability to improve seed germination, enhance seedling growth, development and overall plant productivity. Increased growth and yield following vermicompost application are attributed to the improved soil or media properties such as increased nutrient availability for plant uptake, microbial activity and plant growth regulators which are produced by activity microorganisms such as fungi, bacteria and actinomycetes (Singh et al. 2008; Azarmi et al. 2008). Two studies looked at the possibility of incorporating vermicomposts produced at UFH into pine bark, a traditional horticultural medium in South Africa, with the intention of improving the media properties and subsequent seedling growth. A third study examined the possibility of using fly ash vermicompost in the reclamation of degraded mine land soils.

Mupambwa et al. (2017a) evaluated the influence of substituting pine bark compost (PB) with chicken manure vermicompost (CMV) at 0, 25, 50, 75 and 100%, on media properties and early tomato and cabbage seedling growth. The results showed that substitution of PB with CMV at between 25 and 50% resulted in the most appropriate media physical properties which also translated to seedling emergence above 90% for both vegetables. The same media substitution of between 25 and 50% resulted in a leaf area that was 200% more compared to 100% PB and final plant height that was 100% more than PB alone. The study demonstrated the potential of chicken manure vermicompost in enhancing pine bark growing media's chemical and physical properties. It showed that for effective seedling emergence

1 A Decade of Vermicomposting Research at the University of Fort Hare:... 21

and growth, chicken manure vermicompost can be effective when used as a substitute for pine bark at levels between 25 and 50% (Mupambwa et al. (2017a).

In the second study Mupambwa et al. (2017b) evaluated the possibility of mixing fly ash vermicompost (FA) with pine bark (PB) compost to produce a horticultural growing medium for ornamental plants using ornamental marigolds (*Tagetes* spp.) as the test crop. Fly ash vermicompost was mixed with pine bark compost at 0, 25, 50, 75 and 100% and marigolds seeds were sown with or without fertilizer in the resultant media. The results showed that FA substitution up to 50% significantly improved water holding capacity, total porosity and air filled porosity of the media. After 4 weeks of growth, seedlings in the 25 and 50% FA substituted media had higher plant height and leaf area. The 25% FA treatment resulted in significantly higher number of flowers and buds compared to the 50 and 75%. The study concluded that for effective marigold seedling germination and growth, a 50% FA: 50% PB growing medium is recommended whilst for maturity and flower production, the 25% FA: 75% PB combination is to be preferred.

In the third study, Lukashe et al. (2020) evaluated the potential of fly ash enriched vermicompost (FV) to improve the growth of *C. gayana* in a gold mine waste affected soil as a revegetation strategy. The results showed that the fly ash enriched vermicompost greatly improved the growth of *C. gayana* grown on the gold mine waste affected soil especially at the 40 mg-P/kg FV amendment rate which supplied optimal levels of available P. The improved *C. gayana* growth in vermicompost amended soils was attributed to improved nutrients supply and reduced solubility of potentially phytotoxic trace elements as a result of precipitation due to an increase in soil pH and metal complexation by the organic matter in the vermicompost. Therefore, the revegetation of mine waste contaminated soils can be achieved with addition of fly ash enriched vermicompost at the optimized rate of 40 mg-P/kg FV. Whilst these results require confirmation under field conditions, they have nevertheless demonstrated the potential of fly ash enriched vermicompost in the phyto-stabilization of gold mine waste affected soils.

1.5 Conclusions

Vermicomposting is an earthworm-microbe driven process whose optimal performance depends on some critical parameters. The work done at UFH for approximately 10 years has led to some conclusions regarding the optimization of the vermicomposting process such that it results in stabilized and nutrient rich organic fertilizer free of pathogens over the shortest period. These conclusions are:

1. The carbon and nitrogen in the feedstock mixture must be present at an appropriate ratio for proper nutrition of earthworms and microorganisms participating in the process. The optimal C: N ratio for cow dung-waste paper mixtures feedstock was found to be 30 whilst for chicken manure-waste paper mixtures it was 40. The fact that cattle and chicken manure had different optimal

feedstock C: N ratios for vermicomposting underscored the need for optimizing these parameters for different feedstock materials.

2. Because animal manures, the main organic wastes used in vermicomposting, are laden with pathogens it is important that measures are taken to ensure that vermicomposts prepared from these manures are free of these pathogens. Our work has demonstrated that in the case of cattle manure-waste paper feedstock mixtures a precomposting period of 1 week is an effective measure as it enabled the elimination of over 95% of faecal coliforms, *E. coli* and of *E. coli* O157 which were used as indicator organisms in our study. A precomposting period of at least 1 week is thus recommended as an effective sanitization measure in vermicomposting.

3. Earthworm stocking density for each target waste/waste mixture also needs to be optimized. Our work has shown that for cattle manure-waste paper mixtures enriched with phosphate rock or coal fly ash an earthworm stocking density of up to 22.5 g-worm kg^{-1} feedstock was ideal for producing well-humified vermicomposts with high extractable levels of phosphorus.

4. Incorporation of phosphate rock and coal fly ash in cattle manure-waste paper feedstock mixtures improved the biodegradation of the mixtures and nutrient value of the resulting vermicomposts. Rock phosphate can be incorporated at rates as low as 2% whilst fly ash incorporation at the 1: 2 ratio was found to be most appropriate.

5. Incorporation of *Eisenia fetida* plus *Pseudomonas fluorescens* accelerated the biodegradation process as indicated by significant decreases in C: N ratio resulting in a final C: N ratio of 11 compared to the control which had a much higher C: N ratio of 18.

6. The integrated system involving a combination of thermophilic composting and vermicomposting using *Eisenia fetida* was effective in substantially reducing the concentrations of oxytetracycline and its metabolites 4-epi-oxytetracycline, a-apo-oxytetracycline and b-apooxytetracycline in chicken manures mixed with shredded waste paper at a C: N ratio of 40. Therefore, combined thermophilic composting and vermicomposting not only enhances the bioconversion of organic wastes but also helps to reduce the veterinary antibiotic footprint of livestock manures.

7. Some of the optimized vermicomposts produced through these studies proved to be good components of vegetable and ornamental plants growing media.

    (a) Substitution of pine bark (PB) with chicken manure vermicompost at between 25 and 50% resulted in growing media with appropriate physical properties which supported effective seedling emergence and growth for cabbage and tomato.

    (b) Substitution of pine bark with fly ash vermicompost (FV) up to 50% resulted in growing medium with improved water holding capacity, total porosity and air filled porosity which was ideal for marigold seed germination and growth. However, a 25% FV: 75% PB combination was found to be ideal for marigold maturity and flower production.

8. Fly ash enriched vermicompost applied at 40 mg-P/kg FV proved to be an effective amendment for the growth of *Chloris gayana* grown on a gold mine waste affected soil. This was attributed to improved nutrients supply and reduced solubility of potentially phytotoxic trace elements. Therefore, optimized fly ash vermicompost could be used in the revegetation of mine waste contaminated soils. These findings, however, require confirmation under field conditions.

It is hoped that the findings summarized herein act as a stimulus for further vermicomposting work in Africa that will see vermiculture reach a stage where it is an accepted technology that is widely practised on the continent.

References

Ackers ML, Mahon BE, Leahy E, Goode B, Damrow T, Hayes PS, Bibb WF, Rice DH, Barrett TJ, Hutwagner L, Griffin PM, Slutsker L (1998) An outbreak of *Escherichia coli* 0157:H7 infections associated with leaf lettuce consumption. J Infect Dis 177:1188–1593

Aira M, Monroy F, Dominguez J (2006) C to N ratio strongly affects population structure of Eisenia fetida in vermicomposting systems. Eur J Soil Biol 42:S127–S131

Arancon NQ, Edwards CA, Atiyeh R, Metzger JD (2004) Effects of vermicomposts produced from food waste on the growth and yields of greenhouse peppers. Bioresour Technol 93:139–144

Arancon NQ, Edwards CA, Babenko A, Cannon J, Galvis P, Metzger JD (2008) Influences of vermicomposts, produced by earthworms and microorganisms from cattle manure, food waste and paper waste, on the germination, growth and flowering of petunias in the greenhouse. Appl Soil Ecol 39:91–99

Azarmi R, Sharifi ZP, Reza Satari M (2008) Effect of vermicompost on growth, yield and nutrition status of tomato (Lycopersicum esculentum). Pak J Biol Sci 11(14):1797–1802. https://doi.org/10.3923/pjbs.2008.1797.1802

Bernal MP, Alburquerque JA, Moral R (2009) Composting of animal manures and chemical criteria for compost maturity assessment. A review. Bioresour Technol 100:5444–5453

Biswas DR, Narayanasamy G (2006) Rock phosphate enriched compost, an approach to improve low-grade Indian rock phosphate. Bioresour Technol 97:2243–2251

Busato JG, Lima LS, Aguiar NO, Canellas LP, Olivares FL (2012) Changes in labile phosphorus forms during maturation of vermicompost enriched with phosphorus-solubilizing and diazotrophic bacteria. Bioresour Technol 110:390–395

Centers for Disease Control and Prevention (2006) Ongoing multistate outbreak of *Escherichia coli* serotype 0157:H7 infections associated with consumption of fresh spinach—United States, September 2006. Morb Mortal Wkly Rep 55:1045–1046

Chang JI, Chen YJ (2010) Effects of bulking agents on food waste composting. Bioresour Technol 101:5917–5924

Cieslak PR, Barret TJ, Griffin PM (1993) *Escherichia coli* 0157:H7 infection from a manured garden. Lancet 342:367

Edwards CA, Walker RL, Maskell P, Watson CA, Rees RM, Stockdale EA, Knox OGG (2010) Improving bioavailability of phosphate rock for organic farming. In: Lichtfouse E (ed) Genetic engineering, biofertilisation, soil quality and organic farming, Sustainable agriculture reviews, vol 4. Springer, Dordrecht. https://doi.org/10.1007/978-90-481-8741-6_4

Frederickson J, Howell G, Hobson AM (2007) Effect of pre-composting and vermicomposting on compost characteristics. Eur J Soil Biol 43:S320–S326

Gichangi EM, Mnkeni PNS, Muchaonyerwa P (2008) Phosphate sorption characteristics and external P requirements of selected South Africa soils. J Agric Rural Dev Trop Subtrop 109 (2):139–149

Gunadi B, Blount C, Edwards CA (2002) The growth and fecundity of *Eisenia fetida* (Savigny) in cattle solids precomposted for different periods. Pedobiologia 46:15–23

Gupta R, Garg VK (2009) Vermiremediation and nutrient recovery of non-recyclable paper waste employing *Eisenia fetida*. J Hazard Mater 162:430–439

Lazcano C, Gomez-Brandon M, Dominguez J (2008) Comparison of the effectiveness of composting and vermicomposting for the biological stabilization of cattle manure. Chemosphere 72:1013–1019

Lukashe SN, Mupambwa H, Green E, Mnkeni PNS (2019) Inoculation of fly ash amended vermicompost with phosphate solubilizing bacteria (*Pseudomonas fluorescens*) and its influence on vermi-degradation, nutrient release and biological activity. Waste Manag 83:14–22

Lukashe SN, Mnkeni PNS, Mupambwa HA (2020) Growth and elemental uptake of Rhodes grass (*Chloris gayana*) grown in a mine waste-contaminated soil amended with fly ash-enriched vermicompost. Environ Sci Pollut Res. https://doi.org/10.1007/s11356-020-08354-7

Maboeta MS, Van Rensburg L (2003) Bioconversion of sewage sludge and industrially produced woodchips. Water Air Soil Pollut 150:219–233

Malherbe IDV (1964) Soil fertility, 5th edn. Oxford University Press, London

Mupambwa HA, Mnkeni PNS (2015) Optimization of fly ash incorporation into cow dung – waste paper mixtures for enhanced vermi-degradation and nutrient release. J Environ Qual 44:972–981. https://doi.org/10.2134/jeq2014.10.0446

Mupambwa HA, Mnkeni PNS (2016) *Eisenia fetida* stocking density optimization for enhanced biodegradation and nutrient release in fly ash-cow dung waste paper vermicompost. J Environ Qual 45:1087–1095. https://doi.org/10.2134/jeq2015.07.0357

Mupambwa HA, Mnkeni PNS (2018) Optimizing the vermicomposting of organic waste materials for production of nutrient rich organic fertilizers: a review. Environ Sci Pollut Res 25:10577–10595. https://doi.org/10.1007/s11356-018-1328-

Mupambwa HA, Ravindran B, Mnkeni PNS (2016) Potential of Effective micro-organisms and *Eisenia fetida* in enhancing vermi-degradation and nutrient release of fly ash incorporated into cow dung-paper waste mixture. Waste Manag 48:165–173

Mupambwa HA, Ncoyi K, Mnkeni PNS (2017a) Potential of chicken manure vermicompost as a substitute for pine bark based growing media for vegetables. Int J Agric Biol 19(5):1007–1011. https://doi.org/10.17957/IJAB/15.0375

Mupambwa HA, Lukashe SN, Mnkeni PNS (2017b) Suitability of fly ash vermicompost as a component of pine bark growing media: effects on media physicochemical properties and ornamental marigold (*Tagetes* spp.) growth and flowering. Compost Sci Util 25(1):48–61. https://doi.org/10.1080/1065657X.2016.1180270

Mupondi LT (2010) Improving sanitization and fertilizer value of dairy manure and waste paper mixtures enriched with rock phosphate through combined thermophilic composting and vermicomposting. A Ph.D. thesis, University of Fort Hare, Alice

Mupondi LT, Mnkeni PNS, Muchaonyerwa P (2010) Effectiveness of combined thermophilic composting and vermicomposting on biodegradation and sanitization of mixtures of dairy manure and waste paper. Afr J Biotechnol 9(30):4754–4763

Mupondi LT, Mnkeni PNS, Muchaonyerwa P (2011) Effects of a precomposting step on the vermicomposting of dairy manure waste paper mixtures. Waste Manag Res J 29(2):219–228. https://doi.org/10.1177/0734242X10363142

Mupondi LT, Mnkeni PNS, Muchaonyerwa P, Hupenyu Allan Mupambwa HA (2018) Vermicomposting manure-paper mixture with igneous rock phosphate enhances biodegradation, phosphorus bioavailability and reduces heavy metal concentrations. Heliyon 4(2018): e00749. https://doi.org/10.1016/j.heliyon.2018.e00749

Natvig EE, Ingham SC, Ingham BH, Cooperband LR, Roper TR (2002) *Salmonella* enteric serovar *Typhimurium* and *Escherichia coli* contamination of root and leaf vegetables grown in soils with incorporated bovine manure. Appl Environ Microbiol 68:2737–2744

Ndegwa PM, Thompson SA (2000) Effects of C-to-N ratio on vermicomposting of biosolids. Bioresour Technol 75:7–12

Nelson KL, Brözel VS, Gibson SA, Thaler R, Clay SA (2011) Influence of manure from pigs fed chlortetracycline as growth promotant on soil microbial community structure. World J Microbiol Biotechnol 27:659–668

O'Connor DR (2002) Part 1. Report of the Walkerton Inquiry: the events of May 2000 and related issues. Ontario Ministry of the Attorney General, Queen's Printer of Ontario, Toronto

Peyvast G, Sedghi Moghaddam M, Olfati JA (2007) Effect of municipal solid waste compost on weed control, yield and some quality indices of green pepper (Capsicum annuum L.). Biosci Biotech Res Asia 4(2):449–456

Rangel JM, Sparling PH, Crowe C, Griffin PM, Swerdlow DH (2005) Epidemiology of *Escherichia coli* 0157:H7 outbreaks, United States, 1982–2002. Emerg Infect Dis 11:603–660

Ravindran B, Mnkeni PNS (2016) Bio-Optimization of the carbon to nitrogen ratio for the vermicomposting of chicken manure and waste paper using *Eisenia fetida*. Environ Sci Pollut Res 23(17):16965–16976. https://doi.org/10.1007/s11356-016-6873-0

Ravindran B, Mnkeni PNS (2017) Identification and fate of antibiotic residues degradation during composting and vermicomposting of chicken manure. Int J Environ Sci Technol 14(2):263–270. http://link.springer.com/article/10.1007/s13762-016-1131-z. https://doi.org/10.1007/s13762-016-1131-z

Selvam A, Xu D, Zhao Z, Wong JWC (2012) Fate of tetracycline, sulfonamide and fluoroquinolone resistance genes and the changes in bacterial diversity during composting of swine manure. Bioresour Technol 126:383–390

Singh R, Sharma RR, Kumar S, Gupta RK, Patil RT (2008) Vermicompost substitution influences growth, physiological disorders, fruit yield and quality of strawberry (*Fragaria ananassa* Duch.). Bioresour Technol 99:8507–8511

Tuomela M, Vikman M, Hatakka A, Itavaara M (2000) Biodegradation of lignin in a compost environment: a review. Bioresour Technol 72:169–183

Tzeng TW, Liu YT, Deng Y, Hsieh YC, Tan CC, Wang SL, Huang ST, Tzou YM (2016) Removal of sulfamethazine antibiotics using cow manure-based carbon adsorbents. Int J Environ Sci Technol 13:973–984

Unuofin FO, Mnkeni PNS (2014) Optimization of *Eisenia fetida* stocking density for the bioconversion of phosphate rock enriched cow dung-waste paper mixtures. Waste Manag 34(11):2000–2006. https://doi.org/10.1016/j.wasman.2014.05.018

Chapter 2
State-of-the-Art and New Perspectives on Vermicomposting Research: 18 Years of Progress

Jorge Domínguez

Abstract Vermiculture and vermicomposting are well-established technologies and nowadays constitute a thriving industry that is becoming increasingly important throughout the world. Members of the Soil Ecology Laboratory at the University of Vigo have been studying a wide range of scientific aspects of this discipline and have developed a comprehensive vermicomposting research programme over the past 30 years. This research has included many different aspects of earthworm biology and ecology, the vermicomposting process, and the use of vermicompost for improving plant growth and health. This chapter summarizes the research on vermicomposting conducted in my laboratory, and it represents an up-date of the original text entitled "State-of-the-Art and New Perspectives on Vermicomposting Research", written in 2004 and included in the book Earthworm Ecology. Here, I synopsize the main advances and current state of the art after 18 years of continuous progress in scientific, technical, and industrial-commercial vermicomposting endeavours, illustrating the coming of age of this discipline.

Keywords Enzyme activity · Bioremediation · Earthworm biology · Gut-associated processes · Industrial waste

2.1 Introduction

At the end of the twentieth century, vermicomposting—the transformation of organic waste into a humus-like material called vermicompost mediated by the synergistic actions of earthworms and microorganisms—began to be considered an attractive alternative to thermophilic composting for recycling and conversion of organic matter into organic fertilizers and soil-improving amendments that can be used in horticulture and agriculture. Vermicomposting is now a well-established

---

J. Domínguez (✉)
Grupo de Ecoloxía Animal (GEA), Universidade de Vigo, Vigo, Spain
e-mail: jdguez@uvigo.es

© The Author(s), under exclusive license to Springer Nature Singapore Pte Ltd. 2023
H. A. Mupambwa et al. (eds.), *Vermicomposting for Sustainable Food Systems in Africa*, Sustainable Agriculture and Food Security,
https://doi.org/10.1007/978-981-19-8080-0_2

technology and constitutes a thriving industry that is becoming increasingly important throughout the world.

The Soil Ecology Laboratory (Group of Animal Ecology, GEA) at the University of Vigo (Spain) has conducted comprehensive research on vermicomposting over the past 30 years, studying a wide range of scientific aspects of this discipline, including earthworm biology and ecology, the characterization, functioning and ecology of the vermicomposting process, and the effects of vermicompost on soil and plants. This chapter aims to summarize the research on vermicomposting conducted in my laboratory, and it represents an up-date of the original chapter entitled "State-of-the-Art and New Perspectives on Vermicomposting Research", written in 2004 and included in the book Earthworm Ecology. In this new chapter, I aim to synopsize the main advances and current state of the art after 18 years of continuous progress in scientific, technical, and industrial-commercial aspects of vermicomposting, illustrating the coming of age of this discipline.

Results of an internet survey conducted using three bibliographic search engines (Scopus®, Web of Knowledge®, Google Scholar) and one general search engine (Google Search) revealed that major developments in vermicomposting occurred during the first quarter of the twenty-first century (Fig. 2.1). The main figure shows the number of scientific reports published every year on vermicomposting, with the first document found in the bibliographic search of papers published in 1980. The table shows the output of Google Scholar searches (also academic results but with more wide-reaching information, with 1490 results before 2004, and 24,300 results in total) and the total output of Google Search (1580 results before 2004, and 1,060,000 results in total).

In short, vermicomposting has expanded enormously at various different levels: home application for the management of family or individual household waste, application on farms and in different industrial facilities associated with the agri-food sector, and industrial implementation with the development of large, important companies worldwide.

2.2 Vermicomposting Technologies

Vermicomposting technologies vary depending on the scale of application, ranging from very simple, low technology methods, such as containers, waste heaps and windrows, to medium-scale, moderately complex systems and to automated continuous flow vermireactors. Today, vermicomposting is basically applied at three scales: household or educational; farms and primary sector companies; and industrial level.

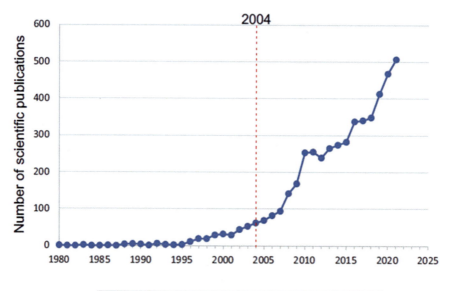

Fig. 2.1 Development of vermicomposting during the last quarter of the twentieth century and the first quarter of the twenty-first century. The search was conducted (March 2022) using three bibliographic engines (Scopus®, Web of Knowledge®, Google Scholar) and one general search engine (Google Search). The dotted line indicates the year (2004) in which the original chapter "State-of-the-Art and New Perspectives on Vermicomposting Research" was published in the book Earthworm Ecology

2.3 Vermicomposting: A Brief Definition

Vermicomposting is the transformation of solid organic waste into vermicompost, which is an excellent biofertilizer. The transformation involves an enhanced bio-oxidation process in which the interactions between epigeic earthworms and microorganisms accelerate the decomposition and stabilization of organic matter and substantially modify the physical, chemical and biological properties of the organic waste. Although earthworms are crucial in the process and drastically alter the biological activity, microorganisms produce the enzymes responsible for the biochemical decomposition of organic matter. The continuous grazing of earthworms on fungi and bacteria reduces microbial biomass and greatly modifies the structure of microbial communities. Decomposition is accelerated during vermicomposting by the huge increase in microbial activity induced by earthworm movements.

2.4 Vermicomposting Earthworms

Earthworms are hermaphroditic, iteroparous soil-living animals with indeterminate growth. Different earthworm species have different life histories and ecological strategies, occupying different soil niches. Among the more than 7000 species described to date, very few (~6) are suitable for vermicomposting. In vermicomposting systems, pure organic matter acts as both the habitat and food, and as soil is not involved, only epigeic earthworms can be used in the process. The earthworm species used in vermicomposting are generally those mentioned in the original chapter (from 2004), and there have been few changes in this respect since then. The earthworms used in vermicomposting are the temperate species *Eisenia andrei* (Bouché 1972) and *Eisenia fetida* (Savigny 1826) and the tropical species *Eudrilus eugeniae* (Kinberg 1867) and *Perionyx excavatus* (Perrier 1872). The biology, life cycle and growth and reproduction rates of these earthworms have been widely reported. Among these species, *E. andrei* and *E. fetida* are the most frequently used, with *E. andrei* being the most widespread and common in vermicomposting companies and farms worldwide. Although it has not yet been generally accepted that *E. fetida* and *E. andrei* are different species, this has been demonstrated and is now recognized by most specialists (Fig. 2.2).

In 2004, it was emphasized that *Eisenia* spp. are temperate organisms, and that the use of tropical species, in particular *P. excavatus* and the African nightcrawler *Eudrilus eugeniae*, would potentially be useful in tropical regions. Although experimental data indicate their aptitude and good performance in vermicomposting, neither of these species has yet replaced *Eisenia* as the queen of vermicomposting worldwide, in all climates, ranging from continental, cold, mid-latitude and temperate climates to equatorial and tropical zones. Some interesting initiatives have been made to find other alternatives, with local species that would at least closely match the quality and aptitude of *Eisenia*. *Dichogaster annae* has been found in some vermicomposting facilities in Brazil, but little is known regarding its life cycle and distribution. *Eudrilus eugeniae* is an African species that, because of its size and life cycle, is more suitable for vermiculture than for vermicomposting. Other epigeic earthworm belonging to the family Eudrilidae, *Hyperiodrilus africanus*, has been also found in compost heaps and litter layers of high organic soils in western and central Africa. However, to date no earthworm species that is more suitable for vermicomposting than *Eisenia* has been found.

2.5 Materials Used in Vermicomposting

In 2004, vermicomposting was basically considered an alternative method of waste treatment. In fact, the first scientific studies focused on the treatment of sewage sludge and animal manures. Many types of waste required pretreatment or conditioning, sometimes by composting, before being acceptable to earthworms. In the

Fig. 2.2 *Eisenia andrei* (above) and *Eisenia fetida* (below) are different earthworm species

intervening years, we have learned that vermicomposting is effective and can be applied to any degrading organic material in a solid or semi-solid state (up to 90% moisture), and that acclimatization or pretreatment of the materials is not necessary.

Of course, thousands of different situations can exist depending on the type of farm and scope, clearly differentiating at least the three levels mentioned above (home vermicomposting for processing domestic waste, medium-scale vermicomposting for management of waste from agro-livestock or agri-food farms, and large-scale, industrial vermicomposting, mainly conducted by waste treatment companies). In medium and large scale operations, the materials processed are usually homogeneous, and mixtures of different materials are generally only processed in domestic vermicomposting. This is important, as we have learned that earthworms become adapted to certain diets, or in other words process homogeneous physical, chemical and microbiological materials, and they do not grow or reproduce as well when these characteristics vary widely and unpredictably.

2.6 Environmental Conditions of Vermicomposting

For maximum efficiency in vermicomposting, very dense earthworm populations must be maintained. This can only be achieved by maintaining ideal conditions for earthworm growth and reproduction. In general, the moisture content of the substrate is a critical factor. The metabolic rates of both earthworms and microorganisms are determined by water availability, so that if the moisture content is too low, both earthworm and microbial activity will be greatly reduced. By contrast, excessively high moisture levels (above 90%) can lead to anaerobic decomposition, and in the case of materials that are dense or have small particle size, the vermicompost will be sludgy and remain wet, forming rocky material that will be difficult to handle, preserve, transport and apply. The optimal moisture content for vermicomposting ranges between 75 and 85%.

Temperature is also important and clearly influences the vermicomposting process. *Eisenia* spp., which as mentioned are the most widely used species in vermicomposting, are temperate species from mid latitudes and develop well at between 15 and 25 °C. At higher temperatures, growth of these species is greatly affected by loss of water, which slows down the process or even leads to death of the earthworms. Temperature must therefore be carefully monitored and controlled. The worms tolerate temperatures below 10 °C, although their metabolism, growth, and reproduction decrease, and vermicomposting thus slows down considerably.

Other environmental factors or physical and chemical conditions will also influence the life cycle of the worms, i.e. their growth and reproduction, as in the case of animals that breathe through their skin and are therefore very sensitive to gas exposure. In 2004, information was already available about the optimum ranges and tolerance to these important variables, including pH, ammonia nitrogen and ammonium gas, and the salt content of the substrate. More data have since been published in relation to the ranges of tolerance of earthworms to other environmental parameters, mainly physical-chemical but also biological. However, I do not think it relevant to report this information here and I believe that the integral management of the vermicomposting system and the care and maintenance of the matrix structure are more important in the present context. As already mentioned, vermicomposting earthworms live in the same substrate that serves as food. The ideal conditions for vermicomposting systems include a healthy matrix and appropriate physical and chemical conditions. The fresh material, i.e. the waste or by-products, to be processed is added to this matrix (or beside it in lateral systems), so that earthworms will process it on demand. When the conditions are suitable and the food is supplied ad libitum or in excess, the population begins to increase and will do so almost exponentially until the carrying capacity or maximum level (more than 20,000 worms per square metre in our pilot studies) is reached. The environmental conditions of the system should then be maintained and the food material should be added continuously for processing by the worms. Maximal rates of waste processing, which are very high, can thus be reached.

The key steps in effective vermicomposting are as follows: (1) addition of the parent material to the surface in thin layers, at frequent intervals, allowing earthworms to move up and concentrate in the upper 15–20 cm of the vermicomposting matrix and to continue to move upwards with the addition of new layers. Vermicomposting operations can be mechanized with a suitable balance between mechanization costs and labour savings; (2) maintenance of aerobic conditions and optimal moisture and temperature ranges in the substrate, avoiding addition of materials with excessive amounts of salts. The addition of fresh materials in thin layers prevents overheating through thermophilic composting, although the heat generated is usually sufficient to maintain suitable temperatures for earthworm growth during cold winter periods. Thus, to maintain a reasonable temperature, vermicomposting should be done under cover; additional heating is not necessary if the waste addition is managed well with application of thicker layers during cold periods to provide some heat derived from thermophilic composting. The temperature can be reduced in summer by using fans or water misting cooling systems.

2.7 How Vermicomposting Works: Stages of the Process

In the chapter written in 2004, a move from the classical approach of considering vermicomposting as the mere addition of earthworm to organic waste, to an approach differentiating two different processes, gut-associated processes (GAP) and cast-associated processes (CAP), was proposed. Vermicomposting involves an active phase, in which earthworms are critical, and a maturation phase, which takes place once the worms leave the substrate, and where microorganisms are the key players. The active phase comprises all of the processes associated with the passage of the waste through the earthworm intestines (GAPs: gut-associated processes). In the maturation phase, earthworm casts begin ageing, while the associated microbial communities experience further turnover (i.e. cast-associated processes, CAPs) (Fig. 2.3). In a prospective study, we recently found that (1) most of the bacterial (96%) and fungal (91%) taxa were eliminated during vermicomposting of sewage sludge, mainly through gut-associated processes (GAP), and (2) that the modified microbial communities in the earthworm casts (faeces) later undergo a process leading to more diverse microbiota than those found in the original sewage sludge (Fig. 2.3).

2.8 Fate of Human Pathogens During Vermicomposting

Animal manures and slurries, sewage sludge and biosolids are generally recycled and applied to agricultural land as the most economical and environmentally sustainable means of treatment and reuse. These materials are valuable as fertilizers and can help maintain soil quality and fertility. However, as they often contain

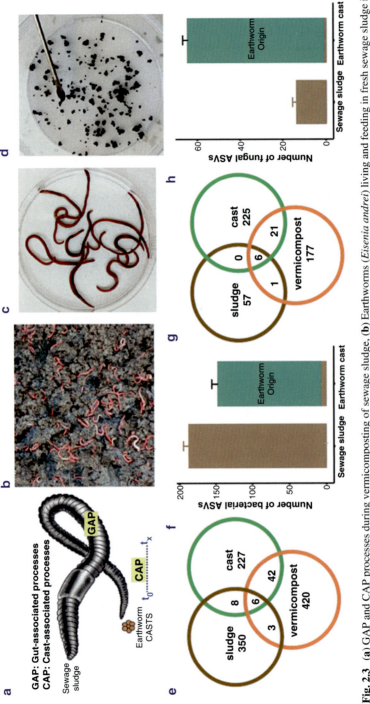

Fig. 2.3 (**a**) GAP and CAP processes during vermicomposting of sewage sludge in the vermireactor, (**c**) Sampling dishes used to collect fresh earthworm cast samples. (**d**) Fresh earthworm casts collected from sampling dishes. Changes in richness and diversity of bacteria and fungi during vermicomposting of sewage sludge. Venn diagrams showing the absolute number of (**e**) bacterial and (**g**) fungal ASVs found in sewage sludge, fresh earthworms casts and vermicompost (3 months old). Effect of GAP (gut-associated processes) on the richness and diversity of (**f**) bacteria and (**g**) fungi. Modified from Domínguez et al. (2021)

pathogenic microorganisms, spreading them on land can lead to pathogens entering the food chain. Therefore, controlling and reducing the levels of human pathogens in animal waste before this is applied to agricultural land will reduce contamination of soil and the surrounding water by pathogens.

Earthworms can reduce the levels of different human pathogens in animal waste via gut-associated mechanisms. The degree of reduction largely depends on the earthworm species and/or the pathogen considered. However, the mechanism and magnitude of this selective elimination during the processing of large amounts of sewage sludge in short vermicomposting processes are currently not very well known.

Elimination of human pathogens during the vermicomposting process is a key part of the process and crucial for its industrial application. Studies in our laboratory have revealed drastic reductions in coliform bacteria (including *E. coli*) during the vermicomposting process. There is also some evidence, which must be demonstrated at larger scales, that the vermicomposting process eliminates or drastically reduces the presence of enteric viruses and eggs of intestinal parasites. However, the biological mechanisms linking the presence of worms to the elimination of pathogens are not fully known. In addition to confirming this elimination, it is essential to determine how long it takes for pathogens to be eliminated from sewage sludge and animal wastes and to determine the underlying mechanisms, the triggering factors and when and how and under what circumstances of the process these occur.

Several studies have reported important reductions in microbial human pathogens during vermicomposting of sewage sludge and animal manures. We have found that earthworm activity, mainly during the gut-associated processes, is a critical factor leading to the rapid reduction of pathogens during vermicomposting (see previous section). The mechanisms involved in reducing or eliminating microbial pathogens may include direct effects of physical disruption during grinding in the earthworm gizzard, microbial inhibition by antimicrobial substances or microbial antagonists produced by the earthworms themselves, and destruction of microorganisms by enzymatic digestion and assimilation.

2.9 Vermicompost Properties

Vermicompost, the final product of vermicomposting, is known commercially as earthworm humus, earthworm castings or vermicasts. It is very fine, porous, biologically stable material with a high water retention capacity, low C/N ratio, high content of nutrients in forms that are easily assimilated by plants. It also includes a rich, complex microbial community with a wide range of beneficial effects to soil-plant systems.

Vermicompost improves soil health and fertility, increases the nutrient content and microbial content of soils, improves water retention and reduces the need for fertilizers and pesticides.

Vermicompost and liquid derivatives have been shown to increase the growth and productivity of many crops. These effects seem to be independent of the chemical nutrients provided and could be the outcome of biological mechanisms derived from microbial activity. The effects of vermicompost on plant growth vary depending on the plant species (and even the variety) considered, as well as the starting material, production process, storage time, and type of soil and potting medium. In studies by our group using next-generation sequencing (NGS) approaches, we have found that different microbial communities emerge during vermicomposting depending on the earthworm species and the starting material, which may imply different biological properties and functional capacities for stimulation of plant growth. The research carried out to date highlights the enormous complexity of vermicompost-plant interactions, and more detailed studies will make important contributions to organic and ecological agriculture and to soil ecology, allowing us to further unravel the complex web of relationships between plants and soil microbial communities.

2.10 Vermicomposting and Enzymatic Activity

The breakdown of macromolecules (e.g. cellulose, hemicellulose, lignin and tannins) during vermicomposting requires the action of various extracellular enzymes, most of which are produced by microorganisms. Changes in enzyme activities during vermicomposting have been extensively studied with the aim of elucidating the biochemical interactions between earthworms and microorganisms during the decomposition of organic matter. Vermicompost contains high amounts of extracellular enzymes involved in nutrient cycling (e.g. phosphatases, glucosidases, cellulase, protease and ureases) and in the degradation of organic pollutants (e.g. laccases, peroxidases and carboxylesterases). We have found that microbial communities are greatly altered by the vermicomposting process, and that the final vermicompost contains a high diversity of bacteria and fungi. Thus, vermicompost has a high load of extracellular enzymes, which are stabilized by organic matter. Recent studies in our laboratory have revealed very high enzymatic activities, including carboxylesterase activity, in vermicompost derived from different types of organic waste. In addition, the gradual increase in the concentrations of humic substances involved in vermicomposting provides some chemical support for the binding of extracellular enzymes, making them more stable and protecting them from proteases or adverse conditions such as temperature changes and desiccation. Therefore, unsurprisingly, the activity of extracellular enzymes remains very high in the final vermicompost. It is important to determine the biological properties of the final vermicompost after drying or ageing, to improve its conservation, management and application.

2.11 Vermicomposting and Bioremediation

The capacity of organisms to accumulate, adsorb, and/or degrade pollutants has led to their potential use for remediation or treatment of contaminated environments. Soil bioremediation involves any process in which a biological system is used to remove environmental pollutants from soil. Natural bioremediation is mediated by native organisms and is often a slow process. With the aim of speeding up this process and accelerating the biodegradation of environmental contaminants, there has been a tendency to focus on biostimulation, which involves the inoculation of non-native microorganisms and the addition of nutrients and other chemicals to increase microbial growth.

The use of vermicomposting and vermicompost for bioremediation purposes is not widespread, although vermicompost is basically organic matter that is rich in chemical and biological nutrients and microorganisms, and that therefore promotes and enhances microbial growth (biostimulation). The synergistic actions of earthworms and microorganisms during vermicomposting enhances detoxification processes and accelerates the removal of pollutants from all types of organic waste. Vermicompost is a microbiologically active substrate, and thus its addition to soil is a way of inoculating the soil with non-native microorganisms that can degrade organic pollutants.

Bioremediation can be done in situ, to treat contaminated sites directly, or ex situ, to treat substrates that have been removed from contaminated sites. In the case of vermicomposting, bioremediation can be achieved in both ways, by applying vermicompost in situ and by processing contaminated materials ex situ to eliminate or reduce the pollutant loads. The vermicomposting process can be considered an ex situ bioremediation process in the sense that it eliminates or reduces the possible pollutant potential of the treated materials. Although less common, in situ bioremediation can be carried out by applying vermicompost as a source of microorganisms, with the aim of decontaminating soil.

The concentration of heavy metals in the substrate increases during vermicomposting because of the loss of mass. However, the bioavailability of the metals decreases since they are sequestered in organo-mineral complexes with humic acids and other polymerized organic fractions. Vermicompost is therefore a good sorbent and can reduce the bioavailability and toxicity of metals when applied to the soil or used in biofilters to detoxify wastewater.

Characteristics such as the high organic matter content and high microbial abundance and diversity, together with the presence of pollutant-detoxifying exoenzymes, make vermicompost an ideal substrate for the bioremediation of contaminated soils. Vermicompost is an organic carbon-rich substrate, and although the total amount of organic matter in the vermicompost is lower than in the original feedstock, the content of humic substances is higher. These humic substances facilitate the formation of metal-humic complexes and the adsorption of organic contaminants. Vermicompost contains a high diversity of microorganisms and extracellular enzymes, and the rate of biodegradation of pollutants is therefore expected to increase in soils treated with vermicompost. These bioaugmentation properties have been addressed in several studies.

Some studies performed in our laboratory have reported the existence of a high level of carboxylesterase activity in vermicompost derived from different types of organic waste. The enzymatic activity is variable, and the rate largely depends on the substrate used and the earthworm activity. Vermicompost can bind organic pollutants, indicating its strong bioremediation power. The incorporation of vermicompost in topsoil could provide a molecular and biochemical barrier reducing the movement of the contaminants and enhancing their biodegradation. Although significant advances have been made in different research laboratories, the use of vermicompost for the bioremediation of contaminated soils and to prevent the effects of pesticides and other organic contaminants on soil function require further research.

In the case of organophosphorus pesticides, carboxylesterase-mediated detoxification involves the formation of a stable enzyme-pesticide complex by the direct interaction between the organophosphorus molecule and the active site of the enzyme. The findings of experiments conducted in our laboratory to study carboxylesterase activity in vermicompost suggest that this enzyme irreversibly binds the organophosphorus chlorpyrifos-oxon, acting as a molecular scavenger of this type of pesticide.

To assess the enzymatic bioremediation potential of vermicompost, we have explored the stability of the carboxylesterase activity in the vermicompost in response to desiccation and ageing. We have found that the esterase activity is very stable, probably due to strong interactions with the humic substances in the vermicompost. This chemical binding protects and stabilizes the enzymes against physicochemical and biological degradation.

Another promising application is the vermicomposting of sludge derived from wastewater treatment (WWTP sludge) as a tertiary treatment to accelerate the degradation of contaminants, including emerging pollutants, antibiotic resistance genes and microplastics.

The amount of sewage sludge generated in wastewater treatment plants is increasing steadily, with hundreds of million tons produced every year all over the world. Microplastics are small (<5 mm diameter) synthetic polymer wastes, derived from a wide range of sources, including clothing, personal care products, and polymer manufacturing and processing industries. The widespread use of plastics, coupled with their resistance to degradation, leads to their accumulation and adverse impacts on the environment. Plastics can adsorb pollutants, such as polycyclic aromatic hydrocarbons, heavy metals, polybrominated diphenyl ethers, and pharmaceutical substances, and may cause chronic toxicity by bioaccumulation in organisms.

Wastewater treatment plants are main sinks of microplastics derived from daily human activities. These plants are quite effective in removing microplastics from the water preventing them from entering natural aquatic systems at this stage. However, during the treatment, the microplastics are retained in the sewage sludge. Land-applied sludge is thus a main source of terrestrial contamination with microplastics, which subsequently enter natural aquatic systems. Therefore, it is important to establish ways of removing plastics from sewage sludge and of assessing the environmental impact of land-applied sludge in terrestrial and aquatic ecosystems. Regulations on the use of sewage sludge and biosolids in agriculture stipulate limits for the contents of human pathogens, and maximum rates of application of metals and nutrients to the soil. Concentration limits vary for all contaminants, and contamination by microplastics has not yet been addressed.

Rates of degradation of plastics and biopolymers are limited by the hydrolysis of ester bonds, and extracellular microbial enzymes (including carboxylesterases) catalyse the hydrolysis of plastic polymers. During vermicomposting the drastic changes that take place in the structure and function of the microbial communities are concomitant with significant increases in extracellular detoxifying enzymatic activity. This activity should trigger and accelerate the enzymatic degradation of plastics. The same approach could potentially be applied to other families of contaminants, such as antibiotic resistance genes, human pathogens and emerging pollutants, which are not usually monitored but can enter terrestrial ecosystems and ultimately aquatic ecosystems causing environmental and health-related problems.

2.12 Conservation of Vermicompost and Processing Strategies

Vermicompost is very rich in microorganisms, and the main qualities that make it a superior fertilizer are its excellent biological properties. These properties have been related to nutrient cycling in the soil or plant growth substrate, and also to the provision of bioactive substances that act as plant growth enhancers (including precursors of auxins, gibberellins and cytokinins, enzymes, humic acids, and beneficial microorganisms and their metabolic products).

The moisture content of vermicomposting substrates should be between 75 and 85% during the process; however, the final moisture content varies greatly, depending on the starting material from which the vermicompost is made. Very wet vermicompost is difficult to store and very expensive to transport (because of its weight) and is also difficult to handle and apply. The moisture content obviously also has an important impact on the price of vermicompost when sold by weight, and vermicompost is therefore generally sold by volume. One of the most difficult technical obstacles to resolve during vermicompost production is drying the material. In most vermicomposting facilities, vermicompost is dried outdoors to reduce the initial water content, from about 75–85% to about 45–50%; however, this is a costly, slow process. It is commonly argued that dry vermicompost can be difficult to re-wet, with significant loss of its biological properties. However, further studies are necessary to confirm this finding.

2.13 Vermicomposting as an Ecological Engineering Technique for Improving Soil Health and Sustainability in Vineyards: A Case Study

Grape is the largest fruit crop in the world, with more than 7 million ha of harvested area and an annual worldwide production of about 80 million tonnes, around 80% of which is used to make wine. The wine sector is very important in economic, social and environmental terms. Winemaking generates millions of tons of grape marc annually as a by-product. Grape marc (also known as bagasse and pomace) consists of the skin, pulp, and seeds that remain after pressing the grapes to obtain the must for making wine. It is a valuable resource and is used to produce ethanol, grape seed oil, bioactive compounds, and animal feed. It can also be used as a nutrient-rich organic soil amendment; however, when applied to soils without prior treatment, it can damage crops due to its acidity and high content of phytotoxic polyphenols. Excessive accumulation of this waste at specific times and in particular areas is problematical, leading to inadequate management, including uncontrolled disposal, dumping on fields without prior treatment, burning and other environmentally unfriendly solutions.

Vineyards are usually managed intensively to maximize wine production, which reduces soil biodiversity, decreases soil fertility and causes environmental problems such as changes in primary production and nutrient cycling, reduction of above-ground biodiversity, high soil erosion rates, groundwater eutrophication and contamination, and global warming.

Soils are among the most biodiverse habitats on Earth, with extremely diverse and complex biological communities that play a key role in the functioning of natural ecosystems. Although soil fauna has a strong influence on soil ecosystems and regulates many important processes, the key steps in the cycling of the main elements are governed by microorganisms. Intense land use generates serious environmental problems and interferes with soil biological processes. Intensive agriculture, which involves conventional tillage and massive, repeated fertilization and pesticide application—and which is associated with low plant diversity—has very negative effects on soil biodiversity, promoting simpler food webs composed of smaller organisms and fewer functional groups. Less diverse communities have been shown to be less resilient to stress than more diverse communities owing to changes in functional capacities. Therefore, intensification of agriculture and the consequent biodiversity loss together lead to environmental problems, such as changes in primary production and nutrient cycling, reduced surface biodiversity, eutrophication of water bodies, and global warming. However, production levels are lower in most sustainable systems than in intensive systems, and they must therefore be optimized.

The discrepancy between high agricultural yields and ecosystem sustainability can only be overcome by major modifications to ecosystem processes. While "green revolution" approaches focus on external manipulation, the internal manipulation of ecosystems has an enormous potential to improve yields, but with fewer

Fig. 2.4 Summary of the ongoing "EWINE" project, started in 2012, and undertaken to close a cycle in which grape pomace is converted into vermicompost, which is then applied to the vineyards as a biostimulant treatment

environmentally negative consequences. Vermicomposting accelerates the decomposition and stabilization of organic matter, substantially modifying the physical, chemical, and biological properties. Vermicompost, the final product of vermicomposting, is a biologically stable, very fine, porous material with a high water retention capacity, a low C/N ratio and a high nutrient content. It also contains complex microbial communities that have a wide range of beneficial effects on the soil-plant system. Vermicompost and its liquid derivatives have been shown to increase the growth and productivity of many crops. These effects occur independently of the nutrients provided and may be due to biological mechanisms derived from the associated microbial activity.

Based on those premises, in 2012, we started the current "EWINE" project, in which the combined and synergistic activity of earthworms and microorganisms during vermicomposting are studied for development of an integrated cycle to convert the waste generated by the wine industry into vermicompost, which is applied to the vineyards as a biostimulant treatment (Fig. 2.4).

Pilot-scale vermicomposting of the residual by-products generated in the wine industry (mainly bagasse and bagasse distilled from white and red grapes) was applied and studied with the aim of producing a high quality vermicompost which meets the requirements demanded by the legislation regulating the production of fertilizers in general, and sustainable, organic and ecological agriculture in particular. The biofertilizer and biopesticide effects of the vermicompost were studied in vineyards in two different geographical areas and two different appellations of origin.

Raw and distilled grape marc derived from white and red wine varieties were processed in pilot-scale vermireactors, yielding a high quality organic, polyphenol-free fertilizer and grape seeds. Earthworm population density and structure and earthworm biomass were monitored or determined monthly. Samples of the processed grape marc and the vermicompost were collected periodically and analysed to determine their chemical and biological properties.

Vermicomposting substantially reduced (by 50–70%) the biomass of grape marc, and the process yielded a nutrient-rich, microbiologically active and stabilized peat-like material that is easily separated from the seeds by sieving. The separation and removal of the seeds eliminates the residual polyphenol-associated phytotoxicity in the vermicompost. The seeds can then be easily processed to obtain polyphenol-rich extracts and fatty acid-rich seed oil. Moreover, the vermicomposting process produces large numbers of earthworms that can be processed as fish bait and as a source of protein for animal feed.

We have also studied the effects of earthworms on the process and the changes during vermicomposting of important biological parameters in microbial communities, such as their abundance, composition, structure, diversity, and activity.

The efficacy of vermicomposting as a low-cost, environmentally safe solution for the treatment and valorization of raw and distilled grape marc, together with an in-depth characterization of the final vermicompost, including its chemical, bio-chemical and microbiological properties is presented in detail in Chap. 6 of this book.

The vermicompost derived from the different wine varieties was applied in solid and liquid formulations to the grapevines in the vineyards where the grapes were harvested to make wine, and the grape marc was obtained.

The effects of the vermicompost on the vineyard soil and in the grapevines were studied during several crop seasons in different vineyards of two different denominations of origin in two biogeoclimatic areas: Rías Baixas, a Spanish *Denominación de Origen Protegida* (DOP) (Protected Designation of Origin) for wines located in the southwest of Galicia (Spain) and Ribeira Sacra, a DOP for wines from the south of the province of Lugo and in the north of the province of Ourense, Galicia (Spain). The vineyard experiments were conducted in the commercial Albariño vineyard *Terras Gauda*, located in O Rosal, Pontevedra (*Vitis vinifera* cultivar Albariño, DOP Rías Baixas), and in the commercial Mencía vineyard *Adegas Moure*, located in A Cova, O Saviñao, Lugo (*Vitis vinifera* cultivar Mencía, DOP Ribeira Sacra).

Fertilization with vermicompost significantly improved grape production in the two denominations of origin and in the two biogeoclimatic areas. We have collected data on nine vintages or harvest between 2014 and 2022, and although the results vary depending on the year, region, grape variety, and age of the plants, in all cases grapevines treated with vermicompost produced significantly more grapes than untreated grapevines.

In the wineries where the vineyard experiments were conducted, wine was made from grapes fertilized with vermicompost derived from grape marc. *Terras Gauda* (Albariño) and *Adegas Moure* (Mencía) made wines elaborated with grapes from vines fertilized with vermicompost derived from grape marc, and control wines elaborated with grapes from vines from the same experimental plot but not treated with vermicompost. Otherwise, both wines were made following the same standard procedures of the wineries.

Blind wine tastings carried out at both wineries revealed notable differences in organoleptic properties (e.g. overall increased complexity, expression, freshness and balance, with better visual intensity and taste persistency) that led to the wine made using grapes from the treated vines better ratings and reviews.

We used metataxonomic approaches to characterize the bacteriota and mycobiota of the must and finished wine from Albariño and Mencía grapevines treated with vermicompost derived from grape marc and controls (standard fertilization) during 2 consecutive years. We found statistically significant, important differences in composition, structure, and predicted metabolic functions of the microbiota.

Our findings suggest an important beneficial role of vermicompost supplementation in the vineyards, improving grape productivity and quality of the wine.

The research has led to important advances in the application of vermicomposting technologies in the wine sector. Both wineries where the experiments have been performed have built vermireactors and they currently process all of their grape marc by vermicomposting and apply the vermicompost to the vineyards.

2.14 Conclusion

This chapter summarizes research on vermicomposting conducted in the Soil Ecology Laboratory, University of Vigo, and it represents an up-date of the original text entitled "State-of-the-Art and New Perspectives on Vermicomposting Research", written in 2004 and included in the book Earthworm ecology. The detailed research has been published in scientific articles that can be consulted in the research group's webpage (http://jdguez.webs.uvigo.es/).

Acknowledgements For their contributions to the vermicomposting research programme developed by the Soil Ecology Laboratory (University of Vigo), I thank Manuel Aira, Fernando Monroy, Cristina Lazcano, María Gómez Brandón, Luis Sampedro, Alberto Velando, Marta Lores, Robert W. Parmelee, Patrick J. Bohlen, Scott Subler, Rola M. Atiyeh, Juan Carlos Sánchez Hernández, Alfredo Ferreiro, Julio Eiroa, Andrea Tato, Domingo Pérez, Jose Carlos Noguera, Sim Yeon Kim, Salustiano Mato, Fuencisla Mariño, Xavier Freire, Ramón Eixo, David Cereijo, Lucía Senra, Aida Costas, Sara Rodríguez, Salomé Álvarez, Pablo G. Porto, Lorena San Miguel, Belen Longa, Julia Arnold, Hans Zaller, Rafael Zas, Michelle Webster, Jose Angel Gago, Cristóbal Pérez, Anxo Conde, Gonzalo Mucientes, Hugo Martínez, Alberto Da Silva, Daniel Fernández Marchán, Luis A. Mendes, Celestino Quintela Sabarís, Gustavo Schiedeck, Daniel Pazzini, Natielo Almeida Santana, Nariane de Andrade, Zaida Inês Antoniolli, Afrânio Guimaraӗs, Mario Domínguez-Gutiérrez, Isabelle Barois, Marcos Pérez Losada, Keith A. Crandall, and all other temporary collaborators in vermicomposting science and technology.

I also thank the vintners at Terras Gauda and Abadía da Cova, particularly Emilio Canas and Evaristo Rodríguez, for providing access to their vineyards and wineries and for their continued partnership and interest in the application of vermicomposting to the wineries in the long-term EWINE project.

This chapter is dedicated to the memory of Clive A. Edwards, a brilliant scientist, friend and mentor who inspired my dedication to earthworm ecology and vermicomposting.

Funding This study was partly co-funded by the Spanish Ministerio de Ciencia e Innovación (PID2021-124265OB-100 and TED2021-129437B-I00), the Xunta de Galicia (ED431C 2022/07) and the European Union's Horizon 2020 research and innovation programme (LABPLAS_101003954).

Chapter 3
Experiences on Methods of Vermicompost Analysis for Plant and Soil Nutrition

Hupenyu A. Mupambwa and Pearson Nyari Stephano Mnkeni

Abstract Vermicomposts and composts are currently being promoted as sources of nutrients in soil fertility following the realization that inorganic synthetic fertilizers have their own limitations in managing soil quality. Most of the composts that are marketed these days are not clearly labeled in terms of their chemical composition, which is important in guiding the application rates of these fertilizers. Various researchers have reported different methods of determining chemical properties of vermicomposts which makes the comparison of results not possible. This chapter shares experiences on various methods that can be used to characterize compost chemical qualities. This includes methods on total carbon content where other researchers have used the dichromate oxidation or ashing method, total versus extractable elemental composition in evaluating compost maturity, among other properties. This chapter intends to encourage further research and discussion on the optimization and standardization of methods of compost chemical characterization.

Keywords Vermicompost · Heavy metals · Cations · Humification parameters · Exchangeable nitrogen · ICP-OES

3.1 Introduction

Agriculture during the twentieth century saw the introduction of the green revolution which resulted in increased use of industrially manufactured fertilizers like urea, ammonium nitrate among others. The use of these fertilizers saw a huge increase in

---

H. A. Mupambwa (✉)
Sam Nujoma Marine and Coastal Resources Research Center, University of Namibia, Henties Bay, Namibia
e-mail: hmupambwa@unam.na

P. N. S. Mnkeni
Faculty of Science and Technology, University of Arusha, Usa River, Arusha, Tanzania

© The Author(s), under exclusive license to Springer Nature Singapore Pte Ltd. 2023
H. A. Mupambwa et al. (eds.), *Vermicomposting for Sustainable Food Systems in Africa*, Sustainable Agriculture and Food Security,
https://doi.org/10.1007/978-981-19-8080-0_3

crop yields and agricultural productivity, as the fertilizers were used to elevate the nutrient levels in soils. However, though the use of industrial fertilizers resulted in increased crop yields, this resulted in agriculture that focused on feeding the crop rather than feeding the soil (Mupambwa et al. 2022). The focus on feeding the crop and ignoring the soil as an important resource resulted in continuous soil degradation which has resulted in higher levels of nutrients being required to maintain yields each season. This has seen most farmers especially smallholder resource poor farmers reporting reduced crop yields due to increased soil degradation such as reduced soil pH, decrease in soil organic carbon, decreased aggregate stability, poor soil structure, among others.

There is growing realization that soil is a resource that is not only physical but also biologically active, which thus needs to be fed in order for it to be able to maintain consistent productivity. This has led to the growing realization that nutrients need to be supplied together with organic materials that are important in feeding the soil microbiome that drives all biological processes in the soil. This saw recent agricultural cropping systems focusing more on organic soil fertility management as a critical driver in making agriculture sustainable. Recently, vermicomposts are now being promoted as effective nutrient rich organic soil fertilizers, and several researchers have been undertaking studies that focus on improving these organic soil fertilizers (Mupambwa and Mnkeni 2019; Unuofin and Mnkeni 2014). For organic fertilizers to be effective in agriculture, there is a need for the evaluation of their physico-chemical properties, and there are various methods that have been recommended for this. Though composts are treated as soils by other researchers, some of the methods recommended for soils are not effective or applicable for composts. This chapter presents some insights on the various methods that have been evaluated on determining key parameters relevant for evaluating the maturity and chemistry of composts.

3.2 Vermicomposts Compared to Soils

Soil is generally defined as the mixture of unconsolidated usually weathered mineral materials with organic matter, water and pore spaces, with the mineral phase contributing up to 45% of this mixture, while organic matter constitutes about 5% (Brady and Weil 2008). On the other hand, composts are defined as mixtures of mainly organic materials with water, minerals and pore spaces, with the organic material constituting the highest concentration and limited mineral material content. Due to the variation in the concentration of the different elements mainly minerals and organic matter, the bulk density of soil, i.e., the weight of soil per unit volume, is generally higher compared to that of composts, which can be an important factor to consider during compost chemical composition compared to that of soils.

3.3 Parameters Critical in Vermicompost Quality

Composts are growing as sources of nutrients in organic agriculture, though unlike inorganic, synthetic fertilizers, these have certain quality parameters that have been determined to be critical in determining the maturity for soil application of these composts. According to Bernal et al. (2009), the process of preparing a compost involves bio-oxidative processes that involve mineralization and partial humification of organic matter which results in the production of a stabilized final product which possess certain qualities. Due to the composting process being biologically driven, it is difficult to have a single parameter that can be used to determine compost maturity (Raj and Antil 2011). The various parameters that are critical in compost or vermicompost quality are presented in Table 3.1.

Interestingly, the subject of compost maturity is littered with diverse values which are influenced by various properties and thus at times a single parameter may have different critical maturity thresholds. There is, however, a need to determine these parameters in each compost in order to determine these maturity parameters and this chapter attempts to list these.

Table 3.1 A summary of the various parameters that have been identified in literature as critical in determining compost or vermicompost maturity

Maturity parameter	Chemical properties	References
Total C to N ratio	Less than 20 and preferably less than 10	Bernal et al. (2009), Meena et al. (2021), and Raj and Antil (2011)
Humic acid to fulvic acid ratio	Greater than 1.9	Raj and Antil (2011)
Water soluble organic-C	Less than or equal to 4 g/kg	Bernal et al. (2009)
WSC/Org-N ratio	Less than 0.55	Meena et al. (2021)
Germination index	Between 50 and 70%	Meena et al. (2021), Raj and Antil (2011), and Bernal et al. (2009)
Organic matter loss	Greater than 42%	Raj and Antil (2011)
Humification index	Greater than 30%	Raj and Antil (2011) and Bernal et al. (2009)
NH_4- N to NO_3- N ratio	Less than 0.16	Bernal et al. (2009)
Polymerization index	Greater than 1	Bernal et al. (2009)
Humification ratio	Greater than 7	Bernal et al. (2009)

3.4 Analysis Important in Vermicompost Quality

3.4.1 Total Carbon Analysis

Carbon is an important element in organic soil fertility management and its determination in composts is critical in the calculation of the C/N ratio, which is a critical compost maturity parameter. In soil, carbon exists in both organic and inorganic forms, with the inorganic component being made up of mainly the carbonates while the wide range of organic molecules constitute the organic carbon (Skjemstad and Baldock 2006). Therefore, in the determination of carbon in soils, there is usually a slight difference between total carbon and organic carbon, since the total carbon is the sum of both carbonate-carbon and the organic carbon. In composts, by contrast, since these are derived from organic matter, almost all the carbon is organic carbon hence it's safe to say the total carbon is equal to organic carbon.

In soils, the organic carbon is generally determined by the dichromate oxidation method where organic C is oxidized by $Cr_2O_7^{2-}$ and then the unreduced $Cr_2O_7^{2-}$ is determined by oxidation-reduction titration with Fe^{2+} (Nelson and Sommers 1996). Other authors have, however, also reported the measurement of organic carbon on composts using this dichromate oxidation method. The total carbon values obtained by this approach are, however, likely to be erroneous as the organic carbon in composts is very high and the oxidation with dichromate does not result in complete oxidation of the C.

The determination of total carbon in composts using the dichromate oxidation method also has the challenge during titration where color change is very difficult to determine using manual titration as color changes often occur after addition of unmeasurable quantities of the Fe^{2+}. Therefore, the dry combustion automated method of determination is the most reliable method of determining total or organic C in vermicomposts. This method is based on the oxidation of carbon under high temperatures in the presence of a stream of oxygen and the carbon dioxide liberated in then quantified using various methods (Nelson and Sommers 1996). The dry combustion method can be done by auto-analyzers that do not require any chemical pre-treatment of the vermicompost sample other than the drying of the sample and grinding it for homogeneity. The ground sample is usually measured in very small quantities usually less than 1 g, and this sample is run through the calibrated machine. As indicated by Nelson and Sommers (1996), the instruments that are widely used for the automatic dry combustion determination of total carbon include the LECO Instruments like the TruMac LECO CNHS auto-analyzer, Carbo-Erba Instruments like the Carbo-Erba NA 1500, and the Perkin Elmer Corp Instruments like the Perkin-Elmer CHN2400.

3 Experiences on Methods of Vermicompost Analysis for Plant and Soil Nutrition 49

3.4.2 Dissolved Organic Carbon

In compost science, at times there is the need to determine the concentration of dissolved organic carbon, which is that carbon that is extractable using various solutions which include distilled water (Jones and Willett 2006), potassium sulfate (Joergensen and Brookes 2005), and sodium hydroxide during the quantification of humification parameters in composts (Sanchez-Monedero et al. 1996). The best method for the determination of this extracted organic composts was described by Anderson and Ingram (1993) and also by Joergensen and Brookes (2005) which also employs the dichromate oxidation of the dissolved carbon.

3.4.2.1 The Method of Anderson and Ingram (1993)

A 4 mL sample is treated with 1 mL of 0.0667 M potassium dichromate (prepared by dissolving 19.622 g of dry potassium dichromate in 1000 mL of deionized water). The mixture is then treated with 5 mL of concentrated sulfuric acid. The solution is then properly mixed and 3–4 drops of an indicator, o-phenanthroline monohydrate, are added. The excess dichromate is then determined by titration using acidified ammonium ferrous sulfate. For this procedure, two blanks are prepared with no sample but having all the reagents, where one blank is heated (HB) while the other is not heated (UB). The organic carbon as a percentage is then calculated using the formula below.

$$\text{Organic carbon } (\%) = \frac{A \times M \times 0.003}{g} \times \frac{E}{S} \times 100 \tag{3.1}$$

$M = 0.4/\text{Standardization titer}$
$M = \text{molarity of ferrous ammonium sulfate (approx. 0.033 M)}$
$A = [\text{mL}_{HB} - \text{mL}_{sample}] \times [(\text{mL}_{UB} - \text{mL}_{HB})/\text{mL}_{UB}] + [\text{mL}_{HB} - \text{mL}_{sample}]$
$g = \text{dry soil/compost mass}$
$E = \text{extraction volume (mL)}$
$S = \text{digest sample volume (mL)}$

The method of Anderson and Ingram (1993) works very well with soils where the amount of organic carbon is less; however, the titration challenge is also observed with compost sample organic carbon extracts, as the end point is very difficult to establish. With composts, very low quantities of the titrating solution are required to observe a color change and at times this happens simply by adding the indicator solution and none of the titrating solution. However, based on experience, it has been observed that if the quantity of compost to be extracted is reduced to less than 2 g and the volume of the extract also reduced, the titration becomes feasible, with clear titer volumes. Research on the analysis of carbon in organic material rich in carbon like composts for both extractable dissolved carbon and actual carbon content, using the dichromate oxidation methods is required for optimization of the procedures.

Current procedures such as those described by Anderson and Ingram (1993) seem to be optimized for soils.

3.4.3 Humification Parameters

In compost maturity, the determination of the various humification parameters is critical in the use of organic soil fertility conditioners. The critical properties that are important in compost include humification ratio, humification index, polymerization index, among others. For the determination of these parameters, it involves the determination of total organic carbon, humic acid fraction, and fulvic acid fraction. The humic and fulvic acid fractions in composts are extracted using a method clearly described by Sanchez-Monedero et al. (1996). Extraction is done using 0.1 M NaOH at the ratio of 1: 20 w/v and shaken on a horizontal reciprocating shaker for 4 h. After the shaking, the extracts are centrifuged at 8000 rpm equivalent to 8.2×103 g Relative Centrifugal Force; and half of the extracts are stored for subsequent analysis of total extractable carbon fraction (C_{tEX}) and the remainder acidified to pH 2 using concentrated sulfuric acid. The pH adjusted extracts are then allowed to coagulate over 24 h at 4 °C by placing in a normal fridge, following which the extracts are centrifuged at 8000 rpm (8200 g) to separate the humic acid (HA) fraction from the fulvic acid (FA) fraction. The fulvic acid (C_{FA}) portion in solution after centrifugation and the (C_{tEX}) fraction are then analyzed for extractable carbon using the dichromate oxidation method as outlined for dissolved organic carbon. The extractable carbon in the extracts is then finally calculated using the formula in Eq. (3.1) described by Anderson and Ingram (1993).

Humification ratio (HR), humification index (HI), percentage of humic acids (Pha), and polymerization index (PI) which are indices used for the evaluation of humification level in the compost are then calculated as shown in Eqs. (3.2–3.5) below:

$$HR = \frac{CtEX}{C} \times 100 \tag{3.2}$$

$$HI = \frac{CHA}{C} \times 100 \tag{3.3}$$

$$Pha = \frac{CHA}{CtEX} \times 100 \tag{3.4}$$

$$PI = \frac{CHA}{CFA} \times 100 \tag{3.5}$$

3.4.4 Total Elemental Content

Total elemental composition is very important in understanding the mineralization process during vermicomposting. Unlike in soils, where the total elemental composition does not contribute mainly to plant nutrition, in composts, this fraction of nutrients contributes to plant nutrition as it is highly labile and can be easily mineralized into plant available forms. In vermicomposts, the processes that are responsible for total elemental composition conversion to plant available forms are driven by the earthworms and micro-organisms (Mupambwa and Mnkeni 2018). Interestingly, though the total elemental composition in composts is very labile, determining this fraction only gives an indication of the elements that are potentially available for plant uptake and cannot substitute the determination of the plant available fractions. However, several authors have reported results of their vermicompost based on the total elemental composition alone, which is misleading as this fraction is always higher than the fraction that plants can actually extract from the compost.

Similar to soils, there are standard methods that are available for both soils and composts that enable for the determination of total elemental composition. These methods involve the use of strong solvents like acids to digest and make available in solution the organic and inorganic elements. Due to composts being mainly organic, the digestion of such materials is fairly simple as it's the inorganically bound elements in soil samples that are more difficult to digest and may require stronger acids like hydrofluoric acid to bring them into solution. We present easy methods that can be effectively used for the determination of total elemental composition of composts without the use of complicated equipment which has also been recommended by AgriLASA (2004). Interesting to note, methods for total elemental composition determination for plant samples are suitable for compost samples as well. A single digest in these procedures can allow for the determination of a full spectrum of inorganic elements such as K, P, Mg, Na, Ca, trace and heavy metals.

3.4.4.1 Compost Wet Ashing Digestion (AgriLASA 2004)

This method employs the use of perchloric acid ($HClO_4$) and nitric acid (HNO_3). A known weight of a finely ground homogenous dry sample between 0.5 and 1 g for composts is added into a digestion tube or a conical flask. It's important to note that for sample drying, air drying is the best method and the use of high temperature ovens needs to be avoided as this can affect the final results obtained. To the weighed sample, 10 mL of 65% nitric acid is added and heated on a digestion block or hot plate at 150 °C for about 30 min or until the brown fumes are no longer being liberated, all under a fume extraction hood. The digestion tubes or conical flasks are then removed from the digestion block or hot plate, respectively, and then allowed to cool, before adding 4 mL of perchloric acid 60% (AgriLASA 2004). The mixture is then heated again for 30 min at the same temperature before being removed from the heating source, cooled and deionized water is added to bring it to 50 mL or any other preferred known volume. During digestion, a few inorganic residues may remain behind and thus the digested and diluted sample may need to be filtered using Whatman number 2 filter paper or the equivalent. Following the digestion process, a blank should also be prepared using the same reagents which are also heated as for the samples. These digestion samples will also need to be used for preparation of the calibration standards to allow for the sample and standard matrix to be the same.

After digestion, the total elemental concentration is determined using various machines that include the Inductively Coupled Plasma-Optical Emission Spectrometer (ICP-OES), Atomic Absorption Spectrophotometer (AAS), and other colorimetric methods. The final sample elemental concentration is then calculated or converted into milligrams per kilogram of compost using the following formula:

$$\text{Elemental composition (mg/kg)} = \frac{c \times v \times f}{w} \tag{3.6}$$

where C = concentration in milligrams per liter.
v = total volume of extraction after dilution.
f = dilution factor.
w = dry weight of sample used.

3.4.5 Total Nitrogen

Unlike the other elements, total nitrogen is determined essentially through two methods i.e., the wet digestion Kjeldahl method and the dry oxidation (combustion) method (Bremer 1996). During the dry combustion method, nitrogen is easily determined simultaneously with total carbon, using an auto analyzer such as the TruMac LECO CNHS auto-analyzer. Of the two methods, the dry combustion method is the simplest method though it requires high capital cost equipment, while the wet digestion method is less capital intensive involving only the use of digestion equipment and the Kjeldahl apparatus to determine the N content in the digest.

3.4.6 Exchangeable Nitrogen Analysis

Nitrogen is the most important element required for plant growth as it is involved in the synthesis of proteins which are critical in the function of all plant processes in plant cells. Nitrogen is the most abundant gas in the atmosphere but plants cannot absorb this as they do with carbon, and have to take this nitrogen as inorganic nitrogen from the soil. In agriculture, the green revolution was driven mainly by the ability to industrially harvest the atmospheric nitrogen through the Haber-Bosch process to generate a fertilizer that is rich in nitrogen. Unfortunately, the major limitation of organic soil fertility management is in the inability of composts to supply high levels of nitrogen equivalent to inorganic synthetic fertilizers. It is therefore important to effectively quantify the amount of extractable nitrogen present in organic fertilizer sources, which allows for the determination of how much extra fertilizer will need to be added. Though samples for extractable nitrogen analysis are expected to be analyzed instantly after collecting, without any drying, this is usually impractical, and thus drying at room temperatures can be done and not at elevated temperatures above 35 °C, as this may lead to considerable loss of exchangeable nitrogen (Mulvaney 1996).

In soils, inorganic nitrogen is mainly found in two forms, i.e., nitrate (NO_3^-) and ammonium (NH_4^+), with very negligible amounts of nitrite (NO_2^-) being also available. Various methods are available for the determination of inorganic nitrogen in composts and these involve the use of extraction and then colorimetric, potentiometric, ultraviolet spectrometric, among others (Mulvaney 1996).

The extraction of the exchangeable inorganic nitrogen forms can be done using 2 M KCl (potassium chloride) solution or with 0.5 M (K_2SO_4) potassium sulfate, with the 2 M KCl solution being the most preferred as the 0.5 M K_2SO_4 solution may result in precipitates forming during refrigeration (Mulvaney 1996; Maynard et al. 2006). The experience from our research has been that from an economics perspective, a 2 M KCl solution will require approximately 149.2 g of potassium chloride per liter of extracting solution, while a 0.5 M K_2SO_4 solution will require approximately 87.13 per liter of extracting solution. Therefore, the use of the 0.5 M K_2SO_4 is much cheaper as less salt is required especially when large quantities of samples are to be processed. For extraction a 1: 10 weight to volume ratio, usually 5 g in 50 mL of extracting solution, is used and this is extracted by shaking on a reciprocating shaker at 160 rpm for 30 min. The solution can then be filtered using Whatmans filter paper number 42 or number 2 or equivalent before subsequent analysis using the colorimetric method as outlined by Okalebo et al. (2002).

3.4.6.1 Colorimetric Determination of Exchangeable Ammonium and Nitrate (Okalebo et al. 2002)

This method involves the use of sodium hydroxide, sodium hypochlorite, sodium nitroprusside, sodium salicylate, and sodium tartrate. The reagents are used to prepare two solutions N1 and N2, and these need to be made at least 24 h before use and stored in the dark. Important to observe is that your sodium hypochlorite should be reagent grade and very fresh as the use of old sodium hypochlorite will result in poor color development. The standard stock solution is prepared using ammonium sulfate, and for the calibrating standards, these are made to volume using the extracting solution which is 0.5 K_2SO_4. In this procedure, 0.2 mL of the sample extract is then mixed with the reagents to develop color, the samples are let to stand for 1 h before measurement using a spectrophotometer at 655 nm. The results of the measurements are in mg per liter of extract, and will need to be converted into milligrams per kilogram of compost using the formula provided in Eq. (3.6).

The method involves the use of sodium hydroxide, salicylic acid, sulfuric acid, and potassium nitrate which is used in preparing the standard stock solution. After adding the reagents to the sample aliquot of 0.5 mL, the mixtures are let to stand for 1 h and color depth is then measured using a spectrophotometer set at 419 nm. It's important to note that since the extraction is done with potassium sulfate, one should only prepare samples that can be analyzed each day so as to avoid keeping the samples in a fridge as they can precipitate as earlier highlighted. For each day, fresh standards need to be prepared so that a new calibration equation can be developed.

3.4.7 Extractable Phosphorus

Phosphorus is the second most important element required for plant growth as it is involved in the energy transfer and nucleic acid synthesis. In soil, phosphorus exists in soil as organic and inorganic P forms and the inorganic fraction is important for plant nutrition and this exists as orthophosphate (HPO_4^{2-} and $H_2PO_4^-$). Various methods have been developed for the quantification of phosphorus in soils and these also apply to composts which includes Mehlich 1 and Mehlich 3; Olsen method; Bray 1, among others. Each of the methods has been observed to be suited for various soil conditions such as pH, levels of calcium carbonate among others, and thus choice of extraction method should be informed by such information as indicated by Sims (2000).

Bray 1 soil test is not suitable for:

- clay soils with a moderately high degree of base saturation
- silty clay loam or finer-textured soils that are calcareous or have a high pH
- value (pH > 6.8) or have a high degree of base saturation
- soils with a calcium carbonate equivalent >7% of the base saturation, or
- soils with large amounts of lime (>2% $CaCO_3$)

Mehlich 1 is best suited for:

- acid soils (pH < 6.5)
- low cation exchange capacities (<10 cmol/kg)
- organic matter contents (<5%).

3 Experiences on Methods of Vermicompost Analysis for Plant and Soil Nutrition 55

- Not suitable for soils that have been recently amended with rock phosphate, and soils with high cation exchange capacity (CEC) or high base saturation

Mehlich 3 is well suited for:

- A wide range of soils, both acidic and basic in reaction

Olsen method is best suited for:

- calcareous soils, particularly those with >2% calcium carbonate
- acidic soils
- Important to note is that after shaking with Olsen extracting solution, the filtered suspensions are then adjusted to pH 5 with 2.5 M sulfuric acid (H_2SO_4) before analysis.

Following the various methods of extraction, the most reliable method for determining the P content in the extracts which requires limited equipment investment is the colorimetric ascorbic acid method. The ascorbic acid method described by Kuo (1996) is a very simple procedure to execute. It involves the use of a desktop spectrophotometer that can be set at 880 nm. It is important to note that all standards must be made to volume with the extracting solution and not with water to ensure that the matrix is the same as that of the samples. If the level of phosphorus is high in the samples, it is important to dilute the samples before color development and the dilution factor (d.f.) should be used to as a factor in Eq. (3.6).

3.4.8 Extractable Cations

Among the most important elements required for crop growth are the cations, i.e., K, Ca, Mg and Na, and monitoring these in compost is important as excess levels of these salts in some soils can be detrimental to plants. Sodium, for example, can have both detrimental effects to soil physical and chemical properties, which can be responsible for soil degradation (Brady and Weil 2008). It is therefore important to monitor the levels of these elements in composts, as high levels of these elements are not preferred at all times. This challenge can be more pronounced if amendments like fly ash, rock phosphate, or seaweed based biochar are used to improve the quality of composts. Elevated levels of cations can influence soil parameters like EC, pH, salinity, exchangeable sodium percentage, and sodium adsorption ratio (Brady and Weil 2008).

For composts, the determination of the concentration of cations is fairly simple and uses procedures used in soil analysis. The procedure involves extraction and then quantification of concentration of these cations in the extracts using the Inductively Coupled Plasma-Optical Emission Spectrometry (ICP-OES) or Atomic Adsorption or Flame Emission Spectrometry. Though there are older references that recommend colorimetric methods for the determination of cations, these methods are outdated as they do not produce results that can be competitively and confidently presented in science.

The extraction of cations involves the use of ammonium acetate extracting solution at a ratio of 1:10 (w/v) which is usually 5 g to 50 mL of extracting solution. The extracts and samples are shaken in 100–150 mL bottles on a reciprocating shaker for 30 min at 180 rpm and filtered using Whatman No. 2 or equivalent filter papers. The extracting solution is prepared by mixing 68 mL of ammonia solution and 57 mL of glacial acetic acid and made to 900 mL with deionized water. This solution is then adjusted to a pH of 7.0 using either ammonia solution or glacial acetic acid, and then made up to the mark of 1000 mL using deionized water (Sumner and Miller 1996). Similar to all other methods, the calibrating standards are prepared using standard stock solutions and these are made to volume using the extracting solution to ensure that the matrix of the samples and the standards is the same. Following the machine determination of the concentration of the various cations in the extracts, the concentration on mg per L are then converted to mg per kg using the formula given in Eq. (3.6).

3.4.9 Extractable Heavy Metals

Trace metals and heavy metals is a class of micro-nutrients that are required by plants for nutrition in very small quantities and when available at higher concentrations in soil or composts, can be detrimental to crop development. Therefore, the determination of these elements which include, Cd, Cr, Ni, Hg, Mn, Pb, Cu, Zn, Se, Fe, Mo, among others is important. Critical threshold levels that are acceptable in soils, composts, and plants have been established in literature. The most common method of extracting heavy metals in soils is that involving the use of diethylenetriaminepentaacetic acid (DTPA) described by Amacher (1996). This DTPA extracting solution is prepared using DTPA, triethanolamine, and calcium chloride dehydrate, adjusted to a pH of 7.3 using 6 M hydrochloric acid. With compost, the main challenge is the extracting sample weight to solution ratio, which is 1:2, i.e., 10 g to 20 mL of extracting solution, which means that the extracts are very small quantities. It is therefore advisable to use higher quantities of sample and extracting solution to allow for larger quantities of extracts for analysis especially if AAS method of analysis is used.

3.4.10 pH and Electrical Conductivity

The most common chemical property that is measured in soils and composts is pH and electrical conductivity (EC). In soils, pH and EC is usually measured in 1: 2.5 weigh over volume ratio, though this ratio does not work with composts as at that ratio composts absorb all the water leaving no solution to measure the pH and

EC. The most ideal ratio is 1: 10, where 10 g of compost is extracted using 100 mL of water, after shaking for 1 min and letting the mixture stand for 1 h. In these solutions, pH and EC are then determined using potentiometric methods that employ digital measuring equipment.

3.5 Conclusions

This chapter presents insights into methods that are recommended for the analysis of various chemical properties of composts that include pH, EC, total carbon and nitrogen, total elemental composition, exchangeable nitrogen, extractable P, Ca, Mg, Ca and K, and extractable heavy metals. Though clear methods of soil analysis exist, this is not the case for composts, and as a result some researchers have used methods that are not applicable for composts. This chapter has provided guidance on which popular methods can be adopted for composts analysis as well as highlights on challenges that can be associated with soil based methods like the dichromate oxidation methods for carbon in composts; the extraction of heavy metals and the amount of extracts generated using the DTPA method, among others. It is hoped that the information shared herein will encourage more research on compost specific methods of chemical characterization, allowing for the adoption of more optimized and standardized methods, rather than simply using soil analysis based methods.

References

Agri Laboratory Association of Southern Africa (AgriLASA) (2004) Soil handbook. Agri Laboratory Association of Southern Africa, Pretoria

Amacher MC (1996) Nickel, cadmium and lead. In: Sparks DL et al (eds) Methods of soil analysis part 3—chemical analysis. SSSA book series, American Society of Agronomy, Madison, pp 739–768

Anderson JM, Ingram JSI (1993) Tropical soil biology and fertility: a handbook of methods. ACB International, Wallingford

Bernal MP, Alburquerque JA, Moral R (2009) Composting of animal manures and chemical criteria for compost maturity assessment. Rev Bioresour Technol 100:5444–5453

Brady NC, Weil RR (2008) The nature and properties of soil, 14th edn. Prentice-Hall, Upper Saddle River, NJ

Bremer JM (1996) Nitrogen-total. In: Sparks DL et al (eds) Methods of soil analysis part 3—chemical analysis, SSSA book series. American Society of Agronomy, Madison, pp 1085–1121

Joergensen GR, Brookes PC (2005) Quantification of soil microbial biomass by fumigation-extraction. In: Margesin R et al (eds) Manual for soil analysis: monitoring and assessing soil bioremediation. Springer, Berlin, pp 281–296

Jones DL, Willett VB (2006) Experimental evaluation of methods to quantify dissolved organic nitrogen (DON) and dissolved organic carbon (DOC) in soil. Soil Biol Biochem 38:991–999. https://doi.org/10.1016/j.soilbio.2005.08.012

Kuo S (1996) Phosphorus. In: Sparks DL et al (eds) Methods of soil analysis part 3—chemical analysis, SSSA book series. American Society of Agronomy, Madison, pp 869–919

Maynard DG, Kalra YP, Crumbaugh JA (2006) Nitrate and exchangeable ammonium nitrogen. In: Carter MR, Gregorich EG (eds) Soil sampling and methods of analysis, 2nd edn. CRC Press, Boca Raton, FL, pp 71–80

Meena MD, Dotaniya ML, Meena MK, Meena BL, Meena KN, Doutaniya RK, Meena HS, Moharana PC, Rai PK (2021) Maturity indices as an index to evaluate the quality of sulphur enriched municipal solid waste compost using variable byproduct of Sulphur. Waste Manag 126:180–190

Mulvaney RL (1996) Nitrogen—inorganic forms. In: Sparks DL et al (eds) Methods of soil analysis part 3—chemical analysis, SSSA book series. American Society of Agronomy, Madison, pp 1123–1184

Mupambwa HA, Mnkeni PNS (2018) Optimizing the vermicomposting of organic wastes amended with inorganic materials for production of nutrient—rich organic fertilizers. A review. Environ Sci Pollut Res Int 25:10577–10595

Mupambwa HA, Nciizah AD, Nyambo P (2022) Can organic soil fertility management sustain farming and increase food security among African smallholder farmers? In: Mupambwa HA, Nciizah AD, Nyambo P, Muchara B, Gabriel NN (eds) Food security for African smallholder farmers. Sustainability sciences in Asia and Africa. Springer, Singapore. https://doi.org/10.1007/978-981-16-6771-8_6

Nelson DW, Sommers LE (1996) Total carbon, organic carbon and organic matter. In: Sparks DL et al (eds) Methods of soil analysis part 3—chemical analysis, SSSA book series. American Society of Agronomy, Madison, pp 961–1010

Okalebo JR, Gathua KW, Woomer PL (2002) Laboratory methods of soil and plant analysis: a working manual. TSBF-KARI-UNESCO, Nairobi

Raj D, Antil RS (2011) Evaluation of maturity and stability parameters of composts prepared from agro-industrial wastes. Bioresour Technol 102:2868–2873. https://doi.org/10.1016/j.biortech.2010.10.077

Sanchez-Monedero MA, Roig A, Martinez-Pardo C, Cegarra J, Paredes C (1996) A microanalysis method for determining total organic carbon in extracts of humic substances: relationships between total organic carbon and oxidisable carbon. Bioresour Technol 57:291–295. https://doi.org/10.1016/S0960-8524(96)00078-8

Sims TJ (2000) Soil test phosphorus: Bray and Kurtz P-1. In: Pierzynski GM (ed) Methods of phosphorus analysis for soils, sediments, residuals, and waters southern cooperative series bulletin no. # 396

Skjemstad JO, Baldock JA (2006) Total and organic carbon. In: Carter MR et al (eds) Soil sampling and methods of analysis, 2nd edn. Canadian Society of Soil Science, Pinawa, MB, pp 225–238

Sumner ME, Miller WP (1996) Cation exchange capacity and exchange coefficients. In: Sparks DL et al (eds) Methods of soil analysis part 3—chemical analysis, SSSA book series. Soil Science Society of America, Madison, pp 1201–1229

Unuofin FO, Mnkeni PNS (2014) Optimization of Eisenia fetida stocking density for the bioconversion of rock phosphate enriched cow dung–waste paper mixtures. Waste Manage 34:2000–2006

Chapter 4
An Outstanding Perspective on Biological Dynamics in Vermicomposting Matrices

Jerikias Marumure, Zakio Makuvara, Claudious Gufe, Richwell Alufasi, Ngavaite Chigede, and Rangarirayi Karidzagundi

Abstract Vermicomposting is the decomposition of organic waste by earthworms and microorganisms such as bacteria and fungi. When various organic wastes are vermicomposted, a nutrient-rich product that can be used as a plant biofertiliser is produced. However, to optimise vermicomposting, a better understanding of the underlying biological dynamics in the vermicomposting community is required. This chapter seeks to explore the biological dynamics during vermicomposting. Firstly, critical organisms involved in vermicomposting, their roles, and the bio-transformation processes involved are critically examined. The yields of vermicompost by different vermicomposting substrates and earthworm species are summarised. Methods for identifying vermicomposting organismal drivers are highlighted. Bacterial succession during vermicomposting is highlighted, as well as various benefits associated with applications of vermicomposting. Lastly, as part of the biological dynamics in vermicomposting, the effects of bacteria and earthworms on plant growth and health are summarised. Understanding biological dynamics in vermicomposting, as a result, can pave the way for novel techniques

---

J. Marumure (✉) · Z. Makuvara
School of Natural Sciences, Great Zimbabwe University, Masvingo, Zimbabwe
e-mail: jmarumure@gzu.ac.zw; zmakuvara@gzu.ac.zw

C. Gufe
Department of Veterinary Technical Services, Central Veterinary Laboratories, Harare, Zimbabwe

R. Alufasi
Biological Sciences Department, Bindura University of Science Education, Bindura, Zimbabwe
e-mail: ralufasi@buse.ac.zw

N. Chigede
Gary Magadzire School of Agriculture, Great Zimbabwe University, Masvingo, Zimbabwe
e-mail: nchigede@gzu.ac.zw

R. Karidzagundi
Materials Development Unit, Zimbabwe Open University, Harare, Zimbabwe
e-mail: karidzagundir@zou.ac.zw

© The Author(s), under exclusive license to Springer Nature Singapore Pte Ltd. 2023
H. A. Mupambwa et al. (eds.), *Vermicomposting for Sustainable Food Systems in Africa*, Sustainable Agriculture and Food Security,
https://doi.org/10.1007/978-981-19-8080-0_4

that have the potential to improve vermicompost quality, thereby improving plant growth and health.

Keywords Bacterial succession · Biotransformation · Earthworms · Microorganisms · Nutrients · Organic wastes · Vermicomposting

4.1 Introduction

The rapid increase of the world population has exerted unprecedented demand for agricultural produces. Farmers began to use chemical fertilisers to increase crop production in response to the rising food demand. However, the practice is costly and harmful to the environment (Aslam et al. 2019a, b). Notably, recent research has linked significant soil degradation and fertility loss to soil toxicity and nutrient imbalance caused by agrochemicals use (Mazur-p et al. 2021; Aslam et al. 2019a, b). Furthermore, the increase in human population, industrialisation, and agricultural activities has resulted in the production and accumulation of large volumes of waste materials, posing a challenge to waste management (Liyarna et al. 2020).

Moreover, agricultural activities generate organic waste in the form of crop residues, which are a source of plant nutrients but are frequently burned without regard for the negative effects on the soil's physical, chemical, and biological properties (Bajal et al. 2019). The burning of crop residues also contributes to greenhouse gas emissions (nitrous oxide (N_2O) and methane (CH_4)), which cause global warming (Jamali et al. 2021). Such waste management methods result in the loss of large quantities of valuable nutrients contained in the organic wastes. Extraction of organic fertilisers from the waste and its application to the soil will go a long way in preventing the degradation of soil ecosystems and improving plant productivity. As a result, developing low-cost, environmentally friendly technologies capable of converting organic waste into a valuable product is required. Waste biodegradation via vermicomposting has emerged as a promising and long-term solution for both solid waste management and sustainable agriculture (Manaig 2016).

Vermicomposting is a biotechnology that uses the complex symbiotic interaction of earthworms (e.g. *Eisenia fetida*, *Eudrilus eugeniae*, and *Perionyx excavatus*) and microorganisms to convert organic waste into a useful nutrient-rich product, vermicompost (Vukovic et al. 2021; Ansari et al. 2020; Indrani et al. 2019; Domínguez et al. 2019; Phukan et al. 2013; Sim and Wu 2010). Though the exact mechanisms are not fully known, the transformation occurs as soon as organic matter enters the earthworm gut (Adhikary 2012). However, the transformations involve organic matter fragmentation, bio-oxidation, and stabilisation (Domínguez et al. 2010). Although microorganisms are responsible for the biochemical degradation of organic wastes, earthworms drive the process by conditioning the substrate and changing the biological activity. This is accomplished primarily by (1) the earthworm directly feeding on the microorganisms (Gómez-Brandón and Domínguez 2014) and (2) making the vermicomposted material more granular (thus increasing surface area) as a result of the earthworms' digestion and fragmentation (Lim et al. 2014). Even after the organic matter has been ejected from the worm's gut, the enzymes will continue to bioconvert it (Olle 2019). When compared to a biodegradation system that does not include earthworms, these activities are reported to increase the rate of organic matter turnover and productivity of microbial communities, thereby increasing the rate of decomposition (Lim et al. 2015; Blouin et al. 2013). The produced vermicompost improves soil fertility on both a physical and biological level, as well as soil conditioning (Lim et al. 2015). The technology has grown in popularity due to its low cost and environmental sustainability (Mazur-p et al. 2021; Purnawanto et al. 2020).

However, apart from earthworms, vermicomposting involves microorganisms (for example, *Actinomycetes, Azotobacter, Nitrobacter, Nitrosomonas*, and *Aspergillus*) that can undergo succession during the process. Notably, as the vermicompost matures, there is a decrease in Gram-negative cocci bacteria and a corresponding increase in Gram-positive bacilli (Ansari and Hanief 2013). In addition, the composting process occurs under aerobic conditions with temperatures ranging from 20 to 25 °C, a pH of 6.8–8.0, and a substrate humidity of 70–75% (Mazur-p et al. 2021; Amaravathi and Reddy Mallikarjuna 2015). Contrary to conventional composting, vermicomposting results in two useful products, namely the earthworm biomass and the vermicompost (Greco et al. 2021; Coulibaly and Bi 2010). Worm castings in vermicompost contain nutrients that plants can use, and the castings have a mucous coating that allows the nutrients to "time-release". Vermicompost produces fine stable granular organic matter that aids in aeration, while hygroscopic mucus absorbs water, prevents waterlogging, and increases water holding capacity (Ansari and Hanief 2013). Studies have reported plant growth promotion of 50–100% by vermicompost over conventional compost and 30–40% over chemical fertilisers (Manaig 2016). Besides being produced 75% faster than conventional compost, vermicomposts are also cleansed of harmful microorganisms and toxic substances and enriched with beneficial soil microbes, which are critical in promoting plant growth, soil health, and crop yield (Manaig 2016).

In addressing the biological dynamics in vermicomposting matrices, the chapter seeks to clarify the following: (1) the organismal key drivers and substrates for vermicomposting, (2) biotransformation processes in vermicomposting, (3) identification of species diversity during vermicomposting, (4) biological succession during vermicomposting, (5) impacts of macro- and microorganisms in vermicompost on plant health, and (6) other benefits and novel techniques for improving the quality of vermicompost. The knowledge gained from the underlying biological dynamics of vermicomposting could lead to process optimisation, resulting in new technologies focusing on increasing plant productivity and the introduction of other novel vermicomposting applications. Such breakthroughs could be viewed as the most critical efforts to improve our understanding of vermicomposting, resulting in improved food security and the provision of goods and services.

4.2 Organisms and the Substrates in Vermicompost

The vermicomposting process involves the interactions between earthworms and microorganisms (Suthar 2009). Earthworms have a critical role in breaking down and degrading the waste by ingesting the organic materials and mineral soil particles (Purnawanto et al. 2020). In addition, earthworms physically mix and loosen, grind and maintain aerobic conditions of the waste materials (Manaig 2016), thereby speeding up the biodegradation process (Greco et al. 2021). The physical action of earthworms produces vermicasts, which provide a larger surface area for microbial colonisation and biochemical activity (Domínguez et al. 2010). Inside the earthworm gut are various earthworm and microbial origin enzymes, mucus, calcium carbonate, and bacteriostatic and microbicidal substances involved in the substrate's biochemical decomposition (Manaig 2016; Domínguez et al. 2010; Suthar, 2009). Among the microorganisms residing in the earthworm gut are phosphate solubilising bacteria, nitrogen-fixing bacteria, and fungi (Liyarna et al. 2020). Therefore, earthworms are the critical drivers of the process as they are involved in the indirect stimulation of microbial communities within the drilosphere through fragmentation and ingestion of fresh organic matter (Domínguez et al. 2010; Domínguez 2004). To a greater extent, the earthworm communities influence the spatial variability of resources, altering microbial communities' density, diversity, structure, and activity (Suthar 2009). Various researchers have reported the effectiveness of different species of earthworms on vermicomposting and results show that vermicompost quality depends on earthworm species and the substrate (Purnawanto et al. 2020).

Although earthworms are known to degrade organic waste, not all species exhibit the same potential (Bhardwaj and Sharma 2015). Several earthworm species have been evaluated for their potential use in vermicomposting, and out of 3000 species, 8–10 species of earthworms have been considered suitable (Sharma et al. 2021), including *Eisenia fetida, Eisenia andrei, Dendrobaena veneta, Dendrobaena hortensis, Eudrilus eugeni*, and *Perionyx excavatus* (Amouei and Yousefi 2017). The most common species used in vermicomposting and vermiculture worldwide are *Eisenia andrei* and *Eisenia fetida* (Sharma et al. 2021; Mousavi et al. 2020; Bajal et al. 2019). Earthworms are categorised into three major groups based on their ecological strategies, feeding habits, and the part of their soil profile (Domínguez et al. 2018). The three categories are epigeic, anecic, and endogeic earthworms.

Epigeic earthworms live in soil and within the organic horizons near the surface. They primarily consume organic waste and convert it into faecal pellets (Domínguez et al. 2010). Furthermore, they exhibit tolerance to many environmental factors and have short life cycles, high reproductive rates, endurance, and resistance to handling (Bajal et al. 2019). Epigeic species of earthworms are the commonly used earthworms for vermicomposting. Due to their natural ability to feed on organic wastes, high reproduction rate, and short life cycle, epigeic species are considered suitable for vermicomposting. Epigeic lumbricids include the species *Dendrobaena veneta, Dendrobaena hortensis, Dendrobaena octaedra, Eisenia fetida, Eisenia andrei, Dendrodrilus rubidus, Eiseniella tetraedra*, and *Allolobophoridellaeiseni*

(Domínguez et al. 2018). However, among these species, *Eisenia fetida* and *Eisenia andrei* have been used extensively in vermicomposting because of their ubiquitous nature, short life cycles, high reproductive rates, and tolerance to a wide temperature and humidity ranges (Domínguez and Edwards 2011; Bajal et al. 2019).

Anecic earthworms include those species that create vertical burrows/gallaries in the soil, extending for several metres throughout the soil profile (Vukovic et al. 2021; Domínguez 2018; Suthar 2009). The burrows are cemented with mucus and other nitrogen-rich body secretions making the burrow attractive to a decomposer community, especially bacteria associated with nitrogen mineralisation (Suthar 2009). The burrowing earthworms surface during the night feed on leaf litter, faeces, and decomposing organic matter and transport them to their gallaries (Domínguez et al. 2018). They excrete organo-mineral faeces at the opening of their burrows, forming conspicuous earthworm casts (Domínguez et al. 2018; Domínguez et al. 2010). Species *Lampito mauritii* and *Lumbricus terrestris* have been identified as anecic earthworms that can potentially decompose organic waste materials (Suthar 2009; Domínguez 2004).

Although both endogeic and anecic earthworms live in burrows and feed on soil and organics, endogeic earthworms prefer horizontal and highly branched burrows to vertical ones. These burrows are created partly by ingestion of soil particles as they make their way and partly by pushing the soil to the sides (Kale and Karmegam 2010). The ingested materials pass through the digestive system, and castings of undigested matter are released along the burrows and at the openings of the burrow on the surface (Kale and Karmegam 2010). However, endogeic and anecic earthworms have lower reproductive rates and longer life cycles than epigeic earthworms (Domínguez et al. 2018).

Several studies have reported on the efficiency of earthworm species in vermicomposting. It is crucial to highlight that the efficiency of these organisms is dependent on a variety of factors, including a favourable living environment, a food source, proper substrate moisture, sufficient ventilation, and protection from extreme hot or cold temperatures (Manaig 2016). Different organic waste materials have been evaluated as vermicomposting substrates, and different nutrition qualities of vermicompost have been reported. Waste materials already evaluated include mushroom waste (Purnawanto et al. 2020; Song et al. 2014), coffee pulp (Velmourougane 2011), rice straw (*Oryza sativa*) (Manaig 2016), paper waste (Arumugam et al. 2015; Velmourougane 2011; Maleki et al. 2016), animal wastes (cow, sheep, pig, and chicken) (Manaig 2016; Coulibaly and Bi 2010), leaf litter (Sharma et al. 2021; Manaig 2016), sewage sludge of wood and paper industry (Amouei and Yousefi 2017), tanning sludge (Malafaia et al. 2015), beverage industry sludge (Singh et al. 2010), paddy straw (Das et al. 2016), sugarcane waste, sheep manure, paper waste, and sewage sludge (Liyarna et al. 2020), vegetable market waste and floral waste (Pattnaik and Reddy 2010), grass and newspaper wastes and animal dung (Mousavi et al. 2020). These substrates have different characteristics, and the nutritional

Table 4.1 Earthworm species used in vermicomposting and their productivity rates in different substrates over 60 days

Earthworm Species used	Type of Substrate used	Yield—Earthworm multiplication (%)[a] or number of earthworms (per m^2)[b]	Comment(s)	Reference
Eudrilus eugeniae and *Perionyx ceylanesis*	Coffee pulp	280[a], 134.1[a] respectively	*E. eugenia* was reproduced at a higher rate than *P. ceylanesis* in coffee pulp.	Velmourougane (2011)
Eisenia fetida	Rice straw Rice straw + grass Dry Grass Clippings	−25 to 875[b] ⁻25 to 1025[b] ⁻25 to 1150[b]	Multiplication of earthworms was higher in vermicompost from dry grass clippings	Indrani et al. (2019)
Eisenia fetida, Eudrilus eugeniae, L. rubellus	Mushroom waste	338.69[a], 336.90[a], 346.4[a] respectively	All the three earthworm species were able to produce appropriate vermicompost, with *E. fetida* showing the best results	Purnawanto et al. (2020)
Eudrilus eugeniae	Sawdust Rice straw (*Oryza sativa*) G. sepium leaves Cattle manure Chicken manure Hog manure	343.8[a] 216.19[a] 81.73[a] 217.30[a] 129.52[a] 294.29[a]	Substrates containing sawdust and rice straw are more conducive to increasing the number and total weight gain of earthworms, whereas *G. sepium* leaves are less conducive.	Manaig (2016)

[a]yield by percentage
[b]yield by number of earthworms

quality of vermicompost, so produced, as well as reproduction and growth of earthworms, depends on the nature of the substrate (Sharma et al. 2021).

A study by Pattnaik and Reddy (2010) on the nutrient content of vermicompost produced from vegetable market waste and floral waste using *Eudrilus eugeniae, Eisenia fetida,* and *Perionyx excavatus,* showed that vegetable waste produces higher nutrient contents than the floral waste. Furthermore, the quality of food source for the earthworms can also influence their reproduction and growth, hence the amount of substrate that can be converted into vermicompost (Manaig 2016), as shown in Table 4.1. A study by Coulibaly and Bi (2010) reported significant differences in the growth and reproduction of *Eudrilus eugeniae* in four animal wastes (cow, sheep, pig, and chicken waste) under laboratory conditions. Maximum weight gain per worm and the highest growth rate per worm per week were obtained in the reactor with chicken waste. However, vermicomposting conditions play a significant role in the process. Factors such as bedding and substrate moisture content, temperature, and decomposition duration, all impact the activity of earthworms and hence the quality of the vermicompost.

4.3 Procedure for Vermicomposting

Vermicomposting involves culturing earthworms outdoors in beds or confined chambers (pit method) in the presence of agricultural waste materials, which can be a mixture of organic materials (Pilli et al. 2019). The pit method has limitations of aeration and increased chances of water logging at the bottom, among others. The steps followed in vermicompost preparation are as follows:

- Waste materials, e.g. cow dung and chopped dried leafy materials, are mixed in a ratio of 3:1 and are kept for partial decomposition for 15–20 days to reduce excessive heating when worms are added.
- A layer of 15–20 cm of chopped dried leaves/grasses is kept as a bedding material at the bottom of the bed.
- Beds of partially decomposed material of size 6 × 2 × 2 feet should be made.
- Earthworms, a population range of 1500–2000, are released on the upper layer of the bed. Earthworms consume biomass and excrete it in a digested form called worm casts.
- Using a can, water is sprinkled on top of beds just after the release of the worms.
- Beds must be kept moist by spraying water daily and by covering them with polythene material.
- The bed must be turned after 30 days to maintain aeration, good worm growth and for proper decomposition.
- Compost gets ready in 45–50 days. The casts are rich in nutrients, growth-promoting substances, beneficial soil micro flora, and inhibiting pathogenic microbes.

NB. The vermicomposting pit must be in a cool, moist, and shady site.

The waste material is reduced in volume and carbon-nitrogen ratio as they are processed by earthworms and decomposed by enhanced microbial activity within the earthworms and their castings. When the vermicompost is completely decomposed, it appears black and granular (Chanu et al. 2018), unlike compost which is dark brown, smelling like earth and crumbling in hands. Watering must be stopped as the vermicompost gets ready, about five (5) days before the harvesting. The vermicompost should be kept over a heap of partially decomposed cow dung so that earthworms could migrate to cow dung from compost. After 48 h, the compost can be sieved to separate worms from vermicast for use. The pit floor must be compact to prevent earthworms from migrating into the soil. The organic material used should be free from plastics, chemicals, pesticides, and metals.

4.4 Earthworm Enemies that Can Reduce Population

(i) ***Predators***: A few predators like ants, birds, and lizards can feed on the worms. Gunny bags must cover the pit or beds of waste being digested. Increasing moisture content on the bed will make the environment un-conducive for ants.

(ii) ***Sunlight and rain***: A thatched roof can protect the worms from these adverse effects, compromising worm multiplication and growth.

(iii) ***Pest and diseases***: Flies are commonly attracted to the decomposing organic material. This can best be dealt with by choosing bedding material and composting material adequately. Also, maintenance of moisture and temperature in the composting unit helps control pests and diseases (Chanu et al. 2018).

4.5 Macro- and Microbial Biotransformation Processes in Vermicomposting

Vermicomposting is based on biotransformation processes in which earthworms and associated microorganisms are exploited in the transformation of organic materials into two major products (1) earthworm biomass and (2) organic fertiliser (vermicompost) (Lee et al. 2018; Sim and Wu 2010). Biotransformations in vermicomposting processes are achieved through the interaction between the earthworms, which drive vermicompost substrate conditioning and biological activity modulation, and microorganisms critical in the biochemical degradation of vermicompost substrates (Cardoso et al. 2008). Generally, processes linked to biotransformation occur within the gut of earthworms and include breaking and splitting of vermicompost substrates, enzyme synthesis and enrichment, and microbial activity promotion and enrichment (Lee et al. 2018; Cardoso et al. 2008). Furthermore, earthworm gut microflora are central in the bioconversion of all organic vermicompost substrates. Therefore, vermicomposting is a biotechnological technique in which vermicompost substrates are transformed into nutrient-rich organic fertiliser known as vermicompost (Benitez et al. 2000).

In vermicomposting processes, earthworms facilitate biotransformation through mixing, consumption, aeration, grinding, and digesting vermicomposting substrates (Mupambwa and Mnkeni 2018; Singh and Kalamdhad 2016). These physical and biochemical processes allow speedy mineralisation and humification of vermicomposting substrates (organic wastes) into stable, nutrient-rich organic fertiliser. However, for the transformation of vermicompost substrates into stable vermicompost, chelating and phytohormonal substances and enzymes are released by worms (Bhat et al. 2017). After consuming vermicompost substrates (organic matter), the organic matter is deposited in the earthworm's gizzard where it is grounded, thereby increasing the surface area for microorganisms. Intestinal substances (antibiotics and mucus) and enzymes (including amylase, protease, urease, chitinase, cellulase, and lipase) further transform grounded matter into vermicasts

(Sulaiman and Mohamad 2020). Biotransformations of vermicompost substrates occur in the earthworm's gut which acts as a bioreactor, and the biotransformation processes involve bioconversion of organic matter and mineralisation of diverse substrate elements. Even after the organic matter has been ejected from the worm's gut, the enzymes will continue to bioconvert it (Olle 2019), resulting in a nutrient-rich product. Notably, the biotransformation of organic matter during vermicomposting occurs during the (1) active phase (earthworms modify vermicompost substrates and associated microbes) and (2) maturation phase (earthworms focus on a new layer of vermicompost substrate and microorganisms take over biotransformation of substrates) (Aira et al. 2007).

Earthworms secrete a consortium of essential enzymes, especially in two regions: (1) the gizzard and (2) the intestine. These enzymes are mainly responsible for the rapid biodegradation of cellulosic vermicompost substrates and protein-containing organic substances (Hand et al. 1988). The most prominent enzymes are cellulases that depolymerise cellulose to glucose, amidohydrolase, proteases and urease, known for nitrogen mineralisation, b-glucosidases and phosphatases are critical in the hydrolysis of glucosides and removal of phosphate groups from a diverse of vermicompost substrates, respectively (Piotrowska-Długosz 2020). Generally, the diverse enzymes secreted in the gut of earthworms have been associated with an enhanced concentration of N, P, and K in vermicomposts, and this occurs through mineralisation and mobilisation (Gusain and Suthar 2020; Edwards and Lofty 1977). Notably, during the biotransformation, the mineralisation of substrate nitrogen and phosphorus results in the formation of nitrites or nitrates and orthophosphates, respectively (Suthar 2009). Earthworms' intestines usually secrete enzymes, including chitinases, proteases, lipases, cellulases, and amylases, which are central in the digestion of vermicompost substrates into nutrients absorbed by the earthworms. However, undigested organic substances from earthworms' guts form vermicastings (Ali et al. 2015).

Due to the action of earthworms (specifically their gut enzymes) and associated microorganisms, the concentrations of phosphorus, nitrogen, and potassium increase (Ndegwa and Thompson 2001). Earthworm gut microflora typically produces acids (including carbonic and sulphuric acids) which solubilise the insoluble mineral nutrients such as potassium. This way, the nutrient content of vermicomposts significantly increases, as indicated by a high content of nitrates, soluble potassium, calcium, and phosphorus (Lim et al. 2015). Biotransformations in vermicomposts include nitrification and denitrification processes facilitated by aerobic and anaerobic microorganisms associated with earthworms (Lv et al. 2020). For example, Thakur and Medhi (2019) discovered facultative anaerobic bacteria that convert nitrates to nitrogen and nitrous oxide, and Sun et al. (2020) found anaerobic denitrifying microorganisms in the gut of earthworms. Biodegradation of vermicompost substrates or organic matter has resulted in pH modulating intermediates, including ammonium and humic acids (Pramanik et al. 2007).

High and speedy mineralisation and microbial decomposition of organic matter have been linked to reduced carbon concentration in vermicomposts, and this is due to the loss of carbon in the form of carbon dioxide during microbial respiration.

However, at the same time, nitrogen is increased during mineralisation, secretion of mucus and other nitrogenous metabolic wastes by earthworms (Sarma et al. 2018). The increase in nitrogen and the decrease in carbon have resulted in a reduced C: N ratio in vermicomposts (Biruntha et al. 2020; Gajalakshmi et al. 2001; Atiyeh et al. 2000). Vermicomposting processes improve the biodegradation of chitin, and it is biotransformed to N-acetyl-D-glucosamine in a chitinolytic system in the presence of hydrolases, chitinase, and N-acetyl b-glucosaminidase (Chattopadhyay 2014; Das et al. 2012). The production of methane gas follows this chitin biotransformation. A study using corn stalks as organic matter reported methane production during vermicomposting processes (Chen et al. 2010). During vermicomposting, metals, including the toxic ones, are taken up by the earthworm skin and intestines, broken down to form complex and nontoxic substances, and removed from the vermicompost substrates (He et al. 2016). The reduction in heavy metals during vermicomposting has been reported in several studies (Wang et al. 2017; Lv et al. 2016; Goswami et al. 2014; Singh and Kalamdhad 2013). Previous studies have indicated that earthworms are involved in the bioaccumulation of heavy metals, which is associated with metallothionein production. The synthesised metallothioneins have been useful in sequestrating heavy metals such as Zn and Mn from vermicomposting substrates (Lv et al. 2016). According to research by Liu et al. (2012) and Lv et al. (2016), metal contaminants (such as Cd, Zn, As, and Cu) were found to decrease during the process while increasing in the tissues of earthworms such as *Eisenia fetida*. The ability of bacteria to precipitate and change the oxidation states of heavy metals during vermicomposting has also been reported. For instance, *Bacillus* sp. and *Arthrobacter* sp. have been shown to reduce Cr (VI) to Cr (III) via bacterial based enzymatic biotransformation (Gutiérrez-Corona et al. 2016). Vermicomposting processes result in vermicompost substrates/organic matter transforming into essential by-products, including humic compounds and soluble salts (Hanc et al. 2019). These humic substances and salts are critical in vermicomposting processes since they significantly improve heavy metal removal from severely contaminated soils (García et al. 2013).

4.6 Methods of Identifying Species Diversity during Vermicomposting

The microbiota in vermicompost is diverse, with most bacteria and only a few fungi. Figure 4.1 shows methods for identifying bacterial and fungal species in vermicompost. The procedures for identifying bacteria differ from those for identifying fungi. Because bacteria and fungi constitute most of the microbiota in vermicompost, this section summarises various methods for identifying the various bacteria and fungi in vermicompost. As observed by studying agar colonies, bacteria and fungi in vermicompost can be identified using macroscopic features such as a microorganism's appearance, size, colour, and smell. The gross morphological/

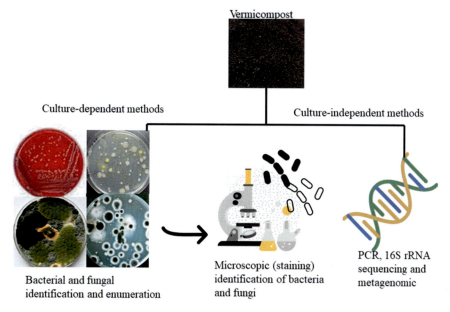

Fig. 4.1 Methods for identifying the diversity of bacterial and fungal species in vermicompost

macroscopic properties of an agar culture can be used to identify the type of microbe (Houpikian and Raoult 2002). Filamentous fungi and moulds, for example, manifest as "hairy" irregularly shaped colonies with visible spores that can be powdery or dusty (Houpikian and Raoult 2002). Fungal colonies can have a variety of colours, with a darker centre (typically raised) and a lighter colour spreading out from it. Filamentous fungi grow radially from the centre of an agar plate, with the youngest on the outside and the older, darker (spore-rich) material on the inside. Filamentous fungi can also produce a single-coloured fuzzy mat with no spores visible. Bacteria typically form distinct and tiny colonies, whereas fungi colonies range in texture from slimy to extremely dry (Houpikian and Raoult 2002).

Bacteria have a distinct odour, whereas filamentous fungi may be odourless or have an earthy odour. Because yeast colonies can resemble bacterial colonies depending on the species and type of agar used, yeast can be the most challenging microorganism to identify using macroscopic features. Depending on the cultural media, the same species may appear differently. Various bacteria and fungi have been discovered using various staining procedures that allow them to be viewed under a microscope. Gram staining is commonly used to identify bacteria and classify them as Gram-positive or Gram-negative. Gram-positive bacteria include *Enterococcus, Streptococcus, Staphylococcus,* and *Clostridium* spp., while Gram-negative bacteria include *Escherichia, Proteus,* and *Shigella* spp. (Houpikian and Raoult 2002). Another staining method is endospore staining, which involves adding a dye to a bacterial sample to check for the presence of spores. Other spore stains are available, but malachite green is probably the most common. There are several types of fungal stains, but they are frequently non-specific. Instead of distinguishing between fungal species, they aid in visualising fungal components for identification. Lactophenol cotton blue, periodic-acid Schiff stain, Grocott's methenamine silver stain, trypan blue, aniline blue, and calcofluor white are all fungal stains (Houpikian and Raoult 2002).

A variety of biochemical tests, such as catalase, oxidase, and sugar fermentation, can be used to detect microorganisms in vermicompost. Bacterial and fungal enumeration assays such as total viable bacteria count and total yeast/mould count can be used to determine species diversity in vermicompost. While culture-dependent methods for identifying microorganisms are still widely used, they have significant limitations, such as the fact that they only apply to in-vitro species and that certain strains have unique biochemical features that do not fit the pattern of any known genus or species (Houpikian and Raoult 2002). Furthermore, culture-based techniques necessitate a substantial investment of time and effort.

Fortunately, many novel methods for detecting microbes are culture-independent, and they frequently reveal minute differences between species that would otherwise go undetected. Molecular identification of bacteria and fungi is becoming more popular in the soil and vermicompost microbiota. The most common molecular methods for identifying bacteria are conventional and real-time PCR (Houpikian and Raoult 2002). Using PCR, one can detect and identify microbial species directly from vermicompost, which speeds up diagnostic procedures (Aira et al. 2019). PCR amplicons are sequenced to identify bacterial/fungal samples using universal PCR primers (Houpikian and Raoult 2002). The 16S rRNA gene is the gold standard for bacterial identification by PCR, whereas the Internal Transcribed Spacer (ITS) region is the critical barcode identifier for fungal species (Houpikian and Raoult 2002). Some genetic fingerprinting methods, such as ARDRA, RFLP, DGGE, and T-RFLP, as well as omics methods, such as metagenomics, meta-transcriptomics, and metabolomics, are critical for detecting and reflecting the entire microbiome and its interactions with the environment (Houpikian and Raoult 2002; Kozich et al. 2013).

Many bioinformatics tools are used to screen the sequence, detect chimaeras, delete amplicon sequence variations (ASVs), identify operational taxonomic units, taxonomic assignment, and identify the core microbiome (Langille et al. 2013; Price et al. 2010). Various bioinformatics softwares are used to quantify taxonomic α-diversity, species diversity (Shannon), species richness (Chao1), phylogenetic diversity (Faith's phylogenetic diversity), and taxonomic β-diversity (principal coordinate analysis—PCoA). Other approaches, such as microarray-based identification, are employed to identify vermicompost bacteria. Microarray-based identification is based on the hybridisation of pre-amplified bacterial DNA sequences to array species-specific oligonucleotide markers. Each probe contains a distinct dye that fluoresces when hybridised.

Earthworm identification is chiefly based on physical traits such as prostomium configuration, orientation, segment number and form of clitellum, spermatheca, and setae configuration (Domínguez et al. 2005). Unfortunately, taxonomic categorisation based on these physical traits is challenging in most species and

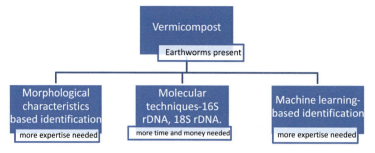

Fig. 4.2 Methods for identifying the diversity of earthworms in vermicompost

needs a high level of knowledge since earthworms have a compact and basic body plan that lacks anatomically sophisticated and highly specialised features (Pop et al. 2003; Pérez-Losada et al. 2009; Boyer and Wratten 2010; Pop et al. 2007). Current molecular-based approaches, such as 18S rDNA, 16S rDNA, and mitochondrial cytochrome oxidase subunit I (COI) sequences, have been employed effectively as an alternate method for earthworm identification (Pop et al. 2003; Boyer and Wratten 2010; Pop et al. 2007). However, these methods necessitate an extensive library of earthworm DNA sequences and require a significant amount of time and money. Recently, computer techniques have been utilised to accurately detect and classify several earthworm species. A machine learning-based earthworm species identification model based on digital photographs of earthworms has been used effectively to categorise distinct animal species in digital photos (Wäldchen and Mäder 2018; Tabak et al. 2018). Figure 4.2 depicts a basic overview of the methods used to identify earthworms in vermicompost.

Following production, the nutritional value of vermicompost may be assessed using standardised chemical analysis procedures similar to those used to assess the nutritional value of soils. The Walkley-Black technique can calculate the total organic matter and organic carbon (%) (Walkey and Black 1934). The Kjeldahl technique may calculate total nitrogen (%), while the phenoldisulphonic acid colourimetric technique can calculate nitrate percentage (%). Bray's approach (Bray and Kurtz 1945) employs the colourimetric method using a spectrophotometer to quantify total phosphorus (%). Total manganese (ppm), total zinc (ppm), total copper (ppm), total potassium (%), and total iron (%) may be determined using the absorption technique utilising the Atomic Absorption Spectrophotometer (AOAC 1995). Ramnarain et al. (2018) found that cow dung and dry grass clippings produced vermicompost with the chemical parameters listed in Table 4.2. Aladesida et al. (2014) also examined the chemical parameters of three varieties of earthworm castings, which are given in Table 4.2.

Table 4.2 Chemical properties of vermicompost produced from cow manure and dry grass clippings (Ramnarain et al. 2018) and three types of earthworm casts (Aladesida et al. 2014)

Parameter	Vermicompost	Earthworms cast (pellet)	Earthworms cast (turret)	Earthworms cast (mass)
Total organic carbon (%)	18.53	0.62	0.42	0.36
Total nitrogen (%)	1.36	0.063	0.048	0.054
C/N ratio	3:1			
Total phosphorous (%)	0.58	11.27	9.50	6.33
Total potassium (%)	0.56	0.62	0.93	0.07
Total manganese (ppm)	544	113	77	19
Total coper (ppm)	26.90	0.45	0.50	0.63
Total zinc (ppm)	611	8.54	6.68	2.04
Total iron (%)	1.56			

4.7 Biological Succession during Vermicomposting

One of the fundamental microbial processes that must be understood as part of biological vermicomposting matrices to optimise the process is bacterial succession during vermicomposting. Several studies have found that vermicomposting increases bacterial abundance and diversity (Rosado et al. 2021; Srivastava et al. 2021; Gómez Brandón et al. 2019). Bacterial succession can be influenced by the type of vermicomposting substrates used and the duration of the vermicomposting process. For example, the vermicomposting of silver wattle during 56 days using the earthworm species *Eisenia andrei* demonstrated the following: (1) significant differences in α- and β-diversity observed, (2) bacterial community more stable at the end of the vermicomposting process, (3) higher number of taxa experienced significant changes in relative abundance starting from the middle of the process, and (4) functional profiles of genes involved in cellulose metabolism, nitrification, and salicylic acid also predicted to change significantly during the vermicomposting process (Rosado et al. 2021). After 14 days of vermicomposting grape marc, significant changes in the bacterial community and bacterial diversity, taxonomic and phylogenetic, were observed in a related study. Furthermore, after 91 days of vermicomposting, twelve bacterial taxa capable of nitrogen fixation and synthesis of plant hormones (salicylic acid) and conferring plant disease suppression were discovered (Gómez Brandón et al. 2019).

The abundance of the dominant phyla can also vary according to earthworm density, biological components of the vermicomposting substrate and where vermicomposting is carried out. Chen et al. (2018) demonstrated the increase in lignocellulolytic microbes during the vermicomposting of medicinal herbal residues, which are known to contain much cellulose. Similarly, the abundance of glycoside hydrolase family 6 (GH_6) cellulase-producing microbial communities during vermicomposting of maise stover was reported by Chen et al. (2022). *Cellulomonas* and *Cellulosimicrobium* are dominant genera harbouring the GH_6 gene and are responsible for cellulose degradation, which is essential for stabilising lignocellulose (Chen et al. 2022). Another study found that composting food waste in a reusable amendment bamboo sphere increased microbial community succession. The bamboo sphere increased bacterial-fungal diversity and microbial community composition by increasing the relative abundance of thermo-tolerant and lignocellulolytic bacteria resulting in a high degree of polymerisation and humification (Wu et al. 2022).

Essentially, the increase in taxonomic and functional diversity of the bacterial community is the foundation of biotransformation during vermicomposting. More importantly, bacterial succession during the vermicomposting process is linked to microbial functions, explaining why vermicompost is beneficial to soil and plants (Domínguez et al. 2019). The vermicompost produced has higher microorganism content, and the activities of microorganisms (*Actinomycetes, Azotobacter, Nitrobacter, Nitrosomonas,* and *Aspergillus*) are responsible for much higher plant productivity than the simple conversion of mineral nutrients into more plant-available forms. During this time, there is also a decrease in Gram-negative cocci bacteria and an increase in Gram-positive bacilli as the vermicompost matures (Ansari and Hanief 2013). The resulting product is beneficial to plant growth and health.

4.8 Impacts of Macro- and Microorganisms in Vermicompost on Plant Health

Earthworms, as previously stated, play a part in a larger microbiota by fragmenting and eating fresh organics, whereas microorganisms produce enzymes that cause the biochemical degradation of organic matter (Vukovic et al. 2021). Furthermore, earthworms interact with other species in the soil and can impact various microflora and microfauna populations (Chaulagain et al. 2017). The activities of these populations make vermicomposts improve the soil's physicochemical properties, which can ultimately have several beneficial effects on plants, including promoting plant growth and health (Lazcano and Domínguez 2014).

4.9 Beneficial Effects of Earthworms and Microorganisms in Vermicomposting

Earthworms and microorganisms involved in vermicomposting have stable and synergistic connections (Vukovic et al. 2021). The amount of available space for microorganisms to live and their properties are all influenced by earthworm physical

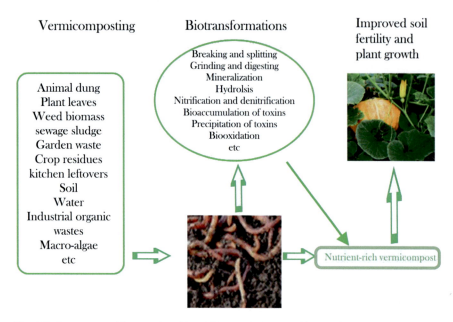

Fig. 4.3 A summary of the organismal key drivers and substrates, biotransformation processes, and the importance of vermicomposting

actions such as substrate aeration, mixing, and grinding. Vermicomposting creates a diverse microbial community with soil fertility-improving properties. This increases the productivity of plants grown with vermicompost (Pilli et al. 2019). In general, bacteria and their actions impact soil characteristics and other soil organisms and contribute to the natural nutrient cycling of carbon, nitrogen, and phosphorus. Genus *Devosia* (Proteobacteria) has been found in cow dung and sawdust vermicompost (Vukovic et al. 2021). This genus has been linked to (1) increased soil fertility, (2) the formation of physicochemical properties that favour the development of aerobic bacteria, and (3) changes in soil microbial communities (Vukovic et al. 2021). An overview of organismal main drivers and substrates, biotransformation processes, and the significance of vermicomposting is shown in Fig. 4.3.

Earthworm numbers, productivity, and dispersion are influenced by the soil habitats in which they live. Because soil texture influences other soil variables such as moisture, nutrients, pH, and temperature, it impacts earthworm populations (Edwards and Bohlen 1996). Light and medium loam soils contain more earthworms than heavy clay, sandy, and alluvial soils (Edwards and Bohlen 1996). Hendrix et al. (1992) discovered a link between soil silt concentration and earthworm richness. In addition, Baker et al. (1992) discovered a modest positive association between soil clay contents and the abundance of *A. trapezoids*, *A. rosea*, and *A. caliginosa*. *A. caliginosa* had the most remarkable positive connection with clay content among these species.

4.10 Effect of Vermicompost on Plant Growth and Yield

Vermicompost has various effects on plants, most of which are positive (Donohoe 2018). Vermicompost promotes plant growth both directly and indirectly by increasing the populations of plant-friendly microorganisms and decreasing soil-borne illnesses (Chaulagain et al. 2017). Vermicompost has several advantages, including increased plant growth and health (Rekha et al. 2018). All these advantages lead to higher plant quality and output.

The inclusion of humic acid (HAs) and several micro-and macro-nutrients in the vermicompost-enriched substrate improves plant growth and development. HAs present in worm casts, even with a modest amount, provide binding sites for nutrients such as phosphorus, potassium, sulphur, iron, and calcium, releasing these elements to plants and stimulating growth (Chaulagain et al. 2017). During vermicomposting, nutrients are transformed into more plant-available forms, resulting in a higher uptake rate by plants (Lazcano and Domínguez 2014). Nitrogen fixers and solubilising phosphate bacteria increase the availability of macro-nutrients such as nitrogen and phosphorus to plants, respectively. Phosphate solubilising bacteria are essential for making phosphorus available for plant uptake. Unlike manufactured nitrogen sources such as those from the Heber-Bosch process, vermicompost supports microorganisms (Donohoe 2018).

The nutrient concentration in vermicompost is significantly higher than in farm manure, resulting in an observed increase in garlic (Allium sativum) growth and output (Donohoe 2018). Vermicompost from sugar mill wastes resulted in a pH drop in various soils (Vukovic et al. 2021). Reduced pH between 6 and 7 promotes nutrient availability and uptake by plants, resulting in improved growth of beans (*Phaseolus vulgaris*). Changes in photosynthetic parameters are strongly related to several factors that reflect plant growth and development. A recent study documented a rise in chlorophyll a, chlorophyll b, and carotenoids with an increase in vermicompost concentration (3%, 6%, 9%, 12%, and 15%) (Vukovic et al. 2021). As a result, photosynthesis rates can increase, boosting growth and productivity (Vukovic et al. 2021).

4.11 Plant Disease Suppressing Potential

Apart from influencing soil fertility and plant growth, vermicomposts are also known for their plant disease suppressing potential. Beneficial bacteria and fungi (Proteobacteria, Bacteroidetes, Verrucomicrobia, Actinomycetes, Aspergillus, Trichoderma, and Firmicutes) are abundant in vermicompost and have antagonistic effects on numerous plant infections such as Fusarium species safeguarding human health. The use of vermicompost and vermicompost extract reduced the severity of *Phytophthora* spp. infections in plants (Chaulagain et al. 2017). The induction of phenolic acids and antifungal activity was linked to aqueous vermicompost extract to manage powdery pea mildew (*Erysiphe cichoracearum*) (Rekha et al. 2018). In addition, disease suppression is primarily due to the secretion of coelomic fluids by earthworms, which kill parasites in the waste. In vitro studies have shown that earthworm coelomic fluid extract has antifungal action (Vukovic et al. 2021). Furthermore, after vermicomposting, an increased salicylic acid and streptomycin production has been noted. Salicylic acid can stimulate plant pathogen resistance mechanisms, and streptomycin has reduced pathogenic bacteria in crop production (Rekha et al. 2018). Vermicompost reduces sporulation, harmful fungal growth, and overall infection in fungal illnesses (Vukovic et al. 2021).

In a related study, vermicompost demonstrated the capacity to protect plants from arthropod and nematode pests by suppressing, killing, repelling, or generating natural resistance in plants to combat them. For example, in tomato, pepper, and cabbage experiments, 20% and 40% vermicompost supplementation significantly reduced arthropods (aphids, buds, mealybug, and spider mite) numbers, as well as a reduction in plant damage (Chaulagain et al. 2017). The use of vermicompost has also been found to control the diversity of soil nematode communities. When compared to the soil from inorganic fertiliser treated plots, soil from vermicompost treated plots had lower plant-parasitic nematodes and increased fungivorous and bacterivorous nematodes (Chaulagain et al. 2017). Vermicompost is the most effective treatment for *Xanthomonas campestris* associated with tomato bacterial spot disease (Chaulagain et al. 2017). Finally, vermicompost is a wonder plant growth enhancer because it contains an enormous population of N_2 fixing bacteria and actinomycetes. Vermicompost dispenses nutrients slowly, making it a slow-release fertiliser that assures a steady supply of nutrients to growing plants. It also binds nutrients due to HAs, thus crucial in dealing with leaching challenges (Chanu et al. 2018).

Several recent studies have identified vermicompost as an essential method in plant disease control (Gudeta et al. 2021; Yatoo et al. 2021; Sulaiman and Mohamad 2020). Vermicompost, with its high nutrient content and plant growth promoters such as auxins, gibberellins, cytokinins, and beneficial microbes, not only improves crop growth and yield but also increases the diversity and activity of antagonistic microbes and nematodes, which aids in the suppression of pests and diseases caused by soil-borne phytopathogens (Yatoo et al. 2021; Machfudz et al. 2020; Reddy et al. 2012; Simsek-Ersahin 2011). Vermicompost tea has also been shown to protect plants from diseases. When applied to plants, it can coat leaf surfaces, reducing the number of available sites for pathogen infection or increasing microbial diversity, killing harmful pathogens (Yatoo et al. 2021).

4.12 Removal of Pollutants and Antimicrobials or Antibiotic Resistance Genes by Vermicomposting

Apart from converting organic waste into a useful nutrient-rich product that is safe to use in soil for crop production (Ibrahim et al. 2016; Phukan et al. 2013; Orozco et al. 1996), several studies have linked vermicomposting to other benefits or applications (Zhao et al. 2022; Gudeta et al. 2021; Yatoo et al. 2021; Shi et al. 2020; Wang et al. 2015). To begin, using earthworms as natural bioreactors for the decomposing organic matter could be used for vermiremediation of specific pollutants such as heavy metals and organic pollutants found in various organic matter sources (Rorat et al. 2017; Olaniran et al. 2013). Although numerous methods exist for removing pollutants from soil and water, less hostile and ecologically friendly methods are required. Vermicomposting has the potential to be an environmentally friendly remediation method (Shi et al. 2020). Vermiremediation is an excellent technique for removing agrochemical contamination from soil. Earthworms have been shown in studies to be capable of dispersing toxins and lowering pesticide concentrations overall (Usmani et al. 2020; Bhat et al. 2018; Suthar 2009). Earthworms are tolerant and can assist in the removal of a wide range of organic and inorganic agrochemicals. Organic contaminants are removed through (1) biodegradation, (2) biotransformation, and (3) physical actions that enhance the natural biodegradation of harmful agrochemicals, converting them into less toxic and valuable stabilised resources (Lacalle et al. 2020; Usmani et al. 2020; Zeb et al. 2020).

Earthworms such as *Eisenia fetida*, *Eisenia tetraedra*, *Lumbricus terrestris*, *Lumbricus rubellus*, and *Allolobophora chlorotica* can remove heavy metals (Cd, Pb, Cu, Hg), pesticides, and lipophilic organic micropollutants such as polycyclic aromatic hydrocarbons (PAH) from soil (Sinha et al. 2008). This is accomplished by absorbing dissolved chemicals via their moist body walls and ingesting them via their mouths as the soil passes through the gut (Ma et al. 2002). Toxic chemicals and heavy metals can bioaccumulate in earthworm tissues without affecting their physiology. Earthworm surface skin tissues secrete an extracellular polymeric substance (EPS) in response to metal stress, responsible for binding with heavy metals, reducing mobility and arresting them on its skin (Alshehrei and Ameen 2021). Earthworm chloragogen cells appear to be primarily responsible for the accumulation of heavy metals absorbed through the gut and their immobilisation in the cells' small spheroidal chloragosomes and debris vesicles (Rajiv et al. 2010).

Vermicomposting has also been linked to reducing antimicrobials and antibiotic resistance genes (ARGs) in some organic matter. Veterinary antimicrobials are poorly absorbed by livestock and are primarily excreted in faeces, where they are then dispersed to the soil when manure is used as fertiliser, increasing the environmental spread of ARGs (Alcock et al. 1999). Before applying manure as a fertiliser to the soil, waste treatment and disposal technologies to reduce antimicrobials and the associated ARGs must be developed. Antimicrobials such as (1) tetracyclines, (2) sulphonamides, (3) quinolones, and (4) macrolides are common in livestock manure, according to research (Cui et al. 2022; Zhao et al. 2022; Wang et al. 2015).

Vermicompost contained many ARGs such as *tet* genes, *sul* genes, *qnr* genes, and *erm* genes (Zhao et al. 2022; Kui et al. 2020). Recent research has shown that applying livestock manure vermicomposting via housefly larvae (*Musca domestica*) and scarab larvae (*Protaetia brevitarsis*) reduces the risk of spreading manure-borne ARGs through the soil-plant system, providing an alternative method for reducing ARGs in organic waste (Zhao et al. 2022; Wang et al. 2015).

4.13 Novel Techniques for Improving Quality of Vermicompost

Although vermicomposting is well-known for its use in the production of nutrient-rich biofertiliser, there are numerous potential techniques for improving its quality. For example, (1) use of selected fungal strains, (2) use of vessels in cold climates, (3) use of slurry method charged with biofertiliser and rock phosphate, and (4) vermicompost of food waste enriched with biochar, cow dung and mangrove fungi, are reported (Ameen and Al-Homaidan 2022; Kumari et al. 2022; Rabari et al. 2020; Maji et al. 2015).

4.13.1 *Use of Fungal Strains*

Enriching the vermicompost waste with selected efficient lignin-degrading fungal strain(s) has the potential to produce qualitatively superior vermicompost with high humic acid (HA) content in a shorter period (Maji et al. 2015). Appropriate fungal strains may improve the efficiency of vermicomposting processes, resulting in higher HA content vermicompost. Humic acid, a component of soil organic matter, is an essential parameter for determining vermicompost quality. It is a complex polymeric organic acid mixture with many molecular weights produced by decomposed plant material, particularly lignin. HA influences the soil's chemical, physical, and biological properties by chelating (ion-exchange and metal-complexing), thus increasing plant bioavailability (Pettit 2006). Fungal strains with celluloxylanolytic and ligninolytic properties are used to accelerate the degradation of the agro-waste substrate, resulting in a higher HA content in the vermicompost. According to research, fungi are effectively involved in the humification and lignin degradation of compost and vermicompost waste (Lopez et al. 2006) and could be used as inocula in a pretreatment process before composting to reduce the substrate's resistance to biodegradation.

4.13.2 Use of Vessels in Cold Climates

Composting suffers greatly in cold climates, resulting in a low-quality product. This is due to the prevailing low temperature, limited microbial activities, and low moisture content. To address this issue, composting technologies that use in-vessel composting instead of open composting systems to prevent energy loss are encouraged. When producing compost for agricultural purposes in cold climates, other factors include substrate pretreatment, insulation techniques, additives, and carrier materials (Kumari et al. 2022).

4.13.3 Use of Slurry Method Charged with Biofertiliser and Rock Phosphate

Enrichment of vermicompost with a consortium of beneficial microorganisms, including solubilising phosphate bacteria and phosphate rock, has been shown in studies to be a viable alternative for supplying soils with phosphatic biofertiliser (PROM) while reducing the use of phosphate fertilisers (Rabari et al. 2020; Phukan et al. 2013; Narayanan 2012). The phosphorus content of phosphatic biofertiliser (PROM) can be as high as 16.5% (as soluble P_2O_5), which is directly assimilable by plants (Narayanan 2012).

4.13.4 Vermicompost of Food Waste Enriched with Biochar, Cow Dung, and Mangrove Fungi

According to a recent study, adding biochar and mangrove fungi improves the efficiency of vermicomposting of polluted food waste (Ameen and Al-Homaidan 2022). The use of mangrove fungi as bio-catalytic actors during vermicomposting proved beneficial in terms of (1) final compost quality (available N, P, and K), (2) composting period shortening, and (3) reduction of heavy metal (Cd, Ni, Pb, Zn, Cu, and Cr) concentrations with the final product being suitable for agricultural use (Ameen and Al-Homaidan 2022).

4.14 Conclusions and Way Forward

The chapter sought to critically examine a unique perspective on biological dynamics in vermicomposting matrices. The primary organismal drivers of vermicomposting have been identified as earthworms and microbes (mostly bacteria). The bioconversion of organic wastes into nutrient-rich biofertiliser is the consequence of the synergistic interaction of earthworms and microorganisms. Organic waste biodegradation occurs in the earthworm's gut through the action of numerous enzymes of the earthworm and microbial origin, mucus, calcium carbonate, bacteriostatic and microbicidal chemicals, albeit the specific paths and processes are unknown. The enzymes continue to biodegrade the organic waste from the worm's intestines even after being ejected. A favourable living environment, a food source, proper substrate moisture, adequate aeration, and protection from adverse effects of temperature are all elements that influence the quality of vermicompost produced and the efficacy of earthworm species in vermicomposting. During vermicomposting, microbial succession occurs, which improves the quality of the vermicompost produced. The primary advantageous qualities of vermicomposting are increased soil fertility, plant development, and disease suppression. Furthermore, vermicomposting has been proven to remove specific pollutants from diverse organic matter sources, including heavy metals, organic pollutants, antibiotics, ARGs, and ARBs. Understanding biological processes in a vermicomposting can, in turn, open the door for new strategies that could increase vermicompost quality. This includes using unique fungal strains, vessels in cold locations, and enriching organic waste with additional enhancers like biochar, cow manure, and mangrove fungi.

References

Adhikary S (2012) Vermicompost, the story of organic gold: a review. Agric Sci 3(7):905–917. https://doi.org/10.4236/as.2012.37110

Aira M, Monroy F, Domínguez J (2007) *Eisenia fetida* (Oligochaeta: Lumbricidae) modifes the structure and physiological capabilities of microbial communities improving carbon mineralization during vermicomposting of pig manure. Microb Ecol 54:662–671

Aira M, Pérez-Losada M, Domínguez J (2019) Microbiome dynamics during cast ageing in the earthworm Aporrectodeacaliginosa. Appl Soil Ecol 139:56–63

Aladesida AA, Dedeke GA, Ademolu K, Museliu F (2014) Nutrient analysis of three earthworm cast-types collected from Ikenne, Ogun State, Nigeria. J Nat Sci Eng Technol 13:36–43

Alcock RE, Sweetman A, Jones KC (1999) Assessment of organic contaminant fate in wastewater treatment plants I: selected compounds and physicochemical properties. Chemosphere 38(10): 2247–2262

Ali U, Sajid N, Khalid A, Riaz L, Rabbani MM, Syed JH, Malik RN (2015) A review on vermicomposting of organic wastes. Environ Prog Sustain Energy 34(4):1050–1062

Alshehrei F, Ameen F (2021) Vermicomposting: a management tool to mitigate solid waste. Saudi J Biol Sci

Amaravathi G, Reddy Mallikarjuna R (2015) Environmental factors affecting vermicomposting of municipal solid waste. Int J Pharm Biol Sci 5(3):81–93

Ameen F, Al-Homaidan AA (2022) Improving the efficiency of vermicomposting of polluted organic food wastes by adding biochar and mangrove fungi. Chemosphere 286:131945

Amouei AI, Yousefi Z (2017) Comparison of vermicompost characteristics produced from sewage sludge of wood and paper industry and household solid wastes:1–6. https://doi.org/10.1186/s40201-017-0269-z

Ansari A, & Hanief A (2013) *Microbial succession during vermicomposting* (No. 537-2016-38589)

Ansari AA, Ori L, &Ramnarain YI (2020) An effective organic waste recycling through vermicompost technology for soil health restoration. In Soil Health Restoration and Management (83–112). Springer, Singapore

AOAC (1995). Association of Official Analytical Chemists. Official Methos of Analysis 20th edn. USA

Arumugam K, Ganesan S, Muthunarayanan V, Vivek S, Sugumar S, Munusamy V (2015) Potentiality of Eisenia fetida to degrade disposable paper cups—an eco-friendly solution to solid waste pollution. Environ Sci Pollut Res 22(4):2868–2876

Aslam Z, Bashir S, Hassan W, Bellitürk K, Ahmad N (2019a) Unveiling the efficiency of vermicompost derived from different biowastes on wheat (Triticum aestivum L.) plant growth and soil health. Agronomy 9(12):791

Aslam Z, Bashir S, Hassan W, Bellitürk K, Ahmad N, Niazi NK et al (2019b) Unveiling the efficiency of vermicompost derived from different biowastes on wheat (*Triticum aestivum* L.) plant growth and soil health. Agronomy 9(12):791

Atiyeh RM, Domínguez J, Subler S, Edwards CA (2000) Changes in biochemical properties of cow manure during processing by earthworms (*Eisenia andrei*, Bouché) and the effects on seedling growth. Pedobiologia 44(6):709–724

Bajal S, Subedi S, Baral S (2019) Utilisation of agricultural wastes as substrates for vermicomposting. IOSR J Agric Vet Sci (IOSR-JAVS) 12(8):79–84

Baker GH, Barrett VJ, Grey-Gardner R, Buckerfield JC (1992) The life history and abundance of the introduced earthworms Aporrectodae trapezoids and Aporrectodaecaliginosain pasture soils in the Mount Lofty Range, South Australia. Aust J Ecol 17(2):177–188

Benıtez E, Nogales R, Masciandaro G, Ceccanti B (2000) Isolation by isoelectric focusing of humic–urease complexes from earthworm (*Eisenia fetida*)-processed sewage sludges. Biol Fertil Soils 31:489–493

Bhardwaj P, & Sharma RK (2015) Vermicomposting efficiency of earthworm species from eastern Haryana Vermicomposting efficiency of earthworm species from eastern Haryana. (July)

Bhat SA, Singh J, Vig AP (2017) Instrumental characterisation of organic wastes for evaluation of vermicompost maturity. J Anal Sci Technol 8(1):1–12

Bhat SA, Singh S, Singh J, Kumar S, Vig AP (2018) Bioremediation and detoxification of industrial wastes by earthworms: vermicompost as powerful crop nutrient in sustainable agriculture. Bioresour Technol 252:172–179

Biruntha M, Karmegam N, Archana J, Selvi BK, Paul JAJ, Balamuralikrishnan B et al (2020) Vermiconversion of biowastes with low-to-high C/N ratio into value added vermicompost. Bioresour Technol 297:122398

Blouin M, Hodson ME, Delgado EA, Baker G, Brussaard L, Butt KR, Brun JJ (2013) A review of earthworm impact on soil function and ecosystem services. Eur J Soil Sci 64(2):161–182

Boyer S, Wratten SD (2010) Using molecular tools to identify New Zealand endemic earthworms in a mine restoration project. Zoology in the Middle East 51(2):31–40. https://doi.org/10.1080/09397140.2010.10638455

Bray RH, Kurtz LT (1945) Determination of total organic and available form of phosphorus in soil. Soil Soc 59:39–45

Cardoso VL, Ramírez CE, Escalante EV (2008) Vermicomposting technology for stabilising the sewage sludge from rural wastewater treatment plants. Water Pract Technol 3:1

Chanu LJ, Hazarika S, Choudhury BU, Ramesh T, Balusamy A, Moirangthem P, Yumnam A, Sinha PK (2018) A guide to vermicomposting-production process and socio-economic aspects. Ext Bull 81:30

Chattopadhyay ST (2014) Biodegradation of novel chitin biocomposites

Chaulagain A, Dhurva P, Lamichhane GJ (2017) Vermicompost and its role in plant growth promotion. Int J Res 4(8):849–864

Chen G, Zheng Z, Yang S, Fang C, Zou X, Luo Y (2010) Experimental co-digestion of corn stalk and vermicompost to improve biogas production. Waste Manag 30(10):1834–1840

Chen Y, Chang SK, Chen J, Zhang Q, Yu H (2018) Characterisation of microbial community succession during vermicomposting of medicinal herbal residues. Bioresour Technol 249:542–549

Chen Y, Zhang Y, Shi X, Xu L, Zhang L, Zhang L (2022) The succession of GH6 cellulase-producing microbial communities and temporal profile of GH6 gene abundance during vermicomposting of maise Stover and cow dung. Bioresour Technol 344:126242

Coulibaly SS, Bi IAZ (2010) Influence of animal wastes on growth and reproduction of the African earthworm species *Eudriluseugeniae* (Oligochaeta). Eur J Soil Biol 46(3–4):225–229. https://doi.org/10.1016/j.ejsobi.2010.03.004

Cui G, Fu X, Bhat SA, Tian W, Lei X, Wei Y, Li F (2022) Temperature impacts fate of antibiotic resistance genes during vermicomposting of domestic excess activated sludge. Environ Res 112654

Das SN, Neeraja C, Sarma PVSRN, Prakash JM, Purushotham P, Kaur M, ... &Podile AR (2012) Microbial chitinases for chitin waste management. In Microorganisms in Environmental Management (pp. 135–150). Springer, Dordrecht

Das S, Deka P, Goswami L, Sahariah B, Hussain N, Bhattacharya SS (2016) Vermiremediation of toxic jute mill waste employing *Metaphire posthuma*. Environ Sci Pollut Res 23:15418–15431

Domínguez J (2004) In: Edwards CA (ed) Earthworm ecology, CRC Press, pp 401–424

Dominguez J, Edwards CA (2011) Relationships between composting and vermicomposting. In: Vermiculture technology earthworms, organic wastes, and environmental management. CRC Press, Boca Raton, FL, USA, pp 11–26

Domínguez J, Aira M, & Gómez-Brandón M (2010) Vermicomposting: earthworms enhance the work of microbes. In Microbes at work. Springer, Berlin, Heidelberg, p 93–114

Domínguez J, Gómez-Brandón M, Martínez-Cordeiro H, Lores M (2018) Bioconversion of Scotch broom into a high-quality organic fertiliser: Vermicomposting as a sustainable option. Waste Manag Res 36:1092–1099

Domínguez J, Aira M, Kolbe AR, Gómez-Brandón M, Pérez-Losada M (2019) Changes in the composition and function of bacterial communities during vermicomposting may explain beneficial properties of vermicompost. Sci Rep 9(1):1–11

Domínguez J, Velando A, Ferreiro A (2005) *Are Eisenia fetida (Savigny, 1826) and Eiseniaandrei (Oligochaeta, Lumbricidae) different biological species?* Pedobiologia 49(1):81–87. https://doi.org/10.1016/j.pedobi.2004.08.005

Donohoe K. (2018) *Chemical and microbial characteristics of vermicompost leachate and their effect on plant growth* (Doctoral dissertation, The University of Sydney). https://ses.library.usyd.edu.au/bitstream/handle/2123/18212/donohoe_kd_thesis.pdf;sequence=1

Edwards CA, Bohlen PJ (1996) Biology and ecology of earthworms, 3rd edn. Chapman and Hall, London

Edwards CA, Lofty JR (1977) Biology of earthworms. Chapman and Hall, London, UK

Gajalakshmi S, Ramasamy EV, Abbasi SA (2001) Potential of two epigeic and two anecic earthworm species in vermicomposting of water hyacinth. Bioresour Technol 76(3):177–181

García AC, Izquierdo FG, de Amaral Sobrinho NMB, Castro RN, Santos LA, de Souza LGA, Berbara RLL (2013) Humified insoluble solid for efficient decontamination of nickel and lead in industrial effluents. J Environ Chem Eng 1(4):916–924

Gómez Brandón M, Aira M, Kolbe AR, De Andrade N, Pérez-Losada M, Domínguez J (2019) Rapid bacterial community changes during vermicomposting of grape marc derived from red winemaking. Microorganisms 7(10):473

Gómez-Brandón M, Domínguez J (2014) Changes of microbial communities during the vermicomposting process and after application to the soil. Crit Rev Environ Sci Technol 44:1289–1312

Goswami L, Sarkar S, Mukherjee S, Das S, Barman S, Raul P, Bhattacharya SS (2014) Vermicomposting of tea factory coal ash: metal accumulation and metallothionein response in *Eisenia fetida* (Savigny) and *Lampitomauritii* (Kinberg). Bioresour Technol 166:96–102

Greco C, Comparetti A, Fascella G, Febo P, La Placa G, Saiano F, Laudicina VA (2021) Effects of vermicompost, compost and digestate as commercial alternative peat-based substrates on qualitative parameters of *Salvia officinalis*. Agronomy 11(1):98

Gudeta K, Julka JM, Kumar A, Bhagat A, Kumari A (2021) Vermiwash: an agent of disease and pest control in soil, a review. Heliyon 7(3):e06434

Gusain R, Suthar S (2020) Vermicomposting of invasive weed Ageratum conyzoids: assessment of nutrient mineralisation, enzymatic activities, and microbial properties. Bioresour Technol 312: 123537

Gutiérrez-Corona JF, Romo-Rodríguez P, Santos-Escobar F, Espino-Saldaña AE, Hernández-Escoto H (2016) Microbial interactions with chromium: basic biological processes and applications in environmental biotechnology. World J Microbiol Biotechnol 32(12):1–9

Hanc A, Enev V, Hrebeckova T, Klucakova M, Pekar M (2019) Characterisation of humic acids in a continuous-feeding vermicomposting system with horse manure. Waste Manag 99:1–11

Hand P, Hayes WA, Frankland JC, Satchell JE (1988) The vermicomposting of cow slurry. Pedobiologia 31:199–209

He X, Zhang Y, Shen M, Zeng G, Zhou M, Li M (2016) Effect of vermicomposting on concentration and speciation of heavy metals in sewage sludge with additive materials. Bioresour Technol 218:867–873

Hendrix PF, Muller BR, Bruce BR, Langdale GW, Parmelee RW (1992) Abundance and distribution of earthworms in relation to landscape factors on the Georgia piedmont, U.S.A. Soil Biol Biochem 24:1357–1361

Houpikian P, Raoult D (2002) Traditional and molecular techniques for the study of emerging bacterial diseases: one laboratory's perspective. Emerg Infect Dis 8(2):122–131. https://doi.org/10.3201/eid0802.010141

Ibrahim MH, Quaik S, & Ismail SA (2016) Vermicompost, its applications and derivatives. In *Prospects of organic waste management and the significance of earthworms* (pp. 201–230). Springer, Cham

Indrani Y, Abdullah R, Ansari A, Ori L (2019) Vermicomposting of different organic materials using the epigeic earthworm *Eisenia foetida*. Int J Recycl Org Waste Agric 8(1):23–36. https://doi.org/10.1007/s40093-018-0225-7

Jamali M, Bakhshandeh E, Khanghahi MY, Crecchio C (2021) Metadata analysis to evaluate environmental impacts of wheat residues burning on soil quality in developing and developed countries. Sustainability 13(11):1–13. https://doi.org/10.3390/su13116356

Kale RD, & Karmegam N (2010) The role of earthworms in tropics with emphasis on indian ecosystems. 2010. https://doi.org/10.1155/2010/414356

Kozich JJ, Westcott SL, Baxter NT, Highlander SK, Schloss PD (2013) Development of a dual-index sequencing strategy and curation pipeline for analysing amplicon sequence data on the MiSeq Illumina sequencing platform. Appl Environ Microbiol 79:5112–5120

Kui H, Jingyang C, Mengxin G, Hui X, Li L (2020) Effects of biochars on the fate of antibiotics and their resistance genes during vermicomposting of dewatered sludge. J Hazard Mater 397: 122767

Kumari S, Manyapu V, Kumar R (2022) Recent advances in composting and vermicomposting techniques in the cold region: resource recovery, challenges, and way forward. Adv Organic Waste Manage:131–154

Lacalle RG, Aparicio JD, Artetxe U, Urionabarrenetxea E, Polti MA, Soto M, Becerril JM (2020) Gentle remediation options for soil with mixed chromium (VI) and lindane pollution: biostimulation, bioaugmentation, phytoremediation and vermiremediation. Heliyon 6(8): e04550

Langille MG et al (2013) Predictive functional profiling of microbial communities using 16S rRNA marker gene sequences. Nat Biotechnol 31:814–821

Lazcano C, Domínguez J (2014) The use of vermicompost in sustainable agriculture: impact on plant growth and soil fertility. In: Miransari M (ed) Soil Nutrients. Nova Science Publishers, Inc., pp 211–233

Lee LH, Wu TY, Shak KPY, Lim SL, Ng KY, Nguyen MN, Teoh WH (2018) Sustainable approach to biotransform industrial sludge into organic fertiliser via vermicomposting: a mini-review. J Chem Technol Biotechnol 93(4):925–935

Lim SL, Wu TY, Clarke C (2014) Treatment and biotransformation of highly polluted agro-industrial wastewater from a palm oil mill into vermicompost using earthworms. J Agric Food Chem 62(3):691–698

Lim SL, Wu TY, Lim PN, Shak KPY (2015) The use of vermicompost in organic farming: overview, effects on soil and economics. J Sci Food Agric 95(6):1143–1156

Liu F, Zhu P, Xue J (2012) Comparative study on physical and chemical characteristics of sludge vermicomposted by *Eisenia fetida*. Procedia Environ Sci 16:418–423

Liyarna R, Krishnaswamy VG, J, M. I. S (2020) Influence of various substrates in vermicomposting its application on different plants (*Abelmoschus esculentus*, *Capsicum annum* and *Trigonella foetum-graceum*). 7(6):6–13

Lopez MJ, del Carmen Vargas-García M, Suárez-Estrella F, Moreno J (2006) Biodelignification and humification of horticultural plant residues by fungi. Int Biodeterior Biodegradation 57(1): 24–30

Lv B, Cui Y, Wei H, Chen Q, Zhang D (2020) Elucidating the role of earthworms in N2O emission and production pathway during vermicomposting of sewage sludge and rice straw. J Hazard Mater 400:123215

Lv B, Xing M, Yang J (2016) Speciation and transformation of heavy metals during vermicomposting of animal manure. Bioresour Technol 209:397–401

Ma Y, Dickson NM, Wong MH (2002) Toxicity of Pb/Zn mine tailings to the earthworm Pheretima and the effects of burrowing on metal availability. Biol Fert Soils 36

Machfudz M, Basit A, & Handoko RNS (2020) Effectiveness of vermicompost with additives of various botanical pesticides in controlling *Plutellaxylostella* and their effects on the yield of cabbage (Brassica oleracea L. var. Capitata)

Maji D, Singh M, Wasnik K, Chanotiya CS, Kalra A (2015) The role of a novel fungal strain *Trichoderma atroviride* RVF 3 in improving humic acid content in mature compost and vermicompost via ligninolytic and celluloxylanolytic activities. J Appl Microbiol 119(6): 1584–1596

Malafaia G, da Costa ED, Guimarães AT, de Araújo FG, Leandro WM, de Lima Rodrigues AS (2015) Vermicomposting of different types of tanning sludge (liming and primary) mixed with cattle dung. Ecol Eng 85:301–306

Maleki S, Hosseini SM, Zare S, Aghyani M, & Pour PG (2016) Effect of vermicompost substrates on amount of organic carbon, total nitrogen and carbon to nitrogen ratio. (September)

Manaig ME (2016) Vermicomposting efficiency and quality of vermicompost with different bedding materials and worm food sources as substrate. 4(1):1–13

Mazur-p A, Garczy M, Hajduk E, Kostecka J, & Butt KR (2021) Use of vermicompost from sugar beet pulp in cultivation of peas (*Pisum sativum* L .). 1–11

Mousavi SA, Sader SR, Farhadi F, Faraji M, Falahi F (2020) Vermicomposting of grass and newspaper waste mixed with cow dung using *Eisenia fetida*: physicochemical changes. 22(1): 8–14

Mupambwa HA, Mnkeni PNS (2018) Optimising the vermicomposting of organic wastes amended with inorganic materials for production of nutrient-rich organic fertilisers: a review. Environ Sci Pollut Res 25(11):10577–10595

Narayanan CM (2012) Production of phosphate-rich biofertiliser using vermicompost and anaerobic digestor sludge—a case study

Ndegwa PM, Thompson SA (2001) Integrating composting and vermicomposting of the treatment and bioconversion of biosolids. Bioresearch Technol 76

Olaniran AO, Balgobind A, Pillay B (2013) Bioavailability of heavy metals in soil: impact on microbial biodegradation of organic compounds and possible improvement strategies. Int J Mol Sci 14(5):10197–10228

Olle M (2019) Vermicompost, its importance and benefit in agriculture. J Agric Sci 2:93–98. https://doi.org/10.15159/jas.19.19

Orozco FH, Cegarra J, Trujillo LM, Roig A (1996) Vermicomposting of coffee pulp using the earthworm Eisenia fetida: effects on C and N contents and the availability of nutrients. Biol Fertil Soils 22(1):162–166

Pattnaik S, Reddy MV (2010) Nutrient status of vermicompost of urban green waste processed by three earthworm species—*Eisenia fetida, Eudriluseugeniae*, and *Perionyx excavatus*. 2010. https://doi.org/10.1155/2010/967526

Pérez-Losada M, Ricoy M, Marshall JC, Domínguez J (2009) Phylogenetic assessment of the earthworm *Aporrectodeacaliginosa* species complex (Oligochaeta: Lumbricidae) based on mitochondrial and nuclear DNA sequences. Mol Phylogenet Evol 52(2):293–302. https://doi.org/10.1016/j.ympev.2009.04.003

Pettit RE (2006) Organic matter, humus, humate, humic acid, fulvic acid and humin. *The Wonderful World of Humus and Carbon*

Phukan IK, Khanikar L, Ahmed CS, Safique S, Jahan A, Baruah A, Phukan I (2013) A novel method for improving the quality of vermicompost. Two a Bud 60:2

Pilli K, Jaison M, Sridhar D (2019) Vermicomposting and its uses in sustainable agriculture. In: Sridhar D, Chandra B, Viswavidyalaya K (eds) Research trends in agriculture sciences. AkiNik Publications, pp 75–88

Piotrowska-Długosz A (2020) Significance of the enzymes associated with soil C and N transformation. In *Carbon and Nitrogen Cycling in Soil* (pp. 399–437). Springer, Singapore

Pop AA, Cech G, Wink M, Csuzdi C, Pop VV (2007) Application of 16S, 18S rDNA and COI sequences in the molecular systematics of the earthworm family Lumbricidae (Annelida, Oligochaeta). Eur J Soil Biol 43:S43–S52. https://doi.org/10.1016/j.ejsobi.2007.08.007

Pop AA, Wink M, Pop VV (2003) Use of 18S, 16S rDNA and cytochrome c oxidase sequences in earthworm taxonomy (Oligochaeta, Lumbricidae). Pedobiologia 47(5–6):428–433. https://doi.org/10.1078/0031-4056-00208

Pramanik P, Ghosh GK, Ghosal PK, Banik P (2007) Changes in organic–C, N, P and K and enzyme activities in vermicompost of biodegradable organic wastes under liming and microbial inoculants. Bioresour Technol 98(13):2485–2494

Price MN, Dehal PS, Arkin AP (2010) FastTree 2—approximately maximum-likelihood trees for large alignments. PLoS One 5. https://doi.org/10.1371/journal.pone.0009490

Purnawanto AM, Ahadiyat YR, Iqbal A (2020) The utilisation of mushroom waste substrate in producing vermicompost: the decomposer capacity of, and. Acta Technol Agric 23(2):99–104

Rabari KV, Patel KM, Chaudhary HL (2020) Effect of rock phosphate enriched different organic manures and chemical fertilisers on growth and yield of wheat. Crop Res 55(1and2):6–9

Rajiv KS, DalsukhV KC, Sunita A (2010) Embarking on a second green revolution for sustainable agriculture by vermiculture biotechnology using earthworms: reviving the dreams of Sir Charles Darwin. J Agric Biotechnol Sustain Dev 2(7):113–128

Ramnarain YI, Ansari AA, Ori L (2018) Vermicomposting of different organic materials using the epigeic earthworm *Eisenia foetida*. Int J Recycl Org Waste Agric 8(1):23–36. https://doi.org/10.1007/s40093-018-0225-7

Reddy SA, Bagyaraj DJ, Kale RD (2012) Management of tomato bacterial spot caused by Xanthomonas campestris using vermicompost. J Biopestic 5(1):10

Rekha GS, Kaleena PK, Elumalai D, Srikumaran MP, Maheswari VN (2018) Effects of vermicompost and plant growth enhancers on the exo-morphological features of Capsicum annum (Linn.) Hepper. Int J Recycl Org Waste Agric 7(1):83–88

Rorat A, Wloka D, Grobelak A, Grosser A, Sosnecka A, Milczarek M, Kacprzak M (2017) Vermiremediation of polycyclic aromatic hydrocarbons and heavy metals in sewage sludge composting process. J Environ Manag 187:347–353

Rosado D, Pérez-Losada M, Aira M, Domínguez J (2021) Bacterial succession during vermicomposting of silver wattle (Acacia dealbata link). Microorganisms 10(1):65

Sarma B, Farooq M, Gogoi N, Borkotoki B, Kataki R, Garg A (2018) Soil organic carbon dynamics in wheat-Green gram crop rotation amended with vermicompost and biochar in combination with inorganic fertilisers: a comparative study. J Clean Prod 201:471–480

Sharma A, Sharma A, Singh S, Vig PA, Nagpal AK (2021) Leaf litter vermi composting: converting waste to resource. IOP Conf Series: Earth Environ Sci 889:2021

Shi Z, Liu J, Tang Z, Zhao Y, Wang C (2020) Vermiremediation of organically contaminated soils: concepts, current status, and future perspectives. Appl Soil Ecol 147:103377

Sim EYS, Wu TY (2010) The potential reuse of biodegradable municipal solid wastes (MSW) as feedstocks in vermicomposting. J Sci Food Agric 90(13):2153–2162

Simsek-Ersahin Y (2011) The use of vermicompost products to control plant diseases and pests. In *Biology of earthworms* (pp. 191–213). Springer, Berlin, Heidelberg

Singh J, Kalamdhad AS (2013) Reduction of bioavailability and leachability of heavy metals during vermicomposting of water hyacinth. Environ Sci Pollut Res 20(12):8974–8985

Singh J, Kaur A, Vig AP, Rup PJ (2010) Role of *Eisenia fetida* in rapid recycling of nutrients from bio sludge of beverage industry. Ecotox Environ Safe 73(3):430–435

Singh WR, Kalamdhad AS (2016) Transformation of nutrients and heavy metals during vermicomposting of the invasive green weed *Salvinia natans* using *Eisenia fetida*. Int J Recycl Org Waste Agric 5(3):205–220

Sinha RK, Bharambe G, Chowdhary U (2008) Sewage treatment by vermi-filtration with synchronous treatment of sludge by earthworms: a low-cost sustainable technology over conventional systems with potential for decentralization. Environmentalist 28:409–420

Song X, Liu M, Wu D, Qi L, Ye C, Jiao J, Hu F (2014) Heavy metal and nutrient changes during vermicomposting animal manure spiked with mushroom residues. Waste Manage 34(11): 1977–1983

Srivastava V, Squartini A, Masi A, Sarkar A, Singh RP (2021) Metabarcoding analysis of the bacterial succession during vermicomposting of municipal solid waste employing the earthworm *Eisenia fetida*. Sci Total Environ 766:144389

Sulaiman ISC, & Mohamad A (2020) The use of vermiwash and vermicompost extract in plant disease and pest control. In *Natural Remedies for Pest, Disease and Weed Control* (pp. 187–201). Academic Press

Sun M, Chao H, Zheng X, Deng S, Ye M, Hu F (2020) Ecological role of earthworm intestinal bacteria in terrestrial environments: a review. Sci Total Environ 740:140008

Suthar S (2009) Vermistabilization of municipal sewage sludge amended with sugarcane trash using epigeic *Eisenia fetida* (Oligochaeta). J Hazard Mater 163(1):199–206

Tabak MA, Norouzzadeh MS, Wolfson DW, Sweeney SJ, Vercauteren KC, Snow NP, Miller RS (2018) Machine learning to classify animal species in camera trap images: applications in ecology. Methods Ecol Evol. https://doi.org/10.1111/2041-210x.13120

Thakur IS, Medhi K (2019) Nitrification and denitrification processes for mitigation of nitrous oxide from waste water treatment plants for biovalorization: challenges and opportunities. Bioresour Technol 282:502–513

Usmani Z, Rani R, Gupta P, & Prasad MNV (2020) Vermiremediation of agrochemicals. In *Agrochemicals Detection, Treatment and Remediation*. Butterworth-Heinemann, p 329–367

Velmourougane KRK (2011) Chemical and microbiological changes during vermicomposting of coffee pulp using exotic (Eudriluseugeniae) and native earthworm (Perionyx ceylanesis) species. 497–507. https://doi.org/10.1007/s10532-010-9422-4

Vukovic A, Velki M, Ecimovic S, Vukovic R, Camagajevac IS, Loncaric Z (2021) Vermicomposting—facts, benefits and knowledge gaps. Agronomy 11(1952):1–15

Wäldchen J, Mäder P (2018) Machine learning for image based species identification. Methods Ecol Evol. https://doi.org/10.1111/2041-210x.13075

Walkey A, Black IA (1934) An examination of Degtjaroff method for determining soil organic matter and proposed modification of the chromic acid titration method. Soil Sci 37:29–38

Wang H, Li H, Gilbert JA, Li H, Wu L, Liu M, Zhang Z (2015) Housefly larva vermicomposting efficiently attenuates antibiotic resistance genes in swine manure, with concomitant bacterial population changes. Appl Environ Microbiol 81(22):7668–7679

Wang Y, Han W, Wang X, Chen H, Zhu F, Wang X, Lei C (2017) Speciation of heavy metals and bacteria in cow dung after vermicomposting by the earthworm, Eisenia fetida. Bioresour Technol 245:411–418

Wu X, Wang J, Yu Z, Amanze C, Shen L, Wu X, Zeng W (2022) Impact of bamboo sphere amendment on composting performance and microbial community succession in food waste composting. J Environ Manag 303:114144

Yatoo AM, Ali M, Baba ZA, Hassan B (2021) Sustainable management of diseases and pests in crops by vermicompost and vermicompost tea. A review. Agron Sustain Dev 41(1):1–26

Zeb A, Li S, Wu J, Lian J, Liu W, Sun Y (2020) Insights into the mechanisms underlying the remediation potential of earthworms in contaminated soil: a critical review of research progress and prospects. Sci Total Environ 740:140145

Zhao X, Shen JP, Shu CL, Jin SS, Di HJ, Zhang LM, He JZ (2022) Attenuation of antibiotic resistance genes in livestock manure through vermicomposting via Protaetiabrevitarsis and its fate in a soil-vegetable system. Sci Total Environ 807:150781

Chapter 5
Insights into Earthworm Biology
for Vermicomposting

Rutendo Nyamusamba, Reagan Mudziwapasi, Fortune Jomane, Unity Mugande, Abigarl Ndudzo, Sicelo Sebata, and Morleen Muteveri

Abstract Vermicomposting refers to the use of earthworms to change organic materials into a humus-like substance called vermicompost. The growth of earthworms with the intention of increasing their numbers is referred to as vermiculture. Vermifiltration is the use of earthworms to treat wastewater. All species of earthworms can be cultured. Common earthworm species include the tiger worm (*Eisenia fetida*), red tiger worm (*E. Andrei*), Indian blue worm (*Perionyx excavatus*), African night crawler (*Eudrilus eugeinae*), and red earthworm (*Lumbricus rubellus*). The conventional names of these earthworms differ from place to place and sometimes among individuals. The success of vermicomposting depends mostly on the soil type and moisture content. Certain soils favor specific species of earthworms. Earthworms are also sensitive to extreme pH and temperature, food type, and stocking density. Extreme conditions can result in earthworm death. Different earthworm culture media compositions can be used to determine the most suitable media and conditions for vermicomposting. Biomass, that is, earthworm population density in

---

R. Nyamusamba
Department of Soil and Crop Sciences, Gary Magadzire School of Agriculture, Great Zimbabwe University, Masvingo, Zimbabwe

R. Mudziwapasi (✉)
Department of Crop and Soil Sciences, Faculty of Agricultural Sciences, Lupane State University, Lupane, Zimbabwe

Research and Innovation, Midlands State University, Gweru, Zimbabwe

F. Jomane
Department of Animal Science and Rangeland Management, Faculty of Agricultural Sciences, Lupane State University, Lupane, Zimbabwe

U. Mugande · A. Ndudzo · S. Sebata
Department of Crop and Soil Sciences, Faculty of Agricultural Sciences, Lupane State University, Lupane, Zimbabwe

M. Muteveri
Department of Physics, Geography and Environmental Science, School of Natural Sciences, Great Zimbabwe University, Masvingo, Zimbabwe

© The Author(s), under exclusive license to Springer Nature Singapore Pte Ltd. 2023
H. A. Mupambwa et al. (eds.), *Vermicomposting for Sustainable Food Systems in Africa*, Sustainable Agriculture and Food Security,
https://doi.org/10.1007/978-981-19-8080-0_5

the soil, maturity, and health of earthworms are significant to vermicomposting. The best conditions for culturing earthworms are the ones under which the earthworms have the highest cocoon production and net reproductive rate. These conditions affect the choice of earthworm species for vermicomposting, vermiculture, and vermifiltration studies. This chapter provides insights into the biology of earthworms in a quest to understand how earthworms work in vermicomposting.

Keywords Earthworms · Net reproductive rate · Specific growth rate · Vermicompost

5.1 Introduction

Vermicomposting is a result of transformation of organic materials into a humus-like material called vermicompost by earthworms (Ramnarain et al. 2018a, b). The worms turn the organic matter into a high nutrient compost that is made up of earthworm casts as well as decomposing organic waste (Devi and Prakash 2015; Ramnarain et al. 2018a, b). Vermicomposting produces a finely divided peat-like material with exceptional moisture holding capacity, aeration, porosity, and drainage (Ismail 2005; Edwards et al. 2011; Ramnarain et al. 2018a, b). This organic fertilizer is also rich in NPK, micronutrients, and important soil microorganisms including phosphate-solubilizing and nitrogen fixing bacteria, actinomycetes, and fungi. Vermicompost acts as a growth promoter and crop plants protect themselves from pathogens (Ramnarain et al. 2018a, b).

The science of vermicomposting is rooted in the knowledge gathered over time on the biology and ecology of earthworms. Queen Cleopatra VII (69 to 31 BC) of Egypt made laws to protect earthworms in that country. A quest to understand earthworms by the early scientists was not born from the need to improve soil health as is the case today but rather to understand observations that were being made on the land. At that time, earthworms were observed to emerge from the earth, especially during the rainy season, and some birds came to eat the worms (Rota 2011). The observation made back then that earthworms were seen on the soil surface more during the rainy season seems to point to the idea of good soil water infiltration compared to the current state of soils where there is more surface runoff than infiltration. This phenomenon could partially explain the lack of observation of earthworms in most of today's crop and vegetable fields. Vermicomposting, as we speak of it today, is trying to turn back the hands of time in improving the soil status in crop fields, gardens, and pastures.

It would not be too far off to hypothesize that the soil was healthy and water infiltration occurred. Documentation of earthworm activity in the environment dates back to the time of Aristotle (384–322 BC), who was a Greek philosopher and scientist. Writings on earthworms by Aristotle were under the heading, "Earth's Intestines," at which time mud, not eggs, was thought to be the origins of earthworms. From his field observations and information from farmers and other observers, Aristotle concluded that earthworms were imperfect animals without blood and other vital organs. Rota (2011) notes how over time, the works of

Aldrovandi and Mouffet, Thomas Willis, Fabriciab Aquapendente, Borelli, and Francesco Redi challenged these Aristotelian views. Discoveries on reproduction and the environments in which earthworms thrive were made.

Earthworms are hermaphrodites that hatch from fertilized eggs. For reproduction to occur, two earthworms mate and exchange sperm, and each earthworm's ova are fertilized by the other earthworm's sperm. This ensures genetic diversity. The ova and sperm are deposited in cocoons outside the earthworm bodies. The ova are fertilized while in the cocoon, and the new earthworms emerge from the cocoon. They reproduce after reaching maturity at 60–90 days. Earthworms can generate new parts of their body if they are cut off. Earthworms reside in dark, moist, and warm environments.

Charles Robert Darwin (1809–1882) is considered one of the first earthworm ecologists with his publication entitled "The Formation of Vegetable Mould, through the Action of Worms, with Observations on their Habitats" (Darwin n.d.). This work points to several aspects of the biology and ecology of earthworms, including habits of worms, denudation (Fowler et al. 2000), and movement of soil. This chapter provides insights into the biology of earthworms in a quest to understand how earthworms work in vermicomposting.

5.2 Earthworms

Earthworms are terrestrial Annelids, belonging to class *Oligochaeta* (Gajalakshmi and Abbasi 2004). According to the earthworm database, over 6000 species of earthworms have been identified (Csuzdi 2012). Earthworms are useful soil quality indicators (Huerta et al. 2009; Bartz et al. 2013) because they are rare in compacted waterlogged soils that have extreme pH values or are very sandy. Agrochemical applications, intensity of tillage and food supply affect earthworms. Earthworms have several benefits that include soil nutrient availability, water and air movement that allow for better drainage, enhancement of soil structure and soil productivity. Therefore, several efforts should be made to increase earthworm populations in soils.

5.3 Earthworm Structure and Biology

Earthworms are tube-shaped and segmented with a digestive system spanning their body length, and they respire through their skin. The earthworms have a double transport system but lack either an exoskeleton or an internal skeleton. Their system consists of a coelom within which circulates a coelomic fluid. This acts as a hydrostatic skeleton. Earthworms have small bristles called setae between the rides of their segments since they do not have tentacles called parapodia like other annelids. They have a simple, closed circulatory system for blood. Earthworms have a peripheral and central nervous system which has 2 ganglia that are above the earthworm's mouth. These are connected to a nerve cord that runs back along the earthworm's length to sensory cells and motor neurons in each segment. A majority of the earthworm chemoreceptors are diffuse near the mouth. These chemoreceptors are mostly used to find food.

The earthworms move using the longitudinal and circumferential muscles, which are on the periphery of each of their segments. Digested food is moved down along the earthworm gut towards the earthworm's anus using similar muscles lining the earthworm gut. Ingested food is moved to the crop and then to the gizzard. After the grinding action of the gizzard, the food is digested by intestinal enzymes while in transit to the anus. Monomers are absorbed into the bloodstream, and undigested food is excreted as feces.

5.4 Classes and Species of Earthworms

Classification is the act or process of classifying organisms according to shared qualities or characteristics. There are several ways in which this can be done, with the traditional more classical approach being the utilization of shared evolutionary ancestry with respect to their phylogeny (Fig. 5.1). Earthworms can be terrestrial or found in fresh water (Table 5.1).

However, a more lexicon approach that is based on the earthworm's functional ecological role is more favorable. This type of grouping focuses on the morphology, ecology, and vertical position (in or above) of the soil. The groupings are commonly known as ecotypes that are very useful in studying earthworms. They are basically three types of earthworms: anecic (deep-burrowing earthworms), endogeic (shallow burrowing earthworms), and epigeic (surface-dwelling earthworms).

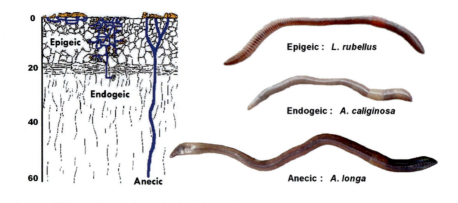

Fig. 5.1 An earthworm cladogram showing evolutionary relationships and developments with the earthworm. (Source http://media1.shmoop.com/images/biology/biobook_taxonomy)

5 Insights into Earthworm Biology for Vermicomposting 93

Table 5.1 Systematic classification of worms

Kingdom	Animalia	(Multicellular, eukaryotic, and heterotrophic)
Phylum	Annelida	Segmented body; segments separated by septa; two body openings and closed circulatory system.
Class	Oligochaeta (oligochaetes)	Lack appendages and few bristles; head absent; fewer setae; no parapodia; landdwelling or found in fresh water, i.e., *Tubifex* earthworm
	Polychaeta (Polychaetes)	Live in salt water (mostly marine); have a distinct head with eyes and tentacles; segmented with parapodia; setae present; lack of a clitellum; pair of bristly fleshy limbs on each segment; some live in tubes including sandworm, fanworm, feather duster worm
	Hirudinea (leeches)	No limbs; carnivores or blood sucking external parasites, inhabit fresh waters including medicinal the leech, *Hirudo medicinalis*

5.4.1 Epigeic Earthworms

Epigeic earthworms, also known as surface-dwelling earthworms, inhabit the topsoil layers (Fig. 5.2). They live in leaf litter, dung, and wood, and they are usually found on the surface and rarely in the soil (<3 cm). They are characterized by the shortest life cycles and are the smallest of the earthworms. They usually feed on decaying plant material such as leaf litter and rotting logs. Their mode of leaf litter processing in natural ecosystems results in greater nutrient leaching into the soil. Since they usually feed mostly on leaf litter and have a short gut retention time, they usually depend on a rapid response of gut microbes to aid in digestion. This therefore translates to a difference in the microorganism composition of soils that have epigeic soils and uningested soils. They usually have pigmented skin that is deep red and dark in color. Only epigeic earthworms are suitable for composting. Examples

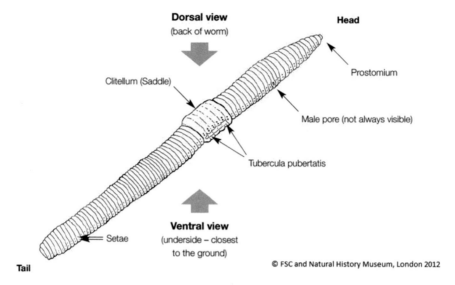

Fig. 5.2 Illustration of epigeic, endogeic, and anecic earthworm

include *Dendrobaena octaedra D. attensi, D. rubidus*, and *Eiseniella tetraedra* (Source: www.mscwbif.org).

5.4.2 Endogeic Worms

Endogeic earthworms, also commonly known as shallow burrowing earthworms, are the most abundant earthworms inhabiting arable land. They usually inhabit the interior of the soil and usually form a network of temporary horizontal burrows in the upper soil layers (Fig. 5.2) that range from 0 to 50 cm. They are larger than epigeic earthworms and are usually medium to large in size. These earthworms rarely come to the soil surface and usually feed on it. They blend the soil, redistributing soil bacteria and producing feces (casts) rich in nutrients (nitrogen, potassium, and sodium). They are often pale in color, as they do not have much skin pigmentation. Examples are *Allolobophora chlorotica, Aporrectodea caliginosa, Aporrectodea icterica,* and *Aporrectodea rosea* (Source: www.mscwbif.org).

5.4.3 Anecic Earthworms

Anecic earthworms are also commonly known as deep-burrowing earthworms. They are the largest earthworms and can live for up to 5 years. They live in deep permanent vertical burrows that are as deep as 180 cm into the ground, and these burrows can be easily identified by the pile of excrement around the surface called a midden. The latter is usually surrounded by a mound of cast material (worm poop) and crowned with fragmented leaf parts or debris. They are also responsible for the creation of galleries that are responsible for depositing cocoons or for shelter from harsh conditions. Anecic earthworms usually forage at night feeding on soil and leaves and usually inhabit their tunnels for the rest of their lives. Morphologically, they are usually dark red dorsally, often darker towards the head.

Porosity and aeration of the soil are improved and water infiltration is enhanced by permanent vertical tunneling through the soil. They are usually absent in fields managed with intensive cultivations and no surface litter (for example, root cropping and plowed soils). Examples of anecic earthworms are *Lumbricus friendi, Lumbricus terrestris, Aporrectodea longa*, and *Aporrectodea nocturna* (Source: www.mscwbif.org).

5.5 Classification of Earthworms

A very important part of recording earthworms is to be able to identify them. Identification is made easy by their various morphological characteristic features (Fig. 5.3). Due to the microstructures on most earthworms, they may require the utilization of a microscope to be able to see certain features. In the absence of a microscope, a strong hand lens can suffice for larger earthworms. Several keys have been developed that can aid with the identification of earthworms, such as Key to the Earthworms of the UK & Ireland by Sherlock (1st edition) (2012); (Key to the Earthworms of the UK & Ireland by Sherlock (2nd edition) (2018), several earthworm identikit as well as the OPAL Earthworm guide.

5.6 The Opal Classification System

The Open-Air Laboratories (OPAL) was a UK-wide citizen science initiative founded and led by Imperial College London (2007–2019) that enabled people to get hands-on interaction with nature while contributing to important scientific research. It was initially funded by the Big Lottery Fund (BLF). Its concept involved taking experts into the heart of communities to explore and investigate the natural world with local people; exchanging knowledge, experience and skills while learning more about local wildlife and their habitats under changing environmental conditions. Citizens who participated brought forth data about their local environment via OPAL's national surveys, which focused on topics such as environmental degradation, loss of ecological diversity, and climate change (Davies et al. 2011; OPAL 2022a, b).

Table 5.2 Activities conducted during an Opal Soil & Earthworm survey

Component	Activities involved
Site characteristics	Recording of various information such as date of collection of survey data, data collectors name, GPS co-ordinates, and description of sampling site (microhabitat/climate/surrounding)
The soil pit and earthworms	Digging ten pits that follow a W shape field sampling pattern, identifying and recording number of caught earthworms (adults or juveniles), approximately 5 min should be taken in each pit (Stroud and Bennet 2018)
Soil properties	Recording of information about soil such as the number of plant roots, the presence of any available objects that is not naturally found within the soil such as construction material cut wood, glass, etc., the hardness of the soil, water retention of the soil, pH and texture of the soil and soil fizz test
Earthworm identification	Utilization of earthworm identification guide to identify and record species of each adult earthworm collected (length and color should always be recorded); characters utilize in identifying earthworms (Fig. 5.3)
Other organisms in your pit	Recording of any other caught in the pit such as beetles and slugs

By facilitating collaborations among scientists, amateur experts, local interest groups, policy makers, and the public closer together, OPAL aimed to form enduring relationships and ensure environmental issues of local and global relevance. It designed a suite of field and desk-based activities that were suitable for all ages, abilities, and backgrounds and, in particular, launched a series of national ecological surveys. These activities increased understanding of the natural environment through data collection covering wildlife and habitat conditions. Each survey consisted of an easy-to-follow set of instructions, a workbook to record results and species identification guides/keys. Partner organizations collaborated in the design, delivery, and adaptation of these surveys (Davies et al. 2011; OPAL 2022a, b).

The Opal Soil and Earthworm survey includes several activities that involve the description of site characteristics, soil pits and earthworms, soil properties, earthworm identification, and an additional search for earthworms in other habitats (OPAL 2022a, b). It is usually best to collect earthworms during autumn and spring. However, timing the sampling during warm, wet conditions usually provides the best earthworm population estimates. The most common tolls to utilize in such a survey include the utilization of spades, pots, and bottles of water, mats and recording sheets (Table 5.2).

The OPAL earthworm guide has been a very successful tool that has been utilized to identify earthworms across the world. However, it also does have several limitations that include

(i) Failure to identify rare earthworms.
(ii) Accuracy rate.

5 Insights into Earthworm Biology for Vermicomposting

(iii) Some of the identification features do not work, for example, the use of a yellow tail to distinguish *Octolasion cyaneum* from other epigeic species of earthworms. *O. lacteum* (a close relative of *O. cyaneum* not included in the key) has been observed with a yellow tail.
(iv) False expectations.

Nevertheless, earthworms with a small amount of practice are a manageable group to master (Entomological Society of Britain 2022).

5.7 Earthworm Species Suitable for Vermicomposting

The scientific name for an earthworm is *Lumbricina*, and the first earthworm species to be named was *Lumbricus terrestris*. The second was named over six decades in 1826 by Savigny 1826. However, by the early 1870s, over 100 species had been described. Currently, according to the species name database, there are over 6000 terrestrial earthworm species, with approximately 150 species that are widely distributed (Csuzdi 2012). There are more than 8300 species in Oligochaeta and half are terrestrial (Reynolds and Wetzel 2004). *Lumbricidae* are the most common earthworms in Europe, north America, and western Asia while *Eudrilidae* are found in West Africa. In South Africa, *Microchaetidae* are abundant in South Africa, *Megascolecidae* in Australia and other parts of Asia, and *Glossoscolecidae* predominate in Central and South America (Edwards 2010).

Epigeic species are more suited for vermicomposting because they can colonize organic wastes naturally, exhibit high rates of consumption, digestibility, and assimilation of organic matter. They can tolerate a wide range of environmental factors, have short life cycles and high rates of reproduction, exhibit tolerance and endurance of handling (Edwards 2010). Suitable species for vermicomposting include temperate and tropical *Eisenia fetida* and *Eisenia andrei*, *Dendrodrilus rubidus*, *Dendrodrilus veneta*, *Lumbricus rubellus*, *Drawida nepalensis*, *Eudrilus eugeniae*, *Perionyx excavates*, and *Polypheretima elongata*. However, only a few of these epigeic species are able to exhibit all the common characteristics that make them suitable for vermicomposting, with only five of them being extensively utilized in vermicomposting.

Earthworms belong to the taxonomic phylum Annelida and the order Oligochaeta. The order Oligochaete comprises 3000 species with 5 families. Earthworms are "ecosystem engineers" and "unheralded soldiers of mankind" due to their activities in the ecosystem. They are found in almost all terrestrial ecosystems and are very versatile. They depend on live and dead organic matter as feed. Earthworms play roles in soil structure improvement, soil improvement and soil-borne plant fungal disease biocontrol. Earthworm body tissues were found to be rich in proteins, carbohydrates, vitamins, minerals, and fats containing 7–10% fat, 60–70% protein, and 8–20% carbohydrates. Earthworms can withstand pH ranges of 4.5 to 9 with a pH optimum of 7. They have an optimum moisture content of 70%, and they tolerate temperatures between 10 and 35 °C (Sinha et al. 2008).

5.8 Earthworm Gut Microbes

Earthworms naturally contain microorganisms in their gut, and they are a host to millions of decomposing and biodegrading microbes in their gut. These come from the food they eat since they feed on decaying and/or dead matter that could be infested with microorganisms. The microorganisms that can be found in the earthworm gut include those from the genera *Bacillus, Pseudomonas,* and *Escherichia.* Earthworms are thus considered saprozoic organisms that are at risk from some of the microorganisms in their drilosphere. Earthworms promote the development of fungi and bacteria that produce antibiotics capable of killing pathogenic microorganisms. In a study carried out by Edward and Fletcher (1988), findings showed that passage of ingesta through the earthworm gut resulted in a 1000-fold increase in the number of bacteria and actinomycetes contained. Earthworm species, reproduction stage, and substrate media determine the enzymatic activity in earthworms. Earthworms act as aerators, grinders, crushers, chemical degraders, and biological stimulators of biomass. To survive, earthworms have a defense mechanism against invading microbes in their food and environment.

5.9 Bacterial Species Associated with Earthworms

The composition and structure of the microbial community in the earthworm's environment, such as in a bioreactor (vermifilter), differs from the gut and casts communities. Consequently, the earthworm gut microbiota can be transient depending on the type of substrate media or type of feed (Medina-Sauza et al. 2019). This is due to competitive interactions between endosymbiotic bacteria residing in the earthworm gut and ingested bacteria (Gómez-Brandón et al. 2011). Biophysical conditions in the earthworm gut, such as the gut's anaerobic environment and the gut fluids' discriminate and repressive activity against specific groups of microorganisms, influence the diversity of microbial communities. Digestive enzyme secretion by earthworms and the inhibitory effects of substances secreted by other bacteria alter microbial populations in the earthworm gut (Gómez-Brandón et al. 2011).

Gammaproteobacteria tolerate extreme physical and chemical conditions (Gómez-Brandón et al. 2011). The proliferation of gammaproteobacteria is of particular importance because medically and scientifically important bacterial species belonging to genera such as *Pseudomonadaceae, Enterobacteriaceae, Vibrionaceae,* and *Xanthomonadaceae* belong to the Gammaproteobacteria group. *Pseudomonas aeruginosa* causes lung infections, *Salmonella* causes enteritis and typhoid fever, while *Klebsiella pneumonia* causes pneumonia. While Firmicutes, Actinobacteria, and Gammaproteobacteria proliferate in the earthworm gut, other species of microbes die and do not increase in number during passage in the earthworm gut. This changes the microbial population in the vermicomposted material, resulting in its biodegradation and detoxification.

Pseudomonads are involved in the degradation of organic pollutants. They have degradative enzymes for more than 70 organic compounds. *Pseudomonads* are responsible for the degradation of some recalcitrant and persistent organic chemicals (Gómez-Brandón et al. 2011; Siles et al. 2021). This can be due to the vast degradative genes encoded on their plasmids and chromosomes (Banta and Kahlon 2007). These genes make bacteria target organisms for xenobiotic degradation in this study. The burrowing action of earthworms will therefore move microbes into their drilosphere, effectively distributing the xenobiotic degrading microbes resulting in improved bioremediation without excavation of contaminated sites or vermicomposted material. Vermicomposting-based systems can therefore be modelled to remove xenobiotic chemicals in solid waste using earthworm gut microorganisms and the actions of earthworms.

5.10 Phosphate-Solubilizing Bacteria in the Earthworm Gut

Bacteria in the earthworm gut have different abilities in the biodegradation of pollutants. Isolated cellulolytic *Aspergillus sp.* and nitrifying *Pseudomonas sp.* from earthworm guts and casts. Some bacterial species have mineralization and solubilization capabilities for organic and inorganic phosphorous, respectively. The ability of microbes to release metabolites such as organic acids determines their phosphate-solubilizing activity. This is achieved through chelation of the cation bound to the phosphate by the hydroxyl and carboxyl groups resulting in formation of soluble forms. Phosphate solubilization can also be achieved through proton extrusion. Some bacteria produce energy through the degradation of polyphosphates releasing phosphates into the wastewater. Phosphate-solubilizing bacteria solubilize insoluble forms of phosphorous to $H_2PO_4^{2-}$ and HPO_4^{2-} which are soluble through chelation, acidification, and exchange reactions. Phosphate-solubilizing bacteria (PSB) have been isolated from the earthworm species *E. fetida, Allolobophora chlorotica,* and *Aporrectodea longa.* The epigeic earthworm *E. fetida* had the highest number of PSB. The phosphate-solubilizing bacteria isolated from the *E. fetida* gut included *Bacillus subtilis, Pseudomonas aeruginosa, Pseudomonas fluorescens,* and *Enterobacter aerogenes.* Microorganisms belong to the genera *Tsukamurella, Klebsiella, Microbacterium, Agromyces, Bacillus, Streptomyces,*

Rhodococcus, and *Pseudomonas.* Species such as *Klebsiella pneumonia* and *Escherichia coli* were isolated at low frequencies. Consequently, enzyme activity is evident and differs in the earthworm gut due to the different profiles obtained in the gut microbes during gut passage. These bacteria were identified using morphological and biochemical methods.

5.11 Earthworm Culture

Earthworm culture is the growth of earthworms with the intention of increasing their numbers. All species of earthworms can be cultured. *Eisenia fetida, E. Andrei, Perionyx excavatus, Eudrilus eugeinae,* and *Lumbricus rubellus* are the common earthworm species used in vermicomposting. The conventional names of these earthworms differ from country to country and sometimes among individuals. The success of earthworm culture depends mostly on the soil type and moisture content. Hence, certain soils favor specific species of earthworms . Earthworms are also sensitive to extreme pH, food type, and stocking density. Different earthworm culture media compositions can be used to determine the most suitable media and conditions for earthworm culture. Population density (biomass) in soil and the maturity and health of earthworms are important factors in earthworm culture. The best conditions for culturing earthworms are the ones under which the earthworms have the highest cocoon production and net reproductive rate. These conditions affect the choice of earthworm species for vermifiltration studies. Earthworms can die in unfavorable soils and under extreme moisture content, pH, and temperature conditions.

An investigation into the growth and reproduction of the deep-burrowing lumbricoides, *Lumbricus terrestris, Aporrectodea longa, and Octolasion cyaneum* revealed reproductive outputs of 18.8, 38.0, and 32.3 cocoons/worm/year for *L. terrestris, A. longa, and O. cyaneum,* respectively. The study showed that earthworms have different durations of life cycles. However, biocides depress earthworm growth when applied alone. Some combinations of biocides can reduce the effect of biocides on earthworm growth. *E. fetida* can change bagasse waste into manure when cocomposted with cow dung at ratios of 0:100, 25:75, 50:50, 75:25, and 100:0 on a dry weight basis. The cow dung helps to improve *E. fetida's* acceptance of the bagasse waste. The highest population increase and minimum mortality of worms was in a 50:50 ratio, making it the best mixture for culture of *E. fetida.* An increase in bagasse waste significantly affects earthworm growth and reproduction. Nitrogen, sodium, and phosphorus increase due to vermicomposting, while the C:N ratio and metal concentrations decrease due to vermicomposting.

5.12 Essentials for Earthworm Culture

The culture of earthworms requires a number of factors that influence their rate of growth and reproduction.

5.12.1 Bedding

Earthworm bedding provides the earthworms with a relatively dependable habitat. The bedding should be relatively highly absorbent since the worms breathe through their skins. It should have a good bulking potential in order to absorb and retain water because the worms die if their skin dries out. Type of bedding material is of critical importance since earthworms are aerobic animals. Thus, different types of bedding materials, particle size range and shape, texture, strength, and rigidity of its structure will affect the porosity of the bedding. Dense or tight packing of bedding material will result in restriction, reduction, or elimination of air flow. This explains the bedding material's bulking potential (Rostami et al. 2010). The bedding should have a low protein/nitrogen composition as well as a high carbon content since earthworms usually feed on their bedding as it undergoes slow degradation. Rapid degradation of bedding due to high protein/nitrogen levels is associated with increased temperatures, creating inhospitable conditions which are fatal (Sadia et al. 2020). The vermicomposting process increases loss of organic carbon thus a high carbon content in bedding is needed to offset and facilitate microbial respiration and degradation of organic wastes by earthworms and microbes in the feed mixtures (Garg and Kaushik 2005).

The floor of the vermiculture unit is supposed to be compacted and hardened in order to discourage earthworm migration into the soil. Vermiculture bedding is prepared by first laying saw dust, then newspaper, straw, and sugarcane waste at the bottom of the container. Newspaper has a high absorbency while sawdust has a characteristic poor to medium absorbency. Moistened fine sand of approximately 3 cm in thickness is then spread as a second layer over the culture bed, followed by a layer of garden soil (3 cm).

5.12.2 Feeding

The growth and reproduction of earthworms as well as the production rate of cocoons is directly influenced by feeding. Factors such as moisture, particle size, organic substrate content, and feed pretreatment affect the feeding rate of earthworms (Sadia et al. 2020). A high organic matter limits the activity of worms, thus resulting in enhanced anaerobic activity of microorganisms, creating anaerobic and foul odor conditions (Hidalgo et al. 2022). The presence of toxic metals in organic feed is fatal to worms. Cow, sheep, horse, and goat dung produce better quality manure which results in better vermicomposting. High levels of ammonia in organic wastes create an inhospitable environment for the earthworms (Gunadi et al. 2002). Many different types of manures have high salt contents, and should be leached first to reduce the salt content before use as bedding since worms are sensitive to salt preferring salt contents of less than 0.5%. Leaching of salts is achieved by running water through the material for a period of time (Gaddie and Donald 1975).

5.12.3 pH

The pH is also an important factor influencing vermicomposting. Epigeic worms can survive in a pH range between 4.5 and 9 (Edwards 1998). The pH of earthworm beds drops over time, that is, if the food source/bedding is originally alkaline, then the pH of the bed drops to neutral or slightly alkaline, but if the food source is acidic, then the pH of the beds can drop well below 7. pH adjustment upwards is accomplished addition of calcium carbonate or peat moss (Gajalakshmi and Abbasi 2004). The recommended pH range is approximately 6.5 to 7.5 although some microorganisms are active at a lower pH of approximately 4. A neutral pH is suitable for the optimal activity of worms, with the favorable range being 4.5–9.0 (Alabi et al. 2022). The pH is dependent upon earthworm sensitivity and physicochemical characteristics of the waste. Differences in physicochemical properties of waste alter the pH during vermicomposting. Activities of the microorganisms change the physicochemical properties of waste during the decomposition as well as mineralization of nitrogen and phosphorus into nitrites/nitrates and orthophosphates. Ammonia and humic acids which are produced as intermediates vermicomposting and types of substrates alter the pH (Cai et al. 2018). The overall pH in the vermicomposting process can drop from alkaline to acidic due to the evolution of CO_2 and the accumulation of organic acids.

5.12.4 Temperature

The optimum temperature range for earthworms is between 25 °C and 37 °C. It promotes optimum activity, growth, metabolism, respiration, reproduction, and cocoon production of earthworms as well as proliferation of microorganisms associated with earthworms (Edwards and Bohlen 1996). The optimum temperature for the stable development of the earthworm population should not exceed 25 °C. *E. fetida* cocoons can be viable for extended periods of deep freezing and remain viable (Sinha et al. 2008). However, but they do not reproduce and do not consume sufficient food at single digit temperatures. Optimal vermicomposting efficiency is achieved at 15 °C for vermicomposting efficiency and 20 °C is for effective reproductive vermiculture and vermicomposting operations. Temperatures above

35 °C will result in either migration of earthworms, or their death (Kaur 2019). Bacterial activity is also optimal at approximately 15–30 °C and it exhibits a twofold increase per 10 °C increase in temperature. Different earthworm species show different responses to temperature. *Dendrobaena veneta* shows optimum growth at lower temperatures and is susceptible to temperature extremes while *Eudrilus eugeniae* and *Perionyx excavatus* show optimum growth at approximately 25 °C. However, they can tolerate temperature ranges between 9 °C and 35 °C (Coulibaly and Irie 2010). Higher temperatures in vermicomposting systems result in loss of nitrogen as ammonia volatilization while lower temperatures fail to destroy pathogenic organisms.

5.12.5 Aeration

Earthworms are aerobic animals and cannot survive in anaerobic conditions thus prefer porous and well aerated compost material. Earthworms also aid aeration by tunneling through their bedding. *E. fetida* has been reported to migrate in high numbers from anoxic water-saturated substrates or in materials in which carbon dioxide or hydrogen sulfide has accumulated (Kaur, 2020). Turning the compost pile is both an effective means of adding oxygen and bringing newly added material into contact with microbes. A pitchfork or a shovel, or an aerator can be used for turning the vermicompost pile. Lack of aeration may produce an odor symptomatic of anaerobic decomposition.

5.12.6 Moisture

The moisture content of organic wastes affects the growth rate of earthworms. In a comparative study by Tchobanoglous et al. (1993), on the vermicomposting process and earthworm growth at different temperature and moisture ranges, it was shown that 65–75% is the most suitable moisture range at all ranges of vermicomposting temperature and this was supported by Ramnarain et al. (2018a, b). The bedding used must be able to hold sufficient moisture as earthworms respire through their skins, since a moisture content of less than 45% can be fatal to the worms. Epigeic species, *E. fetida* and *E. andrei* can survive moisture ranges between 50% and 90%, they exhibit rapid growth at 80% and 90% moisture content (Edwards and Burrows 1988). Bacterial activity decreases in materials with moisture contents lower than 40% and almost stops when the moisture content is lower than 10% (Suthar 2006; Kaur, 2020). Conditions of low moisture delay the sexual development of earthworms (Dominguez and Edwards 1997; Wani et al. 2013). The optimum moisture content for *Eisenia fetida* has been reported to be between 70 and 80%. Some species of earthworms, such as *Lumbricus terrestris,* are adapted to dry conditions, while

Allolobophora chlorotica, Allolobophora caliginosa, and *Aporrectodea rosea,* are susceptible to dry conditions.

5.12.7 Stocking Density

Factors such as initial substrate quality and quantity, temperature, moisture, and soil structure and texture affect the density of earthworms. The copulation frequency of earthworms is high at low population density and it further decreases when the density approaches the carrying capacity of the substrate (Smetak et al. 2007). The optimum stocking density for vermicomposting is 1.60 kg worms/m^2 (Gajalakshmi et al. 2002). It has been reported that *Eisenia andrei* grows slowly in higher population density environments. The earthworms consequently have a lower final body weight.

5.12.8 Carbon: Nitrogen Ratio

The C: N ratio influences cell synthesis, growth, and metabolism of earthworms. The C: N ratio is an indicator of waste stabilization and is widely used as an index for compost maturation. As a result of rapid mineralization and organic matter decomposition, carbon is lost as CO_2 in microbial respiration. However, nitrogen is increased by worms in the form of mucus hence the nitrogenous excretory material results in an overall decrease in the C: N ratio. Initial nitrogen content in the substrate directly influences the final N content of vermicompost and overall, the extent of decomposition. The decrease in pH also plays an important role in nitrogen retention, as at high pH, nitrogen is lost as volatile ammonia (Jiangwe et al. 2019).

5.12.9 Light Sensitivity

Earthworms exhibit negative phototropism (photonegative) and respond to the intensity of light and UV radiation that causes mortality to them (Mishra et al. 2019). Worms are less sensitive to moderate radiant heat than to bright light but very sensitive to contact. Earthworms stop moving on touch, and the response is so strong that it overcomes even light stimulation. With the same senses, they are active and keep on rearing when kept on a surface until they find a suitable burrow.

Studies have shown that vermicompost is up to seven times superior to conventional cattle dung compost in terms of nutrient and plant growth parameters. Earthworms produce a higher content of humus at a faster rate, a process that would otherwise take longer during conventional composting (Suhane 2007). Vital nutrients such as NPK are released in a shorter time and are retained for a relatively longer time in vermicompost in a more bioavailable form compared to conventional compost. As the organic matter passes through the earthworm gut, nutrients are converted to a more bioavailable form through the action of enzymes such as phosphatase, amylase, lipase, cellulase, and chitinase. These enzymes continue to hydrolyze organic matter to release nutrients and make them available to the plant roots. During vermicomposting, phosphatases and phosphate-solubilizing bacteria convert soil phosphate into a bioavailable form for plants (Lazcano et al. 2008). The higher humus content is also key to the growth and survival of plants and gives vermicompost its characteristic higher aeration, porosity, drainage, and water holding capacity compared to traditional composts. The burrowing and feeding activity of earthworms on organic matter promotes the rate of biodegradation by enhancing aerobic bacterial growth, which leads to the production of superior compost that is rich in key nutrients and soil microbes (Suhane 2007). The earthworms also produce coelomic fluid that kills pathogenic organisms.

5.13 Conclusion

Over a period of time, the application of vermicompost results in an increase in the earthworm population in the field as the baby worms grow out of their cocoons. These worms continue to strengthen and restore the soil's natural fertility by improving its physical, chemical, and biological properties in the long term. Vermicompost contains growth-promoting hormones, including auxins, cytokinins, and gibberellins, that are secreted by earthworms. It enhances crop production by stimulating plant growth in cereal crops, fruit crops, vegetables, and herbage. This has been attributed to the greater population of useful soil microbes, such as nitrogen fixers, actinomycetes, yeasts, and fungi, which improve nutrient availability. Farmers have reported improved rates of germination, a higher number of fruits per plant, a reduced incidence of pests and diseases, and improved vegetable and fruit taste and texture, among other enhanced agronomic traits. Although vermicomposting offers substantial benefits, it comes with its own set of challenges. The major drawbacks are that the process requires relatively more care to start and may not kill all pathogens.

References

Alabi DO, Coopoosamy R, Naidoo K, Arthur G (2022) Vermicompost: a sustainable biostimulant and control for agricultural enhancement in Africa. Irish J Agric Food Res 13(1):7. https://doi.org/10.3389/2593-489

Banta G, Kahlon RS (2007) Dehalogenation of 4—Chlorobenzoic acid by *Pseudomonas* isolates. Indian J Microbiol 47:139–143. https://doi.org/10.1007/s12088-007-0027-5

Bartz MLC, Passini A, Brown GG (2013) Earthworms as soil quality indicators in Brazilian no-tillage systems. Appl Soil Ecol 69:39–48. https://doi.org/10.1016/j.apsoil.2013.01.011

Cai L, Gong X, Sun X, Li S, Yu X (2018) Comparison of chemical and microbiological changes during the aerobic composting and vermicomposting of green waste. PLoS One 13(11): e0207494. https://doi.org/10.1371/journal.pone.0207494

Coulibaly SS, Irie ZB (2010) Influence of animal wastes on growth and reproduction of the African earthworm species Eudrilus eugeniae (Oligochaeta). Eur J Soil Biol 46(3–4):225–229. https://doi.org/10.1016/j.ejsobi.2010.03.004

Csuzdi C (2012) Earthworm species, a searchable database. Opusula Zoologica Budapest 43(1): 97–99

Darwin CR (n.d.) *The formation of vegetable mould through the action of worms with observations on their habits*

Davies L, Bell JNB, Bone J, Head M, Hill L, Howard C, Hobbs SJ, Jones DT, Power SA, Rose N, Ryder C, Seed L, Stevens G, Toumi R, Voulvoulis N, White PCL (2011) Open air laboratories (OPAL): a community-driven research programme. Environ Pollut 159:2203–2210

Devi J, Prakash M (2015) Microbial population dynamics during vermicomposting of three different substrates amended with cow dung. Int J Curr Microbiol Appl Sci 4(2):1086–1092

Dominguez J, Edwards CA (1997) Effects of stocking rate and moisture content on the growth and maturation of *Eisenia andrei* (Oligochaeta) in pig manure. Soil Biol Biochem 29:743–746. https://doi.org/10.1016/S0038-0717(96)00276-3

Edwards CA (1998) Earthworm ecology, 2nd edn. CRC Press, Boca Raton, p 389

Edwards CA (2010) Biology and ecology of earthworm species used for vermicomposting. Vermiculture technology: earthworms, organic waste and environmental management. Pp 25–37

Edwards CA, Bohlen PJ (1996) Biology and ecology of earthworm, 3rd edn. Chapman and Hall, London, p 426

Edwards CA, Burrows I (1988) The potential of earthworm composts as plant growth media. In: Edwards CA, Neuhauser E (eds) Earthworms in waste and environmental management. SPB Academic Press, Netherlands, pp 21–32

Edwards CA, Fletcher KE (1988) Interactions between earthworms and microorganisms in organic-matter breakdown. Agric Ecosyst Environ 24(1-3):235–247. https://doi.org/10.1016/0167-8809 (88)90069-2

Edwards CA, Subler S, Arancon N (2011) Quality criteria for vermicomposts. *Vermiculture technology: earthworms, organic waste and environmental management*. CRC Press, Boca Raton, pp 287–301

Entomological society of Britain (2022) Limitations of the Opal Earthworm Guide https://www.earthwormsoc.org.uk/OPALkey

Fowler K, Greenfield H, van Schalkwyk L (2000) The identification and significance of ceramic ecofacts from Ndondondwane, South Africa. Southern African Field Archaeol 9:32–42

Gaddie RE, Donald ED (1975) Earthworms for ecology and profit. 1st vol. scientific earthworm farming. Bookworm Publishing Company, California, p 180

Gajalakshmi S, Abbasi SA (2004) Earthworms and vermicomposting. Indian J Biotechnol 3:486–494

Gajalakshmi S, Ramasamy E, Abbasi S (2002) Vermicomposting of paper waste with the anecic earthworm Lampito mauritii Kinberg. Indian J Chem Technol 9:306–311

Garg VK, Kaushik P (2005) Vermistabilization of textile mill sludge spiked with poultry droppings by epigeic earthworm *Eisenia fetida*. Bioresour Technol 96:1063–1071. https://doi.org/10.1016/j.biortech.2004.09.003

Gómez-Brandón M, Aira M, Lores M, Domínguez J (2011) Epigeic earthworms exert a bottleneck effect on microbial communities through gut associated processes. PLoS One 6(9):e24786. https://doi.org/10.1371/journal.pone.0024786

Gunadi B, Blount C, Edwards CA (2002) The growth and fecundity of *Eisenia fetida* (Savigny) in cattle solids precomposted for different periods. Pedobiologia 46:15–23. https://doi.org/10.1078/0031-4056-00109

Hidalgo D, Corona F, Martín-Marroquín J (2022) Manure biostabilization by effective microorganisms as a way to improve its agronomic value. Biomass Conv Bioref. https://doi.org/10.1007/s13399-022-02428-x

Huerta E, Kampichler C, Geissen V, Ochoa-Gaona S, Jond B, Hernández-Daumás S (2009) Towards an ecological index for tropical soil quality based on soil macrofauna. Pesq Agropec Bras 44:1056–1062

Ismail SA (2005) The earthworm book. Other India Press, Mapusa, Goa, p 92

Jiangwe J, Komakech AJ, Karugi J, Amann A, Wanyama J, Lederer J (2019) Assessment of a cattle manure vermicomposting system using material flow analysis: a case study from Uganda. Sustainability 11:573. https://doi.org/10.3390/su11195173

Kaur T (2019) Vermicomposting: an effective option for recycling organic wastes. Organic Agriculture. IntechOpen. https://doi.org/10.5772/intechopen.91892

Lazcano C, Gomez-Brandon M, Dominguez J (2008) Comparison of the effectiveness of composting and vermicomposting for the biological stabilization of cattle manure. Chemosphere 72:1013–1019

Medina-Sauza RM, Álvarez-Jiménez M, Delhal A, Reverchon F, Blouin M, Guerrero-Analco JA, Cerdán CR, Guevara R, Villain L, Barois I (2019) Earthworms building up soil microbiota, a review. Front Environ Sci 7:81. https://doi.org/10.3389/fenvs.2019.00081

Mishra CS, Nayak S, Samal S (2019) Low intensity light effects on survivability, biomass, tissue protein and enzyme activities of the earthworm *Eudrilus eugeniae* (Kinberg). Invertebr Surviv J 16(1):8–14

OPAL (2022a) The Soil and Earthworm survey. http://www.opalexplorenature.org/surveys/soilsurveys

OPAL (2022b) Open Air Laboratories. Explore Nature. http://www.opalexplorenature.org/. Accessed 2022 March 2. Natural History Museum

Ramnarain YI, Ansari A, Ori L (2018b) Vermicomposting of different organic materials using the epigeic earthworm *Eisenia fetida*. Int J Recycl Org Waste Agric. https://doi.org/10.1007/s40093-018-0225-7

Ramnarain YI, Ori LYDIA, Ansari AA (2018a) Effect of the use of vermicompost on the plant growth parameters of Pak choi (Brassica rapa var. chinensis) and on the soil structure in Suriname. J Global Agric Ecol 8(1):8–15

Reynolds JW, Wetzel MJ (2004) NomenclaturaOligochaetologica. SupplementumQuartum. A catalogue of names, descriptions and type specimens of the Oligochaeta. Natural History Survey Special Publication, Illinois

Rostami R, Najafi SH, Nabaee A, Eslami A (2010) Survey of *E. foetida* population on pH, C/N ratio process rate in vermicompost. J Environ Stud 35:93–98

Rota E (2011) Early oligochaete science, from Aristotle to Francesco Redi. Arch Nat History 38: 136–163. https://doi.org/10.3366/anh.2011.0011

Sadia MA, Hossain MA, Islam MR, Akter T, Shaha DC (2020) Growth and reproduction performances of earthworm (*Perionyx excavatus*) fed with different organic waste materials. J Adv Vet Animal Res 7(2):331–337. https://doi.org/10.5455/javar.2020.g426

Siles JA, García-Sánchez M, Gómez-Brandón M (2021) Studying microbial communities through co-occurrence network analyses during processes of waste treatment and in organically amended soils: a review. Microorganisms 9(6):1165. https://doi.org/10.3390/microorganisms9061165

Sinha RK and Gokul B (2007) Vermiculture revolution: rapid composting of waste organics with improved compost quality for healthy plant growth while reducing ghg emissions. *3rd National Compost Research and Development Forum Organized by COMPOST Australia*, Murdoch University, Perth

Sinha RK, Nair J, Bharambe B, Patil S, Bapat P (2008) Vermiculture revolution: a low-cost and sustainable technology for management of municipal and industrial organic wastes (solid and liquid) by earthworms with significantly low greenhouse gas emissions. In: Daven JI, Klein RN (eds) Progress in waste management research. Nova Science Publishers, New York, pp 158–229

Smetak KM, Jonhson-Maynard J, Lloyd JE (2007) Earthworm population density and diversity in different aged urban systems. Appl Soil Ecol 37(1):161–168. https://doi.org/10.1016/j.apsoil.2007.06.004

Stroud J & Bennet A (2018) How to count earthworms in great soils; fact sheet. Agriculture and Horticulture development Board

Suhane RK (2007) Vermicompost. Publication of Rajendra Agriculture University, Pusa, Bihar, India, p 88

Suthar S (2006) Potential utilization of guargum industrial waste in vermicompost production. Bioresour Technol 97:2474–2477. https://doi.org/10.1016/j.biortech.2005.10.018

Tchobanoglous G, Theisen H, Vigil S (1993) Integrated solid waste management. McGraw-Hill, New York, p 4

Wani KA, Mamta, Rao RJ (2013) Bioconversion of garden waste, kitchen waste and cow dung into value added products using earthworm *Eisenia fetida*. Saudi J Biol Sci 20:149–154

Part II
Vermicompost Production

Chapter 6
Vermicomposting as an Eco-Friendly Approach for Recycling and Valorization Grape Waste

María Gómez-Brandón, Manuel Aira, and Jorge Domínguez

Abstract Finding strategies to treat, dispose of and reuse organic wastes is of utmost need. Biological processes offer the possibility to transform them into safer end products with benefits for both agriculture and the environment. This is of particular interest for the winemaking industry given its increasing activity worldwide with the subsequent generation of a wide variety of waste streams. The purpose of this chapter is to evaluate the effectiveness of the vermicomposting process as a low-cost and environmentally safe solution for the treatment and valorization of raw and distilled grape marc, the major solid by-products derived from the winery and distillery industry. We give an overview of the performance of the vermicomposting trial together with an in-depth characterization of the respective vermicomposts by looking at a combination of physicochemical, biochemical and microbiological indicators.

Keywords Winery waste · Enzymatic activities · Microbial diversity · Vermicompost · Soil amendment

6.1 Introduction

The winemaking industry constitutes one of the most important worldwide, agro-industrial sectors from an economic, social and cultural perspective (Hussain et al. 2008). An estimation carried out by the International Organization of Vine and Wine stated that about 260 million hl were produced globally in 2020 (http://www.oiv.int). The European Union (EU) comprises 44% of world wine-growing areas, with Spain, France and Italy as the three Member States accounting for 76% of EU areas under vines. Other major wine producing countries are the United States, Argentina, Chile, Australia and South Africa.

M. Gómez-Brandón (✉) · M. Aira · J. Domínguez
Grupo de Ecoloxía Animal (GEA), Universidade de Vigo, Vigo, Spain
e-mail: mariagomez@uvigo.es

© The Author(s), under exclusive license to Springer Nature Singapore Pte Ltd. 2023
H. A. Mupambwa et al. (eds.), *Vermicomposting for Sustainable Food Systems in Africa*, Sustainable Agriculture and Food Security,
https://doi.org/10.1007/978-981-19-8080-0_6

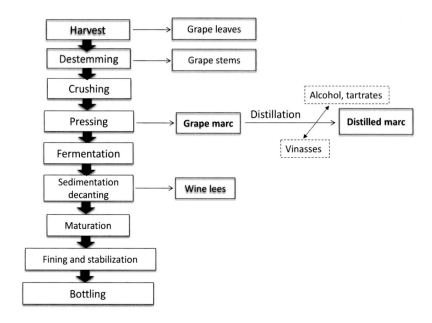

Fig. 6.1 Processing steps and by-products derived from the winery and distillery industry

Vitis vinifera is considered the most commonly cultivated grape variety for wine production (FAO 2015). From grape harvest to bottling, the increasing activity related to the winemaking process generates an ample variety of solid and liquid by-products including vine shoots, grape marc, wine lees, spent filter cakes, vinasses and winery wastewater (Fig. 6.1). Taken together, it raises the necessity to search for profitable and sustainable options to treat, dispose of or to properly reuse these waste streams in order to reduce to a minimum their disposal and avoid negative environmental impacts (Spigno et al. 2017; Muhlack et al. 2018; Gómez-Brandón et al. 2019a). This is of particular interest for grape marc, also known as grape bagasse or grape pomace given that approximately 25% of the grape mass turns into this by-product during wine production (Domínguez et al. 2014).

Grape marc is composed of the seeds, pulp and stalks that remain after pressing the grapes (Domínguez et al. 2016, 2017; Fig. 6.1). Around 1 kg of this by-product results from the production of 5 L of wine, accounting for a worldwide production of 10–13 Mtons per year. A way to economically valorize grape marc is through distillation that permits the recovery of ethanol for its use in the elaboration of alcoholic beverages (Fig. 6.1; Cisneros-Yupanqui et al. 2022). In a simplified way, the sugars present in the raw marc are subjected to alcoholic fermentation under anaerobic conditions, followed by the subsequent distillation that can be carried out in batches or on a continuous basis (Botelho et al. 2020). Nonetheless, the distillation process also involves the generation of distillery effluents including vinasses and distilled grape marc (Zhang et al. 2017; Fig. 6.1) that altogether can represent a challenging waste disposal problem for the winery and distillery industry.

The use of raw and distilled marc as a soil amendment constitutes as an alternative option for their valorization, promoting waste prevention and new recycling goals. Both winery by-products are rich in organic matter and macronutrients that are essential for plant development (Requejo et al. 2016; Domínguez et al. 2016, 2017). However, if not properly treated, their application can cause soil acidification, groundwater pollution and oxygen depletion in soil due to the release of tannins and polyphenols with antimicrobial and phytotoxic effects (Domínguez et al. 2016).

The biological treatment of grape marc, either raw or distilled, through aerobic biodegradation processes offers a promising avenue to handle and process these by-products with the dual purpose of fertilizer production and environment protection (Gómez-Brandón et al. 2019a). The stabilization of raw and distilled grape marc via composting has been explored in this regard (Zhang and Sun 2016; Muhlack et al. 2018; Martínez-Salgado et al. 2019; Paradelo et al. 2019; Pinter et al. 2019; Cortés et al. 2020). Nevertheless, the low pH of grape marc could negatively affect microbial activity and inhibit the transition between mesophilic and thermophilic composting phases (Paradelo et al. 2019). Co-composting with other organic materials has been used as a more effective alternative improving not only the compost quality but also shortening the time required to achieve the stabilization of raw and distilled marc (Zhang and Sun 2016; Costa Barros et al. 2021).

Vermicomposting has also been successfully used to dispose of and treat large quantities of a wide variety of organic wastes (Domínguez and Gómez-Brandón 2012), and can provide a means to overcome the potential limitations of composting with regard to the stabilization of grape marc. The advantage of vermicomposting is twofold: it favours pH neutralization as a result of earthworms' activity (Domínguez et al. 2014; Gómez-Brandón et al. 2020a); and at the same time it offers the possibility to obtain a high-quality fertilizer with beneficial effects on soil and plants (Domínguez et al. 2016, 2017; Gómez-Brandón et al. 2020a, b). The bioconversion of biomass waste into organic fertilizers through the process of vermicomposting is linked to the synergistic actions of earthworms and microbial communities (Domínguez et al. 2010). The underlying biological mechanisms involved largely determine the dynamics of the process and, consequently, the properties of the end product for its further use as a plant growth promoter and in plant disease suppressiveness (Gómez-Brandón and Domínguez 2014).

The purpose of this chapter is to provide an overview of the vermicomposting process as a low-cost and environmentally safe solution for the treatment and valorization of raw and distilled grape marc derived from the winery sector. We will present a case study to illustrate the effectiveness of vermicomposting for the stabilization of these winery by-products together with an in-depth characterization of the final vermicomposts by looking at a combination of physicochemical, biochemical and microbiological indicators.

6.2 How Vermicomposting Works: The Experimental Set-up

Raw and distilled marc of the grape variety *Vitis vinifera* L. cv. Mencía, which represents 95% of the annual red grape harvest in northwestern Spain, were processed in pilot-scale vermireactors (6 m^2; Fig. 6.2a) using the earthworm *Eisenia andrei* (Oligochaeta, Lumbricidae; Fig. 6.2b). This is one of the most widely used earthworm species in vermicomposting facilities because of its high rate of consumption, digestion and assimilation of organic matter, and its tolerance to a wide range of temperature and moisture conditions (Domínguez and Edwards 2011a). The vermireactors were continuously fed for almost a year with grape marc, either raw or distilled, allowing the population density of earthworms to reach its maximum capacity, with an average value of $11{,}115 \pm 2827$ individuals m^{-2} that corresponds with a mean biomass of 1361 ± 415 g m^{-2}. This points to the suitability of raw and

Fig. 6.2 (**a**) Overview of the pilot-scale vermireactors fed with raw and distilled marc; (**b**) Processing of grape marc by the earthworm *Eisenia andrei*; (**c**) Fresh layer of grape marc on top of the plastic mesh for the performance of the vermicomposting trial

distilled marc derived from Mencía grape cultivar as an optimum feedstock for earthworm growth and reproduction providing sufficient energy to sustain large populations in vermicomposting systems.

To evaluate the effectiveness of vermicomposting for the stabilization of grape marc, a new layer of the substrate (raw or distilled marc, 50 kg fresh weight) was placed over a plastic mesh that allowed earthworm migration and prevented the mixing of the processed grape marc and the vermicomposting bedding (Fig. 6.2c). Control reactors containing the same amount of raw or distilled marc in the absence of earthworms were also set up during the vermicomposting trial. All the reactors with and without earthworms were covered with a shade cloth to keep the moisture level of the substrate at approximately 85% during the trial. This is considered an optimum moisture level for the growth and reproduction of epigeic earthworms, as well as for a good performance of the vermicomposting process (Domínguez and Edwards 2011a). After a two-month period, the substrate was completely processed by the earthworms and the raw and distilled grape marc-derived vermicomposts as well as the respective control samples were collected from the surface of the mesh, stored at 4 °C or air-dried depending on the analytical measurement, and further processed for their characterization.

6.3 Physicochemical Characterization and Nutrient Content of Grape Marc Vermicompost

Among the physicochemical parameters, pH is one of the most widely used to study the dynamics of the vermicomposting process (Ali et al. 2015), as it affects environmental conditions that are relevant for microbial growth and survival (Jin and Kirk 2018). Changes in pH may influence not only the solubility and bioavailability of nutrients and trace elements over the course of the process (Domínguez et al. 2016, 2017), but also the synthesis and activity of extracellular enzymes (Sánchez-Hernández and Domínguez 2017). Vermicomposting was effective to neutralizing the acidity of raw and distilled grape marc, reaching a pH value close to neutrality in the respective vermicomposts after the two-month period (Table 6.1). An increase in pH towards neutral values as a result of earthworms´ activity is an important asset for the potential usefulness of vermicompost as a soil amendment, considering that the response of crops to organic amendments is more favourable when soil pH ranges from weak-acidic to weak-alkaline levels (Luo et al. 2018; Cataldo et al. 2021).

The electrical conductivity (EC) is also considered an important physicochemical parameter to evaluate during vermicomposting because it reflects the degree of salinity and can largely affect the survival and reproduction of earthworms if it reaches values higher than 8.0 mS cm^{-1} (Domínguez and Edwards 2011b). The EC levels of the studied vermicomposts (circa 0.3 mS cm^{-1}; Table 6.1) seem optimal for earthworm survival and were in agreement with those reported in previous vermicomposting trials dealing with grape marc (Cástková and Hanc 2019;

Table 6.1 Overview of the physicochemical properties and the nutrient content of the vermicomposts obtained from raw and distilled marc after a two-month period. Values are means $\pm$ standard error. Nutrient data are on a dry weight (dw) basis

	Raw marc		Distilled marc	
	Control	Vermicompost	Control	Vermicompost
pH	3.76 ± 0.05	6.90 ± 0.03	4.30 ± 0.07	7.53 ± 0.03
EC (mS cm^{-2})	0.84 ± 0.02	0.31 ± 0.02	0.89 ± 0.03	0.20 ± 0.01
Ca (mg kg^{-1} dw)	3680 ± 49	4391 ± 196	3257 ± 52	3669 ± 55
K (mg kg^{-1} dw)	$23,654 \pm 568$	$11,956 \pm 351$	$17,721 \pm 748$	$13,161 \pm 415$
P (mg kg^{-1} dw)	3116 ± 61	1864 ± 38	2582 ± 56	1818 ± 96
Mg (mg kg^{-1}dw)	1135 ± 15	988 ± 40	976 ± 31	1047 ± 49
Mn (mg kg^{-1} dw)	40 ± 0.4	37 ± 1	39 ± 0.6	43 ± 2
Fe (mg kg^{-1} dw)	92 ± 4	129 ± 6	98 ± 3	134 ± 33
S (mg kg^{-1} dw)	1528 ± 31	1141 ± 45	1579 ± 32	1650 ± 64

Gómez-Brandón et al. 2020a). The grape marc-derived vermicomposts had EC values lower than those in the control treatment likely due to a reduced presence of mineral salts in available forms such as K, P and NH_4^+ which are known as main contributors to the electrical conductivity.

Epigeic earthworms are known to accelerate the turnover rate of organic matter during vermicomposting (Domínguez and Gómez-Brandón 2012), by grazing directly on microorganisms and/or by increasing the surface area available for microbial attack through comminution. Supporting this, the initial mass of the raw and distilled marc (50 kg fresh mass; 10.75 kg dry mass) was considerably reduced, by approximately 75%, as a result of the earthworm activity, reaching a final fresh mass of 9.26 kg (2.44 kg dry mass) after 2 months of vermicomposting. The pronounced reduction of the initial mass of grape marc leads to an increase in the density of grape seeds (Domínguez et al. 2014, 2016, 2017). Polyphenolic compounds represent around 7% of seeds' composition and in this regard, Domínguez et al. (2014, 2016) reported that separating the seeds from the vermicompost through sieving may offer a twofold advantage: (i) to obtain a polyphenol-free vermicompost for its safe application into soil considering that some phenolic compounds can negatively affect plant growth and/or disrupt the microbial community balance in the rhizosphere (Jilani et al. 2008) and (ii) to further exploit the nutritional, cosmetic and health-promoting properties of polyphenols recovered from the seeds with respect to their antioxidant, antifungal and scavenging activities (Lores et al. 2013; García-Jares et al. 2015; Álvarez-Casas et al. 2016; Barba et al. 2016).

As a consequence of the enhanced mineralization in the presence of earthworms, macro- and micronutrients such as calcium (Ca) and iron (Fe) appeared in higher concentrations in the grape marc-derived vermicomposts when compared to the control without earthworms (Table 6.1). However, this did not hold for potassium (K) and phosphorus (P) whose contents were much lower in the vermicomposts obtained from raw and distilled marc (Table 6.1). A decrease in these nutrients' concentration throughout the process of vermicomposting might occur via leaching, immobilization by microorganisms or earthworms, or precipitation in the form of non-mineral salts (Domínguez et al. 2018). This decreasing trend in K and P was not observed when cattle manure was used as feedstock in previous vermicomposting trials (Domínguez and Gómez-Brandón 2013). Grape marc is of lignocellulosic nature and the resulting vermicompost can be thought to represent the process of a single gut—that is, the starting material passed only through the earthworm gut. Nonetheless, in vermicomposting applications with cattle manure or other types of manure the feedstock has already passed through the vertebrate gut (i.e. cow, pig, horse).

To sum up, the values reported with regard to the physicochemical and the nutrient properties of raw and distilled grape marc-derived vermicomposts generally matched with those considered to have the quality criteria of a good vermicompost (Domínguez and Edwards 2011b; Domínguez et al. 2017).

6.4 Biochemical and Microbiological Characterization of Grape Marc Vermicompost

During vermicomposting, the biochemical decomposition of the organic matter is primarily accomplished through the chemical reactions performed by microorganisms via the production of extracellular enzymes. However, the effects that earthworms have on microbial communities in terms of biomass, activity and/or composition will largely determine the dynamics of the process and the quality of the end product (Gómez-Brandón and Domínguez 2014; Gómez-Brandón et al. 2020b). Bearing this in mind, over the last twenty years our research group has been exploring the relationships between earthworms and microorganisms during vermicomposting of a wide variety of organic wastes by assessing the changes in microbial biomass carbon and nitrogen, as well as in basal respiration as a proxy of microbial activity, and in the activity of extracellular enzymes involved in the major nutrient cycles (Aira et al. 2002, 2006a, b, 2007; Aira and Domínguez 2008, 2011; Domínguez and Gómez-Brandón 2013; Domínguez et al. 2018; Almeida-Santana et al. 2020; Gómez-Brandón et al. 2021). Shifts in community composition over the course of vermicomposting have also been studied by using culture-independent methods such as phospholipid fatty acid analysis (Aira et al. 2011; Gómez-Brandón et al. 2011, 2012, 2013), and more recently by molecular tools based on DNA extraction, followed by amplification (via polymerase chain reaction, PCR) and sequencing of marker genes (Domínguez et al. 2019; Kolbe et al. 2019; Gómez-Brandón et al. 2019b, 2020c; Rosado et al. 2022).

6.4.1 Microbial Biomass and Activity

Assessing vermicompost stability is crucial for its potential usefulness as a soil amendment and plant growth promoter. After incorporation to soils, unstable vermicomposts may induce enhanced microbial activity and cause oxygen deficiency and a variety of phytotoxicity problems to plant roots. Stability is related to the degree to which vermicomposts have been decomposed to more stable organic materials and it is typically evaluated by different respirometric measurements and/or by studying the transformations in the chemical characteristics of organic matter (Bernal et al. 2017).

Both the raw and distilled grape marc-derived vermicomposts were characterized by lower levels of microbial biomass carbon and microbial activity assessed as basal respiration when compared to the control without earthworms (Fig. 6.3a, b). Earthworms' activity can reduce microbial biomass directly by selectively feeding on specific taxa during transit through the earthworm gut (Drake and Horn 2007; Aira et al. 2015), or indirectly by accelerating the depletion of resources available for the microbes. In line with this, the resulting vermicomposts were characterized by lower abundances of both bacteria and fungi after the two-month period (Fig. 6.3c,d).

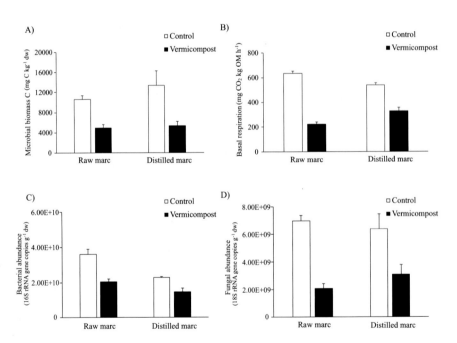

Fig. 6.3 Impact of earthworms (*Eisenia andrei*) on microbial biomass and activity after vermicomposting of raw and distilled marc for a period of 2 months: (**a**) Microbial biomass carbon; (**b**) Basal respiration, a measure of microbial activity; (**c, d**) Bacterial and fungal abundances estimated by real-time PCR. Values are mean ± SE. Control is the same treatment in the absence of earthworms

Altogether, these findings point to the effectiveness of vermicomposting at biologically stabilizing these winery by-products, as reflected by decreases in both microbial biomass and its activity in comparison with the control treatment.

6.4.2 Enzymatic Activities

From a functional perspective, the activity of extracellular enzymes has received considerable attention because of their sensitivity to environmental changes and their role in the breakdown or mineralization of major nutrients like carbon (C), nitrogen (N) and P into inorganic forms that can be used by plants (Acosta-Martínez et al. 2018).

Three enzyme activities involved in the C cycle were measured in the grape marc-derived vermicomposts, namely the β-glucosidase, the cellobiohydrolase and the xylosidase (Fig. 6.4a–c). They have a relevant role in organic matter decomposition processes because they catalyse the hydrolysis of cellulose and hemicellulose, two of the main components in the plant debris (Nannipieri et al. 2012). Both β-glucosidase and cellobiohydrolase activities had lower values in the resulting vermicomposts in comparison with the control without earthworms (Fig. 6.4a, b). It is likely that the accelerated depletion of resources due to earthworm activity leads to reduced enzyme activities towards the end of the process (Benítez et al. 2005). The lower levels of these C-associated enzymes were in line with the reduced microbial biomass and activity in the vermicomposts, reinforcing the use of these enzyme activities as indicators of the suitability of vermicomposting for the biological stabilization of raw and distilled marc.

However, the opposite trend was observed for xylosidase whose activity was higher in the vermicompost than in the control samples (Fig. 6.4c). A plausible explanation relies on the increasing concentrations of humic substances that appear as vermicomposting progresses providing chemical support for binding extracellular enzymes, and protecting them against proteases or adverse environmental conditions (Castillo et al. 2013). Likewise, leucine-aminopeptidase activity responsible for the release of amino acids from polypeptides, as well as alkaline phosphomonoesterase reached higher values in the raw and distilled marc-derived vermicomposts (Fig. 6.4d, f). This latter enzyme is involved in the mineralization of organic P into phosphate by hydrolysing phosphoric (mono) ester bonds under alkaline conditions. In contrast, acid phosphomonoesterase showed lower or similar values than those in the control (Fig. 6.4e). Earthworms can alter nutrient cycling and increase N and P uptake by plants through a combination of biochemical and chemical pathways (Medina-Sauza et al. 2019). Taken together, our findings are consistent with these observations providing evidence of the enhancing effect of the earthworm *E. andrei* on certain enzyme activities involved in the breakdown or mineralization of N and P into inorganic forms that can be used by plants.

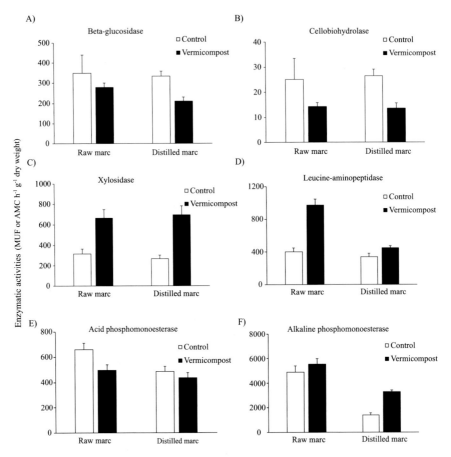

Fig. 6.4 Impact of earthworms (*Eisenia andrei*) on enzymatic activities related to C-, N- and P-cycles after vermicomposting of raw and distilled marc for a period of 2 months: (**a**) Beta-glucosidase, (**b**) Cellobiohydrolase, (**c**) Xylosidase, (**d**) Leucine-aminopeptidase and (**e, f**) Acid and Alkaline phosphomonoesterases. Values are mean ± SE. Control is the same treatment in the absence of earthworms. Units are given as nanomoles of 4-methyl-umbelliferone (MUF) h^{-1} g^{-1} dry weight, except for leucine-aminopeptidase which activity is expressed as nanomoles of 7-amino-4-methyl coumarin (AMC) h^{-1} g^{-1} dry sample

6.4.3 Microbial Richness and Diversity

Abundant research has shown that the use of vermicomposts has beneficial effects when used as an amendment for soil or plant growth media for a wide variety of agronomic and horticultural crops (Lazcano and Domínguez 2011; Blouin et al. 2019). Various factors such as an improved availability of air and water, the presence of plant growth regulating substances, and the mitigation or suppression of plant diseases have been proposed as plausible, albeit not exclusive, mechanisms by which such improvement is achieved (Gómez-Brandón and Domínguez 2014).

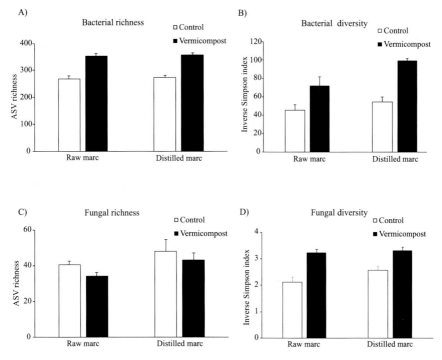

Fig. 6.5 Impact of earthworms (*Eisenia andrei*) on microbial richness calculated as the number of observed amplicon sequence variant (ASVs), and microbial diversity assessed as the inverse Simpson index after vermicomposting of raw and distilled marc for a period of 2 months: (**a**) Bacterial richness, (**b**) Bacterial diversity, (**c**) Fungal richness and (**d**) Fungal diversity. Values are mean ± SE. Control is the same treatment in the absence of earthworms

Vermicomposts are considered bioactive organic materials and there exist evidence about microbial-based mechanisms that may explain the positive influence of vermicompost on soil and plants (Domínguez et al. 2019; Gómez-Brandón et al. 2020b). We observed that the richness and diversity of bacterial communities were higher in the grape marc-derived vermicomposts compared to the control treatment (Fig. 6.5a, b). The same trend was recorded for fungal diversity (Fig. 6.5d), whilst fungal richness had similar values in the presence and the absence of earthworms (Fig. 6.5c). Microbial diversity may benefit soils in two major ways, by safeguarding a steady nutrient supply through decomposition processes that cope with changing environmental conditions, and secondly by supplying a microbiota that is disease suppressive. In this regard, higher abundances of putative genes involved in the biosynthesis of antibiotics or plant hormone synthesis were found in vermicomposts when compared to the starting material (Domínguez et al. 2019; Kolbe et al. 2019; Gómez-Brandón et al. 2020b).

6.5 Conclusions

Vermicomposting has proven to be an effective, environmentally sound management system for processing raw and distilled grape marc derived from the winery industry. Earthworm activity favoured the stabilization of these by-products resulting in a final vermicompost characterized by a higher concentration of macro- and micronutrients essential for plant growth. Moreover, lower values of microbial biomass and microbial activity, indicative of stabilized materials, together with a reduced activity of hydrolytic enzymes involved in cellulose degradation were found after two months of vermicomposting. This was accompanied by an increased microbial richness and diversity in the grape marc-derived vermicomposts reinforcing their potential role as bioactive organic materials for its further use as an amendment for soil or plant growth media.

Acknowledgements This study was supported by the Spanish Ministerio de Ciencia e Innovación [PID2021-124265OB-I00 and TED2021-129437B-I00] and the Xunta de Galicia [grant numbers ED431C 2022/07]. MGB acknowledges support by the Programa Ramón y Cajal [RYC-2016-21231; Ministerio de Economía y Competitividad]. Paul Fraiz is also acknowledged for his valuable help with English wording and editing.

References

Acosta-Martínez V, Cano A, Johnson J (2018) Simultaneous determination of multiple soil enzyme activities for soil health-biogeochemical indices. Appl Soil Ecol 126:121–128

Aira M, Bybee S, Pérez-Losada M, Domínguez J (2015) Feeding on microbiomes: effects of detritivory on the taxonomic and phylogenetic bacterial composition of animal manures. FEMS Microbiol Ecol 91:fiv117

Aira M, Domínguez J (2008) Optimizing vermicomposting of animal wastes: effects of dose of manure application on carbon loss and microbial stabilization. J Environ Manag 88:1525–1529

Aira M, Domínguez J (2011) Earthworm effects without earthworms: inoculation of raw organic matter with worm-worked substrates alters microbial community functioning. PLoS One 6: e16354

Aira M, Gómez-Brandón M, González-Porto P, Domínguez J (2011) Selective reduction of the pathogenic load of cow manure in an industrial-scale continuous-feeding vermireactor. Bioresour Technol 102:9633–9637

Aira M, Monroy F, Domínguez J (2002) How earthworm density affects microbial biomass and activity in pig manure. Eur J Soil Biol 38:7–10

Aira M, Monroy F, Domínguez J (2006a) C to N ratio strongly affects population structure of *Eisenia fetida* in vermicomposting systems. Eur J Soil Biol 42:127–131

Aira M, Monroy F, Domínguez J (2006b) *Eisenia fetida* (Oligochaeta, Lumbricidae) activates fungal growth, triggering cellulose decomposition during vermicomposting. Microb Ecol 52: 738–746

Aira M, Monroy F, Domínguez J (2007) Earthworms strongly modify microbial biomass and activity triggering enzymatic activities during vermicomposting independently of the application rates of pig slurry. Sci Tot Environ 385:252–261

Ali U, Sajid N, Khalid A, Riaz L, Rabbani MM, Syed JH, Malik RN (2015) A review on vermicomposting of organic wastes. Environ Prog Sustain Energy 34:1050–1062

Almeida-Santana N, Seminoti Jacques RJ, Antoniolli ZI, Martínez-Cordero H, Domínguez J (2020) Changes in chemical and biological characteristics of grape marc vermicompost during a two-year production period. Appl Soil Ecol 154:103587

Álvarez-Casas M, Pájaro M, Lores M, García-Jares M (2016) Characterization of grape marcs from native and foreign white varieties grown in northwestern Spain by their polyphenolic composition and antioxidant activity. Eur Food Res Technol 242:655–665

Barba FJ, Zhu Z, Koubaa M, Sant'Ana AS, Orlien V (2016) Green alternative methods for the extraction of antioxidant bioactive compounds from winery wastes and by-products: a review. Trends Food Sci Technol 49:96–109

Benítez E, Sainz H, Nogales R (2005) Hydrolytic enzyme activities of extracted humic substances during the vermicomposting of a lignocellulosic olive waste. Bioresour Technol 96:785–790

Bernal MP, Sommer SG, Chadwick D, Qing C, Guoxue L, Michel FC Jr (2017) Current approaches and future trends in compost quality criteria for agronomic, environmental and human health benefits. Adv Agron 144:143–233

Blouin M, Barrere J, Meyer N, Lartigue S, Barot S, Mathieu J (2019) Vermicompost significantly affects plant growth. A meta-analysis. Agron Sustain Dev 39:34

Botelho G, Anjos O, Estevinho LM, Caldeira I (2020) Methanol in grape derived, fruit and honey spirits: a critical review on source, quality control, and legal limits. PRO 8:1609

Castillo JM, Romero E, Nogales R (2013) Dynamics of microbial communities related to biochemical parameters during vermicomposting and maturation of agroindustrial lignocellulose wastes. Bioresour Technol 146:345–354

Cástková T, Hanc A (2019) Change of the parameters of layers in a large-scale grape marc vermicomposting system with continuous feeding. Waste Manage Res 37:826–832

Cataldo E, Fucile M, Mattii GB (2021) A review: soil management, sustainable strategies and approaches to improve the quality of modern viticulture. Agronomy 11:2359

Cisneros-Yupanqui M, Rizzi C, Mihaylova D, Lante A (2022) Effect of the distillation process on polyphenols content of grape pomace. Eur Food Res Technol 248:929–935

Cortés A, Moreira MT, Domínguez J, Lores M, Feijoo G (2020) Unraveling the environmental impacts of bioactive compounds and organic amendment from grape marc. J Environ Manag 272:111066

Costa Barros ES, de Amorim MCC, Olszevski N, de Sousa e Silva PT (2021) Composting of winery waste and characteristics of the final compost according to Brazilian legislation. J Environ Sci Health B 56:447–457

Domínguez J, Aira M, Gómez-Brandón M (2010) Vermicomposting: earthworms enhance the work of microbes. In: Insam H, Franke-Whittle I, Goberna M (eds) Microbes at work: from wastes to resources. Springer, Berlin, pp 93–114

Domínguez J, Aira M, Kolbe AR, Gómez-Brandón M, Pérez-Losada M (2019) Changes in the composition and function of bacterial communities during vermicomposting may explain beneficial properties of vermicompost. Sci Rep 9:9657

Domínguez J, Edwards E (2011a) Biology and ecology of earthworm species used for vermicomposting. In: Edwards CA, Arancon MQ, Sherman RL (eds) Vermiculture technology: earthworms, organic waste and environmental management. CRC Press, Boca Raton, Florida, pp 25–37

Domínguez J, Edwards E (2011b) Relationships between composting and vermicomposting: relative values of the products. In: Edwards CA, Arancon MQ, Sherman RL (eds) Vermiculture technology: earthworms, organic waste and environmental management. CRC Press, Boca Raton, Florida, pp 1–14

Domínguez J, Gómez-Brandón M (2012) Vermicomposting: composting with earthworms to recycle organic wastes. In: Kumar S, Bharti A (eds) Management of organic waste. Intech Open Access Publisher, pp 29–48

Domínguez J, Gómez-Brandón M (2013) The influence of earthworms on nutrient dynamics during the process of vermicomposting. Waste Manage Res 31:859–868

Domínguez J, Gómez-Brandón M, Martínez-Cordeiro H, Lores M (2018) Bioconversion of Scotch broom into a high-quality organic fertilizers: vermicomposting as sustainable option. Waste Manage Res 36:1092–1099

Domínguez J, Martínez-Cordeiro H, Ávarez-Casas M, Lores M (2014) Vermicomposting grape marc yields high quality organic biofertiliser and bioactive polyphenols. Waste Manage Res 32:1235–1240

Domínguez J, Martínez-Cordeiro H, Lores M (2016) Simultaneous production of a high-quality biofertilizer and bioactive-rich seeds. In: Morata A, Loira I (eds) Grape and wine biotechnology. Intech Open Science, Rijeka, Croatia, pp 167–183

Domínguez J, Sánchez-Hernández JC, Lores M (2017) Vermicomposting of wine-making products. In: Galanakis CM (ed) Handbook of grape processing by-products: sustainable solutions. Academic Press, Elsevier, London, UK, pp 55–78

Drake HL, Horn MA (2007) As the worm turns: the earthworm gut as a transient habitat for soil microbial biomes. Annu Rev Microbiol 61:169–189

Food and Agriculture Organization of the United Nations. FAO STAT 2015. http://faostat.fao.org

García-Jares C, Vázquez A, Lamas JP, Pájaro M, Álvarez-Casas M, Lores M (2015) Antioxidant White grape seed phenolics: pressurized liquid extracts from different varieties. Antioxidants 4:737–749

Gómez-Brandón M, Aira M, Domínguez J (2020b) Vermicomposts are biologically different: microbial and functional diversity of green vermicomposts. In: Bhat SA, Pal Vig A, Li F, Ravindran B (eds) Earthworm assisted remediation of effluents and wastes. Springer Nature, Singapur, pp 150–170

Gómez-Brandón M, Aira M, Kolbe AR, Pérez-Losada M, Domínguez J (2019b) Rapid bacterial community changes during vermicomposting of grape marc derived from red winemaking. Microorganisms 7:473

Gómez-Brandón M, Aira M, Santana N, Pérez-Losada M, Domínguez J (2020c) Temporal dynamics of bacterial communities in a pilot-scale vermireactor fed with distilled grape marc. Microorganisms 8:642

Gómez-Brandón M, Domínguez J (2014) Recycling of solid organic wastes through vermicomposting: microbial community changes throughout the process and use of vermicompost as a soil amendment. Crit Rev Environ Sci Technol 44:1289–1312

Gómez-Brandón M, Lazcano C, Lores M, Domínguez J (2011) Short-term stabilization of grape marc through earthworms. J Hazard Mat 187:291–295

Gómez-Brandón M, Lores M, Domínguez J (2012) Species-specific effects of epigeic earthworms on microbial community structure during first stages of decomposition of organic matter. PLoS One 7:e31895

Gómez-Brandón M, Lores M, Domínguez J (2013) Changes in chemical and microbiological properties of rabbit manure in a continuous-feeding vermicomposting system. Bioresour Technol 128:310–316

Gómez-Brandón M, Lores M, Insam H, Domínguez J (2019a) Strategies for recycling and valorization of grape marc. Crit Rev Biotechnol 39:437–450

Gómez-Brandón M, Lores M, Martínez-Cordeiro H, Domínguez J (2020a) Effectiveness of vermicomposting for bioconversion of grape marc derived from red winemaking into a value-added product. Environ Sci Pollut Res 27:33438–33445

Gómez-Brandón M, Martínez-Cordeiro H, Domínguez J (2021) Changes in the nutrient dynamics and microbiological properties of grape marc in a continuous-feeding vermicomposting system. Waste Manag 135:1–10

Hussain M, Cholette S, Castaldi RM (2008) An analysis of globalization forces in the wine industry: implications and recommendations for wineries. J Global Market 21:33–47

Jilani G, Mahmood S, Chaudhry AN, Hassan I, Akram M (2008) Allelochemicals: sources, toxicity and microbial transformation in soil - a review. Ann Microbiol 58:351–357

Jin Q, Kirk MF (2018) pH as a primary control in environmental microbiology: 1. Thermodynamic perspective. Front Environ Sci 6:21

Kolbe AR, Aira M, Gómez-Brandón M, Pérez-Losada M, Domínguez J (2019) Bacterial sucession and functional diversity during vermicomposting of the white grape marc *Vitis vinifera* v. Albariño. Sci Rep 9:7472

Lazcano C, Domínguez J (2011) The use of vermicompost in sustainable agriculture: impact on plant growth and soil fertility. In: Miransari M (ed) Soil nutrients. Nova Science Publishers, New York, pp 230–254

Lores M, Álvarez-Casas M, Llompart M, García-Jares M (2013) Uvariño: cosmetic power from white grapes. Expression Cosmétique 23:146–149

Luo G, Li L, Friman V-P, Guo J, Guo S, Shen Q, Ling N (2018) Organic amendments increase crop yields by improving microbe-mediated soil functioning of agroecosystems: a meta-analysis. Soil Biol Biochem 124:105–115

Martínez-Salgado MM, Ortega Blu R, Janssens M, Fincheira P (2019) Grape pomace compost as a source of organic matter: evolution of quality parameters to evaluate maturity and stability. J Clean Prod 216:56–63

Medina-Sauza RM, Álvarez-Jiménez M, Delhal A, Reverchon F, Blouin M, Guerrero-Analco JA, Cerdán CR, Guevara R, Villain L, Barois I (2019) Earthworms building up soil microbiota, a review. Front Environ Sci 7:81

Muhlack RA, Potumarthi R, Jeffery DW (2018) Sustainable wineries through waste valorization: a review of grape marc utilization for value-added products. Waste Manag 72:99–118

Nannipieri P, Giagnoni L, Renella G, Puglisi E, Ceccanti B, Masciandaro G, Fornasier F, Moscatelli MC, Marinari S (2012) Soil enzymology: classical and molecular approaches. Biol Fertil Soils 48:743–762

Paradelo R, Moldes AB, Barral MT (2019) Evolution of organic matter during the mesophilic composting of lignocellulosic winery wastes. J Environ Manag 116:18–26

Pinter IF, Fernández AS, Martínez LE, Riera N, Fernández M, Aguado GD, Uliarte EM (2019) Exhausted grape marc and organic residues composting with polyethylene cover: process and quality evaluation as plant substrate. J Environ Manag 246:695–705

Requejo MI, Fernández-Rubín de Felis M, Martínez- Caro R, Castellanos M, Ribas F, Arce A, Cartagena M (2016) Winery and distillery derived materials as phosphorus source in calcareous soils. Catena 141:30–38

Rosado D, Pérez-Losada M, Aira M, Domínguez J (2022) Bacterial succession during vermicomposting of silver wattle (*Acacia dealbata* link). Microorganisms 10:65

Sánchez-Hernández JC, Domínguez J (2017) Vermicompost derived from spent coffee grounds: assessing the potential for enzymatic bioremediation. In: Galanakis CM (ed) Handbook of grape processing by-products: sustainable solutions. Academic Press, Elsevier, London, pp 55–78

Spigno G, Marinoni L, Garrido G (2017) State of the art in grape processing by-products. In: Galanakis CM (ed) Handbook of grape processing by-products: sustainable solutions. Academic Press, Elsevier, London, pp 1–23

Zhang L, Sun X (2016) Improving green waste composting by addition of sugarcane bagasse and exhausted grape marc. Bioresour Technol 218:335–343

Zhang N, Hoadley A, Patel J, Lim S, Li C (2017) Sustainable options for the utilization of solid residues from wine production. Waste Manag 60:173–183

Chapter 7
Vermitechnology: An Underutilised Agro-tool in Africa

Ebenezer Olasunkanmi Dada and Yusuf Olamilekan Balogun

Abstract The increasing human population growth, with its attendant pressure on food supply, has continued to drive the search for sustainable agro-tool practices. The need to improve crop yield and increase food production have resulted in the evolution of various agricultural practices, including manual and mechanical tillage practices, shifting cultivation, chemical fertilisation and weed control. Although many of these conventional agricultural methods have been effective in improving food output, their adverse effects on the biotic and abiotic environment question their efficiency and sustainability. Vermitechnology is a potentially sustainable tool to solving many environmental and agricultural challenges that confront humankind. Africa faces a number of challenges, including the twin problem of rising population size and food shortage. In spite of this, vermitechnology has remained relatively under-researched in the continent, and the opportunities presented by the application of earthworms and their products to improve plant and animal production have not been fully tapped. This chapter describes, among others, the basics of vermitechnology, its applicability as an agro-tool, and the potential it presents as a cost-effective tool to addressing some agricultural challenges in Africa.

Keywords Agricultural challenges · Vermitechnology · Vermiagrotool · Vermirobotics

7.1 Introduction

The increasing human population growth, with its attendant pressure on food supply, has continued to drive the search for sustainable agro-tool practices. The need to improve crop yield and increase food production have resulted in the evolution of

---

E. O. Dada (✉) · Y. O. Balogun
Department of Cell Biology and Genetics, Environmental Biology Unit, Faculty of Science, University of Lagos, Lagos, Nigeria
e-mail: eodada@unilag.edu.ng

© The Author(s), under exclusive license to Springer Nature Singapore Pte Ltd. 2023
H. A. Mupambwa et al. (eds.), *Vermicomposting for Sustainable Food Systems in Africa*, Sustainable Agriculture and Food Security,
https://doi.org/10.1007/978-981-19-8080-0_7

various agricultural practices, including manual and mechanical tillage practices, shifting cultivation, chemical fertilisation and weed control. Although many of these conventional agricultural methods have been effective in improving food output, their adverse effects on the biotic and abiotic environment question their efficiency and sustainability. Vermitechnology has been identified as a potentially sustainable tool to solving many environmental and agricultural challenges that confront mankind (Sinha et al. 2010; Heuzé et al. 2020).

Vermitechnology is the systematic application of knowledge derived from studying the properties and importance of earthworms to solve environmental and human-related problems. Vermitechnology is an important fast-rising biotechnology that has gained and continued to gain attention since the end of the twentieth century, and is now widely used in many developed and developing countries (Li and Li 2010). The ubiquity of earthworms and their natural quality-maintaining and replenishing role in the soil justify the wide application of vermitechnology in many fields, such as sustainable crop production, pharmaceutics, biomonitoring and bioremediation, among others. Vermitechnology is divided into many overlapping parts, including vermiculture, vermimedicine, vermiremediation and vermicomposting. The application of earthworms and their products to improve plant and animal production can be referred to as vermiagrotechnology. Due to the importance attached to food production, vermiagrotechnology has gained more attention worldwide, in terms of researches, discoveries and innovations. Many developed and developing countries, such as China, United States of America, United Kingdom, Canada, India and Malaysia have made a giant stride vermiagrotechnology (Sinha et al. 2010).

In Africa, however, vermitechnology, including vermiagrotechnology, has remained relatively under-researched, and the opportunities presented by the application of earthworms and their products to improve plant and animal production have not been fully tapped. This is in spite of the fact that Africa faces a number of challenges, including the twin problem of exponentially rising population growth and food insecurity. The Food and Agriculture Organisation (FAO) estimated that 281.6 million people are in Africa, of which over one-fifth of the population were hunger stricken in 2020, which is 46.3 million more than reported in 2019 (FAO 2021). The aftermath of Covid 19 outbreak has further worsened the situation, with a 60% increase in the number of people facing hunger crises in Africa, as at 2020 (ACSS 2021). It is a paradox to note that whereas many advanced and developed nations are already challenged with public health concerns of overweight and obesity (FAO 2021), African countries are experiencing the fastest growing number of underfed, compared to other parts of the world (UN SDGs 2022). A good approach to mitigating this problem is to focus attention on vermitechnology as a cost-effective tool to improving food production in the continent.

7.2 Basics of Vermitechnology

Vermitechnology is a wide and integral aspect of biotechnology. There are many perspectives regarding the scope of vermitechnology. To draw a clear line across the extremes and borders of vermitechnology appears almost impossible, owing to its numerous applications and prospects. Some authors merely relate vermitechnology to vermiculture, vermicomposting and vermifiltration. Senapati (1992) and Gopi (2017) defined vermitechnology as the conversion of organic wastes by earthworms into quality manure. Kiyasudeen et al. (2016) described vermitechnology as methods, processes and designs that relate to vermicomposting. However, the application of vermitechnology in waste management goes beyond the degradation of solid organic wastes, to the removal of contaminants, including heavy metals, treatment of wastewater, and even management of greenhouse gases (Dada et al. 2016; Samal et al. 2019; Singh et al. 2020). In addition to vermicomposting, other prevailing aspects of vermitechnology include vermi-agro production (the application of earthworms and their products in sustainable crop production), vermimedicine (an ancient aspect of vermitechnology that explores and exploits the various therapeutic potential of earthworms), and vermi-industrial production (the use of earthworms to produce some raw materials for industrial products) (Sinha et al. 2010).

Boruah (2019) defined vermitechnology as the branch of science that studies the importance and utilisation of different epigeic earthworm species to answer problems related to ecology and environment. As broad and concise this definition is, it restricts the type of earthworm used, and limits the application of vermitechnology to ecological and environmental problems. Based on the ubiquitous nature of earthworms and their vast applications that are still open to discoveries, a proper and befitting definition for vermitechnology should depict its vast applications and be open-ended to accommodate the various aspects that have received less attention, such as mimicking the nature earthworms to mitigate engineering problems (vermiconstruction/vermirobotics). Hence, the definition provided by Dada et al. (2021a) which described vermitechnology as the science of using earthworms to improve food production and tackle environmental and other human challenges seems very much applicable. This definition accounts for the various applications and prospects of vermitechnology, starting from its involvement in the provision of man's most important need (food production), his immediate responsibility (environmental protection), and all other applications and prospects.

7.3 Branches of Vermitechnology

Vermitechnology has been randomly classified depending on what earthworms or their products are deployed to do. These classifications include Vermiculture/ Vermicomposting, Vermiremediation, Vermimedicine, Vermirobotics and

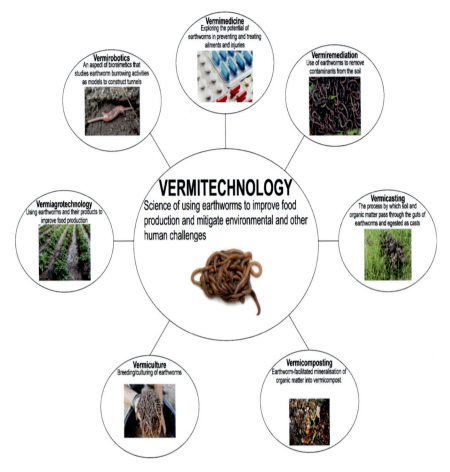

Fig. 7.1 Branches of vermitechnology

Vermiagrotechnology. Vermiculture and vermicomposting are a twin technology whose resultant products are earthworms and vermicompost. Vermiremediation is the use of earthworms to remove contaminants from the soil. Vermimedicine is the science of exploring the potentials of earthworms in preventing and treating ailments and injuries. Vermirobotics is an aspect of biomimetic that studies earthworms' burrowing activities as models to construct tunnels. Vermiagrotechnology is the science of using earthworms and their products to tackle food insecurity. These classifications are overlapping, and the list is emerging. A brief description of some of these and other vermiremediation biotechnology is provided in the following paragraphs and annotated in Fig. 7.1.

7.4 Vermiculture

Vermiculture simply means culture, breeding or rearing of earthworms. Its goal is to increase the number of worms in order to obtain sustainable harvest (Munroe 2012). The worms so obtained may be used for supplementing natural field density, vermicomposting, animal feeds and so on. Like other specialised farming such as aquaculture, vermiculture involves orderly steps, which include procuring of seed stocks, culture of the proper variety, and providing optimum food, moisture, air and temperature conditions (Munroe 2012). Several earthworm species have been identified as suitable vermiculture candidates. These include *Eisenia fetida, Eisenia andrei, Eudrilus eugeniae, Perionyx excavatus, Perionyx sansibaricus, Lumbricus rubellus, Enchytraeids* spp., *Dendrobaena veneta, Perionyx hawayana, Dendrobaena rubida, Amynthas differensis, Lampito mauritii* and *Hyperiodrilus africanus* (Suthar 2007; Munnoli et al. 2010). Most of these species are epigeic (surface litter dwelling) earthworms which are considered the most prolific, when compared with endogeics (soil mineral layer dwelling earthworms) and anecics (deep burrowing earthworms) (Gajalakshmi and Abbasi 2004).

7.5 Vermicomposting

Vermiculture and vermicomposting are, many times, a twin technology whose resultant products are earthworms and vermicompost. Vermicomposting focusses on using earthworms to facilitate the conversion of organic materials into humus-like material, known as vermicompost (Munroe 2012). Although microorganisms are mainly responsible for the biochemical degradation of the organic matter, earthworms play an important role in the process by fragmenting and conditioning the substrate, increasing the surface area for growth of microorganisms and altering its biological activity (Munnoli et al. 2010). The goal of vermicomposting is to process the organic material as quickly and efficiently as possible. In vermicomposting, the desire is to keep the earthworms' density level at optimum all of the time, but in vermiculture, the population density of worms is better kept low enough so that breeding rates are optimised (Munroe 2012).

Earthworms provide suitable conditions (mucus, optimum temperature and pH, moisture, etc.) that help to promote the growth of millions of biodegrader organisms (Dada et al. 2015; Samal et al. 2019). These microorganisms, including species of the genera *Bacillus, Pseudomonas, Enterobacter, Azotobacter, Klebsiella, Proteus and Streptococcus* produce a cocktail of enzymes such as invertases, ureases, and alkaline phosphatases, proteases, lipases that actively degrade organic wastes (Dada et al. 2021a). Organic wastes that can be degraded using this technology include varieties of hydrocarbon and non-hydrocarbon solid and liquid wastes from different sources (Kiyasudeen et al. 2016; Dada et al. 2021a). Vermicomposting is usually preferred in organic waste management owing to its efficacy in waste degradation, environmental suitability, cost-effectiveness and its low or no energy requirement. The product of vermicomposting is vermicompost. Vermicompost is a stabilised/mineralised by-product of earthworms' organic waste degradation that is used in sustainable farming to enhance plant growth and protect them from pest infestation (Kiyasudeen et al. 2016).

7.6 Vermiremediation

Vermiremediation is the use of earthworms and/or their products to remove contaminants from the soil environment. Earthworms play very important roles in maintaining the quality of their natural habitats. They do this naturally in two ways: (1) degradation and mineralisation of organic waste, (2) accumulation of non-degradable wastes from substrates (Dada et al. 2021a). The processes by which earthworms degrade organic wastes are variously referred to as vermidegradation, vermicomposting and vermifiltration. On the other hand, the process whereby earthworms take up contaminants or wastes into their body is referred to as vermiaccumulation (Dada et al. 2021a).

7.7 Vermicasting and Drilodegradation

Vermicasting and drilodegradation are both natural vermidegradation processes with little or no human interference, unlike vermicomposting and vermifiltration that are wholly set up by man. Vermicasting and drilodegradation both occur in soils that have been physicochemically and biologically influenced by earthworms. Vermicasting occurs above the soil surface while drilodegradation occurs below the soil surface around earthworm burrow walls (Shi et al. 2020). Vermicasting is the formation of casts by a series of bioturbated soil deposition made by earthworms in their natural habitats. Vermicasts harbour millions of microorganisms egested by earthworms along with soil during casting (Dada et al. 2021a). Drilosphere, the surface of earthworm burrows, is also a microbial hotspot and, consequently, a site of active organic degradation (Hoang et al. 2016; Shi et al. 2020). These two processes are yet to receive much research attention.

7.8 Vermimedicine

Vermimedicine explores the potentials of earthworms in preventing and treating injuries and ailments. Although vermimedicine is just beginning to get scientific and research attention, its recorded account dates back to over 2300 years (Afreen and Shaikh 2020). Ancient traditional systems that used earthworms to cure various types of diseases include Traditional Chinese Medicine (TCM), Ayurveda (India), Western (Mediterranean origin) and Traditional Arabic and Islamic Medicine (TAIM) (Cooper et al. 2012). These traditional systems reportedly used earthworms to cure ailments like asthma, epilepsy, high blood pressure, schizophrenia, leg ulcers, mumps, eczema, urticaria and anaphylaxis diseases, burns and scald fractures, chronic lumbago, skin crevices, blood deficiency apoplexy, acute injury of soft tissues, vertigo, hematemesis and hematuria, digestive ulcer, vesicle calculus and cancer (Cooper et al. 2012; Sun 2015; Afreen and Shaikh 2020). Scientific studies have confirmed the efficacy of earthworms in curing many ailments owing to some therapeutic compounds present in their body, such as fibrinolytic enzymes, lumbrokinase, collagenase, superoxide dismutase, cholinesterase, catalases, glycosidases, metallothioneins, antitumor protein, glycoprotein extract, gut mobility regulation peptide, antibacterial peptide, carbamidine, lumbritin, lumbrofobrim, terrestrolumbrolysin, purin, vitamin B, tyrosine, succinic acid, lauric acid and unsaturated fatty acid (Sun 2015).

7.9 Vermirobotics

Vermirobotics is an aspect of biomimetic that studies earthworm burrowing activities and their burrows as models to construct tunnels. It has been observed that earthworm burrows are durable and rarely collapse, despite being built within a short period, and their overlapping nature. This is clearly an indication that earthworms are efficient burrowing agents and can serve as models for tunnel construction. Hence, earthworms are imitated, in structure and function, to build durable tunnels within the shortest period possible. In 2020, General Electricity, a United States-based research firm, received a 2.5 million dollars funding, through the Defence Advanced Research Projects Agency (DARPA), the Pentagon's cutting-edge research and development branch, to demonstrate the feasibility of rapidly constructing tactical tunnel networks that enable secure, responsive resupply for ground troops (Mizokami 2020). The project was intended to create secure and communicable underground channels over long distances in the United States' megacities that would be very useful in the events of foreseen and unforeseen fighting. The prototype built, called Robo-worm, mimics the structure and function of earthworms that enable it to easily dig through soil, channelling durable tunnels in the process (Niskayuna 2021).

7.10 Vermiagrotechnology

Vermiagrotechnology, the science of using earthworms and their products to tackle food insecurity, is about the most explored of all branches of vermitechnology. Vermiagrotechnology provides an easy, eco-friendly, sustainable and cost-effective solution to food insecurity. Darwin regarded earthworms as friends of farmers, playing unparalleled, important roles in the history of life on earth (Darwin 1881; Sinha et al. 2010). The contribution of earthworms to global crop yield and soil fertility is enormous (Gopi 2017). Not only are earthworms known as indispensable in soil fertility and crop production, their relevance in animal agriculture also has been known much earlier than plant agriculture. They have long been used as baits for fishes, feedstock for domestic animals, and as food and medicine for man, in many ancient cultures (Heuzé et al. 2020). Now, earthworms are cultured and utilised in various ways, for sustainable food production.

7.11 Agricultural Challenges in Africa and Opportunities Presented by Vermitechnology

Africa confronts the problems of rapid population growth, food insecurity and undernourishment. The Food and Agricultural Organisation report (FAO 2015) showed that one out of five individuals in African population is undernourished. More recent report showed that African countries are experiencing the fastest growing number of undernourished, compared to other parts of the world (UN SDGs 2022). Over the past three decades, the population size of Africa has doubled, from 635 million people in 1990 to 1.2 billion people in 2018 (Dodo 2020). Despite the high amount of labour input into agriculture, African countries still have to rely excessively on food importation to meet the dietary needs of the populace. Between 2016 and 2018, Africa imported 85% of her food from other continents (Seleshie 2021). The increasing growth in population size has resulted in rapid urbanisation, which caused farmlands to be converted to homes, industries, roads and other infrastructures (Gashu et al. 2019). The consequent reduction in the size of farmlands has led to a decline in agricultural output, and a rise in food insecurity and malnutrition. Other causes of food insecurity in Africa include conflicts, failure of agricultural policies, poor governance and health issues (Dodo 2020; ACSS 2021; World Food Programme 2022).

Another major reason attributed to the food security issues in Africa is climate change. Climate change is a global issue, which has long retarded Africa from achieving its sustainable global development goals (SDGs). It causes variation in weather conditions and intense weather events, which increases the spread of pests and diseases, occurrence of natural disasters, and disruption of ecosystem functioning and services (Gashu et al. 2019). It has also been observed that inadequate investments in the agricultural research and development sector has led to a significant drop in agricultural total factor productivity (TFP), that is, the gross amount of crops, livestock and aquatic products produced per unit labour, material and capital (GAP 2021). To mitigate these problems, Africa must refocus attention to sustainable agricultural development, through increased research and investment. Africa, especially Sub-Saharan Africa, is potentially viable for high agricultural productivity growth, owing to availability of vast arable land, uncultivated land, and massive labour force (Shimeles et al. 2018; Nwachukwu 2019). Since earthworms are ubiquitous in nature, vermitechnology presents affordable and sustainable options to increase food output in Africa. Specifically, vermitechnology can be practically employed in agriculture to boost crop and animal production in many ways, some of which are described below.

7.12 Earthworms as Animal Feed

As explained above, some earthworm species, especially soil surface dwelling (epigeic) species, can be cultured in situ or ex situ to increase field density or for other uses like composting (vermicomposting), fishing and animal feed. Early humans used earthworms as baits for fishing and hunting, and in the eighteenth century, earthworms were generally known for fish feedstocks (Dominguez et al. 2004; Dedeke et al. 2013). The viability of earthworms as fish feed then brought about the idea of studying their nutritional properties and extending their use to other aspects of animal agriculture. Studies on earthworm meals reveal that they contain high quantity of protein and other nutritive compounds (Sogbesan et al. 2007). The protein, lipid, carbohydrates and mineral contents of earthworm powder vary from 50% to 70%, 5% to 10%, 5–21% and 2–3% respectively, with varieties of vitamins, including niacin (Mohanta et al. 2016; Sun and Jiang 2017). Studies that compared earthworm meals with conventional fish and poultry meals reported outstanding results for earthworm meals (Sogbesan et al. 2007; Dedeke et al. 2013; Djissou et al. 2017).

Earthworms also contain considerable amounts of essential and non-essential amino acids (Istiqomah et al. 2009; Antonova et al. 2021). For instance, tyrosine, which is essential in the production of several important brain chemicals, was found in earthworms at concentrations much higher than recommended by the Food and Agriculture Organisation of the United Nations (Antonova et al. 2021). Earthworms also has lipid contents naturally high in saturated, monounsaturated, polyunsaturated fatty acids, such as pentadecylic, myristic, lauric, palmitoleic, oleic and linolenic fatty acids (Gunya et al. 2016; Mohanta et al. 2016; Isea-León et al. 2019). Earthworms are therefore strongly recommended as a very viable animal protein source for fish and livestock.

7.13 Earthworms as Soil Fertility Improvers

Earthworms naturally influence the quality of soil in which they inhabit, mainly by bioturbation and geophagy (soil feeding). Bioturbation is the mixing and reworking of soil particles through biological activities like burrowing, soil digging and mole construction (Fattore et al. 2020). Earthworms contribute greatly to soil formation and evolution. They are found in almost all terrestrial regions and their impacts on soil stretch far beyond mere physical mixing of soil, as in the case of many bioturbating animals (Meysman et al. 2006). Earthworm bioturbation influences virtually all properties of soil ranging from physical to biogeochemical. Burrows created by earthworms by vertical and horizontal movement through soil, help in soil aeration and infiltration, which are both important qualities that facilitate the survival and growth of relevant soil microbes and biota, including plants. The displacement of soil particles by the alternating upward and downward movement of earthworms helps to replenish oxides in the subsoil (Meysman et al. 2006). Earthworm bioturbation also moderates and optimises other properties of soil such as soil texture, bulk density, electrical conductivity and pH (Blouin et al. 2013).

Soil particles ingested by earthworms, alongside the constituent organic matter, pass through earthworm guts which contain populations of degrading microorganisms that help to stabilise and mineralise organic matter. These soil particles are egested as casts, which contain abundant minerals and millions of microorganisms that are beneficial for plant growth. Vermicasts also contain enzymes such as proteases, alpha and beta amylases, ureases, phosphatases and nitrogen dehydrogenases that enhance seed germination and promote plant physical and biochemical growth and development (Suthar 2010; Dada et al. 2021b). Earthworm casts also contain ammonia, urea, oxidisable organic matter, some exchangeable forms of plant nutrients and plant growth-promoting hormones (Krishnamoorthy and Vajranabhaian 1986). In practical terms therefore, earthworms can be harvested from the vermiculture media, and seeded into farmlands to optimise soil earthworm density or trigger earthworm activities.

7.14 Use of Vermicompost as Biofertilisers

Vermicompost is an earthworm-conditioned soil, whose constituent organic matter has been decomposed and stabilised as a result of interactions with earthworms and microorganisms (Dominguez 2004). Vermicompost possesses high porosity, density, aeration and water-holding capacity, and low carbon to nitrogen ratio (Kiyasudeen et al. 2016). Earthworms consume soil and organic substances and excrete them as vermicompost or vermicasts, which contain high humus content, macronutrients (nitrogen, phosphorus and potassium), micronutrients, beneficial soil microflora, actinomycetes and plant growth regulators (Adhikary 2012). The most important processes that aid in transformation of ingested soil and organic matters into valuable product constituting essential nutrients and active components of microbial biomass are the activities of endosymbiotic microbes and gut enzymes (cellulase, protease, chitinase acid and phosphatase) of earthworm (Zhang et al. 2000).

Vermicompost can easily be produced by farmers in their farms, as opposed to chemical fertilisers can only be produced in the factories, and with very high economic and environmental consequences (Hussaini 2013). Studies on vermicompost and its process have long gained attention in countries like England, France, the Netherlands, Germany, Italy, Spain, Poland, United States, Cuba, Mexico, Bahamas, China, Japan, Philippines, India and other parts of Southeast Asia (Edwards 2004). Vermicompost is a potent and affordable biofertilisers farmers in Africa should focus attention, in place of chemical fertilisers that are relatively costly and environmentally unfriendly.

7.15 Vermifluid and Vermitea as Liquid Biofertilisers

Vermifluid is a liquid obtained by passing water through the column of live earthworms, after their guts are voided (Suthar 2010), while vermitea or vermicast tea is a liquid that is collected after the passage of water through vermicompost bed or earthworm casts (Litterick et al. 2004). Vermitea and vermifluid are potent organic fertilisers that contain nitrogenous excretory product and growth-promoting hormones, nutrients and essential enzymes (Bhardwaj and Sharma 2016; Senthilmurugan et al. 2018; Dada et al. 2021b). Vermifluid and vermitea are used as foliar spray and soil-drench fertilisers.

Vermifluid may be obtained by heat stress or cold stress methods. Heat stress method involves agitating earthworms in warm distilled water (at 40 °C) for five to six minutes. Cold stress method involves keeping earthworms in a bowl containing ice cubes, at −5 °C, for three to four minutes, with occasional stirring, then rinsing in cold water for seven to eight minutes (Chattopadhyay 2015).

Vermitea can be processed or obtained by allowing water to pass through a vermicompost bed. The product of this method is commonly termed vermicompost tea (VCT), although some authors have referred to it as vermiwash (Suthar 2010; Senthilmurugan et al. 2018). VCT can be aerated or non-aerated. Non-aerated vermitea is prepared at room temperature without stirring for a defined period of time (Scheuerell and Mahaffee 2002), whereas aerated VCT is produced by allowing water to recirculate through loose vermicompost bed in aerobic conditions. The product formed using this method is termed aerated vermicompost tea. Vermitea may also be obtained by soaking earthworm casts in distilled water, stirring, allowing it to stand for twenty-four hours, and thereafter filtering, with the filtrate serving as vermitea or vermicast tea (Musa et al. 2017).

7.16 Vermiagrotechnology in Africa Relative to Other Continents

Vermiagrotechnology as a branch of vermitechnology is the application of knowledge and science of earthworms to increase crop and animal production. Earthworms have been actively involved in plant growth and agricultural production, right from man's existence, through their important roles in soil formation and fertility (Anderson and James 2017; Anderson et al. 2017). Darwin, in 1881, provided the first elaborate description of how earthworms modify soil properties to suit plant growth, by churning and aerating the soil with their burrows, breaking down and burying large fragments of organic matter and making them available for plants' use (Darwin 1881; Anderson et al. 2017). However, the period when Darwin's book was published was also the period of industrial revolution when chemistry was discovered, and that became the major field of interest for scientists. As a result, the findings of Darwin were left unattended until almost after a century when sustainable crop production became expedient. Farmers cultured earthworms on a smaller scale until 1970 when it became a commercial process (Sinha et al. 2010). The first large-scale vermiculture experiment was conducted in Holland in 1970, followed by Canada, England, USA, Italy, Philippines, Thailand, China, Japan, Korea, France, Brazil, Israel, Russia and Australia (Sinha et al. 2010; Bhat et al. 2017).

Earthworms were used as angling bait, fish feed and duck feed for many centuries in many ancient cultures like the Native Americans, Aborigenes in Australia, Maori in New Zealand, Ye'kuana Amerindians of the Alto Orinoco in Venezuela and the Fujian, Guangdong, Taiwan and Henan provinces in China (Sun and Jiang 2017; Heuzé et al. 2020). During the eighteenth century, earthworms were generally regarded as animal feedstock, and in the twentieth century, they were first produced on industrial scale for fish farming in the United States, California in particular (Heuzé et al. 2020).

Vermicomposting started as a practice intended to solve challenges related to wastes management and food production, but has now transcended to a global market, worth USD 101,850 million, and projected to reach USD 234,370 million by 2028, with a compound annual growth rate of 14.9% (Market Reports World 2022). According to the Market Reports World (2022), North America is the largest market of vermicomposting with about 16% market share, followed by Europe and Japan that account for about 30% market share. The remaining 54% market share is accounted for by countries in Asia Pacific, Latin America, Middle East and Africa. The key players are MyNOKE in New Zealand, NutriSoil and Davo's Worm Farms in Australia, Earthworm in New Zealand, Wormpower in North America, Kahariam Farms in Philippines, SAOSIS and Sri Gayathri Biotec in India, Jialiming in China, Dirt Dynasty and SLO County Worm Farm in California, USA, Agrilife and Suman Vermi Compost in India.

With regard to research publications related to vermicompost, reports showed that India has had the highest research productivity for over the past two decades, followed by USA, Spain, Brazil, Mexico, Iran, China, South Africa, Italy, Malaysia and Argentina (Ghorbani and Sabour 2020). These countries are spread mainly across Europe, Asia, North America, with one each from South America and Africa. Ghorbani and Sabour (2020) reported that the number of publications from India was almost 4.5 times more than that from the USA, which ranked second on the list of productivity, and 5 times more than Spain that ranked third.

Customarily, the number of publications in any field is an index of the research progress in the field. It then follows that India made the most significant progress in the number of publications over the period of research (1993–2017). Relative to India, the developed countries made no significant progress in the number of publications during the period, probably owing to their already developed and organised agricultural and waste systems, of which vermicomposting programmes are component part (Ghorbani and Sabour 2020).

While many developed and developing countries in other regions and continents of the world have substantially embraced and applied vermiagrotechnology, vermitechnology remains an underutilised agro-tool in Africa. Ghorbani and Sabour (2020) reported that out of the top twenty countries/territories that were most productive in terms of vermicomposting research, South Africa, eighth on list, was the only African country represented. Up until the twentieth and early twenty-first century, vermicomposting research and application in Africa countries remained generally low, except for few studies from South Africa (Venter and Reinecke 1988; Reinecke et al. 1992; Riley et al. 2008).

In recent times however, vermicomposting is being embraced and applied in some African countries. Our online search revealed the existence of some vermicomposting outfits, like Global Warming, Goodbugs, Worm farm, Quietscapes and Suburban Earthworms in South Africa; Green Cycle Technologies and JV Comsoil in Ghana; HK Global Enterprise in Uganda; and Kenworks Ventures Company Limited in Kenya. In spite of these, vermitagrotechnology remains underutilised in Africa, relative to many developed countries in Europe, Asia, America and Australia. To achieve the United Nations-initiated Sustainable Development Goals' objective of maximising viable, chemical-free agricultural productivity and food resilience, through agroecological processes (UN SDGs 2022), African countries need to increase funding and research in vermitechnology, particularly vermiagrotechnology.

7.17 Conclusion

Africa faces a number of challenges, including the twin problem of rising population size and food insecurity. Whereas many advanced and developed nations are already challenged with public health concerns of overweight and obesity, African countries are experiencing the fastest growing number of underfed, compared to other parts of the world. A potentially sustainable tool to solving agricultural challenges and increasing food production is vermitechnology. While developed countries, who are relatively food secure, have effectively incorporated vermitechnology into their agricultural and environmental programmes, in Africa, vermitechnology has remained relatively under-researched, and the opportunities presented by the application of earthworms and their products to improve plant and animal production have not been fully tapped. It is therefore important for African countries to increase funding and research in vermiagrotechnology and vermitechnology.

References

Adhikary S (2012) Vermicompost, the story of organic gold: a review. Agric Sci 3(7):905–917. https://doi.org/10.4236/as.2012.37110

Afreen S, Shaikh A (2020) Therapeutic uses of earthworm – a review. Int J Adv Ayurveda, Yoga, Unani, Siddha Homeopathy 9(1):571–580. https://doi.org/10.23953/cloud.ijaayush.469

Africa Centre for Strategic Studies (ACSS) (2021) Food insecurity crisis mounting in Africa. https://pdf.printfriendly.com/downloads/pdf_1648712764_b06d96A4.pdf. Accessed 30 Mar 2022

Anderson F, James S (2017) The evolution of earthworms. https://blogs.biomedcentral.com/bmcseriesblog/2017/06/01/the-evolution-of-earthworms/. Accessed 18 Mar 2022

Anderson FE, Williams BW, Horn KM et al (2017) Phylogenomic analyses of Crassiclitellata support major Northern and Southern Hemisphere clades and a Pangaean origin for earthworms. BMC Evol Biol 17(123). https://doi.org/10.1186/s12862-017-0973-4

Antonova E, Titov I, Pashkova I et al (2021) Vermiculture as a source of animal protein. E3S Web of Conferences 254:08006. https://doi.org/10.1051/e3sconf/202125408006

Bhardwaj P, Sharma RK (2016) Effect of vermifluid and vermicompost on the growth and productivity of Moong dal. J Chem Biol Phys Sci 6(4):1381–1388. https://www.citefactor.org/article/index/98185/pdf/effect-of-vermiwash-and-vermicompost-on-the-growth-and-productivity-of-moong-dal

Bhat SA, Singh J, Vig AP (2017) Instrumental characterization of organic wastes for evaluation of vermicompost maturity. J Anal Sci Technol 8(1). https://doi.org/10.1186/s40543-017-0112-2

Blouin M, Hodson ME, Delgado EA et al (2013) A review of earthworm impact on soil function and ecosystem services. Eur J Soil Sci 64(2):161–182. https://doi.org/10.1111/ejss.12025

Boruah T (2019) Vermitechnology: an overview. http://babrone.edu.in/blog/index.php/2019/02/04/vermitechnology-an-overview/. Accessed 25 Mar 2022

Chattopadhyay A (2015) Effect of vermiwash of Eisenia foetida produced by different methods on seed germination of green mung, *Vigna radiate*. Int J Recycl Org Waste Agricult 4(4):233–237. https://doi.org/10.1007/s40093-015-0103-5

Cooper EL, Balamurugan M, Huang C-Y et al (2012) Earthworms dilong: ancient, inexpensive, noncontroversial models may help clarify approaches to integrated medicine emphasizing neuroimmune systems. Evid Based Complement Alternat Med 2012:164152. https://doi.org/10.1155/2012/164152

Dada EO, Abdulganiy T, Owa SO et al (2021b) Tropical wetland earthworm vermifluid promotes mitotic activities and root growth in allium cepa at low concentrations. Chiang Mai Univ J Nat Sci 20(3):e2021064. https://doi.org/10.12982/CMUJNS.2021.064

Dada EO, Akinola MO, Owa SO et al (2021a) Efficacy of vermiremediation to remove contaminants from soil. J Health Pollut 11:29. https://doi.org/10.5696/2156-9614-11.29.210302

Dada EO, Njoku KL, Osuntoki AA et al (2015) A review of current techniques of physico-chemical and biological remediation of heavy metals polluted soil. Ethio J Environ Stud Man 8(5): 606–615. https://doi.org/10.4314/ejesm.v8i5.13

Dada EO, Njoku KL, Osuntoki AA et al (2016) Heavy metal remediation potential of a tropical wetland earthworm, *Libyodrilus violaceus* (beddard). Iran J Energy Environ 7(3):247–254. https://doi.org/10.5829/idosi.ijee.2016.07.03.06

Darwin C (1881) The formation of vegetable mould through the action of worms, with observations on their habits. John Murray, London

Dedeke GA, Owa SO, Olurin KB et al (2013) Partial replacement of fish meal by earthworm meal (*Libyodrilus violaceus*) in diets for African catfish, *Clarias gariepinus*. Int J Fish Aquac 5(9): 229–233. https://doi.org/10.5897/IJFA2013.0354

Djissou ASM, Ochiai A, Koshio S et al (2017) Effect of total replacement of fishmeal by earthworm and *Azolla filiculoides* meals in the diets of Nile tilapia *Oreochromis niloticus* (Linnaeus, 1758) reared in concrete tanks. Indi J Fish 64(1):31–36. https://doi.org/10.21077/ijf.2017.64.1.55317-05

Dodo MK (2020) Understanding Africa's food security challenges. In: Mahmoud B (ed) Food security in Africa. IntechOpen, London. https://doi.org/10.5772/intechopen.91773

Dominguez J (2004) State-of-theart and new perspectives on vermicomposting research. In: Edwards CA (ed) Earthworm ecology, 2nd edn. CRC Press, Boca Raton, pp 401–424. https://doi.org/10.1201/9781420039719.ch2

Edwards CA (ed) (2004) Earthworm ecology. CRC Press, London

Fattore S, Xiao Z, Godschalx AL et al (2020) Bioturbation by endogeic earthworms facilitates entomopathogenic nematode movement toward herbivore-damaged maize roots. Sci Rep 10(1). https://doi.org/10.1038/s41598-020-78307-0

Food and Agriculture Organization of the United Nations (FOA) (2021) Regional overview of food security and nutrition 2021. https://reliefweb.int/sites/reliefweb.int/files/resources/cb7496en.pdf. Accessed 30 Mar 2022

Food and Agriculture Organization of the United Nations (FOA) (2015) State of food insecurity in the world: in brief. https://www.fao.org/3/i4671e/i4671e.pdf. Accessed 30 Mar 2022

Gajalakshmi S, Abbasi SA (2004) Neem leaves as a source of fertilizer-fum-pesticide vermicompost. Bioresour Technol 92:291–296. https://doi.org/10.1016/j.biortech.2003.09.012

Gashu D, Demment MW, Stoecker BJ (2019) Challenges and opportunities to the African agriculture and food systems. Afr J Food Agric Nutr Develop 19(1):14190–14217. https://doi.org/10.18697/ajfand.84.BLFB2000

Ghorbani M, Sabour MR (2020) Global trends and characteristics of vermicompost research over the past 24 years. Environ Sci Pollut Res 28:94–102. https://doi.org/10.1007/s11356-020-11119-x

Global Agricultural Productivity (GAP) (2021) 2021 Global Agricultural Productivity report https://globalagriculturalproductivity.org/2021-gap-report/downloads/. Accessed 25 Mar 2022

Gopi P (2017) Vermitechnology, a scenario of sustainable agriculture - a mini review. VISTAS 6(1):51–56

Gunya B, Masika PJ, Hugo A et al (2016) Nutrient composition and fatty acid profiles of oven-dried and freeze-dried earthworm *Eisenia foetida*. J Food Nutr Res 4(6):343–348. https://doi.org/10.12691/jfnr-4-6-1

Heuzé V, Tran G, Sauvant D et al (2020) Earthworm meal. https://www.feedipedia.org/node/665 Accessed 2 Mar 2021

Hoang DTT, Pausch J, Razavi BS et al (2016) Hotspots of microbial activity induced by earthworm burrows, old root channels, and their combination in subsoil. Biol Fertil Soils 52(8):1105–1119. https://doi.org/10.1007/s00374-016-1148-y

Hussaini A (2013) Vermiculture bio-technology: an effective tool for economic and environmental sustainability. Afr J Environ Sci Technol 7(2):56–60. https://www.ajol.info/index.php/ajest/article/view/88669/78261

Isea-León F, Acosta-Balbás V, Rial-Betancoutd LB et al (2019) Evaluation of the fatty acid composition of earthworm *Eisenia andrei* meal as an alternative lipid source for fish feed. J Food Nutr Res 7(10):696–700. https://doi.org/10.12691/jfnr-7-10-2

Istiqomah L, Sofyan A, Damayanti E et al (2009) Amino acid profile of earthworm and earthworm meal (*Lumbricus rubellus*) for animal feedstuff. J Indones Trop Anim Agric 34(4). https://doi.org/10.14710/jitaa.34.4.253-257

Kiyasudeen K, Ibrahim MH, Quail S (2016) Prospect of organic waste management and the significance of earthworms. Springer, Switzerland. https://doi.org/10.1007/978-3-319-24708-3

Krishnamoorthy RV, Vajranabhaian SN (1986) Biological activity of earthworm cats. An assessment of plant growth promoter or levels in the casts. Proc Indian Acad Sci Anim Sci 95(3): 341–351. https://www.ias.ac.in/article/fulltext/anml/095/03/0341-0351

Li KM, Li PZ (2010) Earthworms helping economy, improving ecology and protecting health. Int J Environ Eng 10:354–365

Litterick A, Harrier L, Wallance P et al (2004) The role of uncomposted materials, composts, manures and compost extracts in reducing pest and disease incidence and severity in sustainable temperate agricultural and horticultural crop production - a review. Plant Sci 23:453–479. https://doi.org/10.1080/07352680490886815

Market Reports World (2022) Vermicompost market size [2022–2028] is estimated to be worth USD 234370 million at a CAGR of 14.9%. https://www.globenewswire.com/news-release/2022/03/11/2401514/0/en/Vermicompost-Market-Size-2022-2028-is-Estimated-to-Be-Worth-USD-234370-Million-at-a-CAGR-of-14-9-Industry-Share-Growth-Demand-Key-Players-Revenue-Gross-Margin-Opportunities-Challen.html. Accessed 17 Mar 2022

Meysman F, Middleburg J, Heip C (2006) Bioturbation: a fresh look at Darwin's last idea. Trends Ecol Evol 21(12):688–695. https://doi.org/10.1016/j.tree.2006.08.002

Mizokami K (2020) Why the military is building a tunnelling earthworm. https://www.popularmechanics.com/military/weapons/a32700671/tactical-tunnel-earthworm-robot/ Accessed 6 Mar 2022

Mohanta KN, Sankaran S, Veeratayya SK (2016) Potential of earthworm (*Eisenia fetida*) as dietary protein source for Rohu (*Labeo rohita*) advanced fry. COG Food Agric 2:1138594. https://doi.org/10.1080/23311932.2016.1138594

Munnoli PM, da Silva JAT, Bhosle S (2010) Dynamics of the soil-earthworm-plant relationship: a review. Dyn Soil Dyn Plant 4(1):1–21

Munroe B (2012) Manual on vermiculture and vermicomposting. Organic Agriculture Centre of Canada 6:1–57. http://oacc.info/vermiculture_farmersmanual_gm.pdf

Musa SI, Njoku KL, Ndiribe CC et al (2017) The effect of vermi tea on the growth parameters of *Spinacia oleracea* L. (Spinach). J Environ Sci Poll Res 3(4):236–238. https://doi.org/10.13140/RG.2.2.25268.71041

Niskayuna NY (2021) The autonomous dig: GE's giant earthworm tunneling robot finds its own way. https://www.ge.com/news/press-releases/the-autonomous-dig-ges-giant-earthworm-tunneling-robot-finds-its-own-way. Accessed 9 Mar 2022

Nwachukwu JU (2019) Agricultural production amid conflict: implications for Africa's regional development. In: Edomah N (ed) Regional development in Africa. IntechOpen, London. https://doi.org/10.5772/intechopen.86613

Reinecke AJ, Viljoen SA, Saayman RJ (1992) The suitability of *Eudrilus eugeniae*, *Perionyx excavatus* and *Eisenia foetida* (Oligochaeta) for vermicomposting in Southern Africa in terms of their temperature requirements. Soil Biol Biochem 24(12):1295–1307

Riley SJ, Panikkar A, Shrestha S (2008) Development of an aerobic vermiculture whole-of-waste treatment plant and potential uses in sub-Saharan Africa. Proceedings of the decentralised water and wastewater systems international conference, pp 59–66

Samal K, Raj-Mohan A, Chaudhary N et al (2019) Application of vermitechnology in waste management: a review on mechanism and performance. J Environ Chem Eng. https://doi.org/10.1016/j.jece.2019.103392

Scheuerell SJ, Mahaffee WF (2002) Compost tea: principals and prospects for plant disease control. Compost Sci Util 10:313–338. https://doi.org/10.1080/1065657X.2002.10702095

Seleshie L (2021) What's behind Africa's skyrocketing imports yet increased production growth? https://www.theafricareport.com/123719/whats-behind-africas-skyrocketing-imports-yet-increased-production-growth/. Accessed 30 Mar 2022

Senapati BK (1992) Vermitechnology as an option for recycling of cellulose waste in India. In: Subbarao NS, Balagopalan C, Raman-Krishna SV (eds) New trends in biotechnology. Oxford & IBH Pub. Co., New Delhi, pp 347–358

Senthilmurugan S, Sattanathan G, Vijayan PPK et al (2018) Evaluation of different concentration of vermiwash on seed germination and biochemical response in Abelmoschus esculentus (L.). Int J Biol Res 3(1):228–231. https://www.biologyjournal.in/download/188/3-1-56-714.pdf

Shi Z, Liu J, Tang Z et al (2020) Vermiremediation of organically contaminated soils: concepts, current status, and future perspectives. Appl Soil Ecol 147. https://doi.org/10.1016/j.apsoil.2019.103377

Shimeles A, Verdier-Chouchane A, Boly A et al (2018) Introduction: understanding the challenges of the agricultural sector in sub-Saharan Africa. In: Shimeles A, Verdier-Chouchane A, Boly A (eds) Building a resilient and sustainable agriculture in sub-Saharan Africa. Palgrave Macmillan, Cham. https://doi.org/10.1007/978-3-319-76222-7_1

Singh R, Srivastava P, Verma P et al (2020) Exploring soil responses to various organic amendments under dry tropical agroecosystems. In: Climate change and soil interactions. Elsevier, Amsterdam, pp 583–611. https://doi.org/10.1016/b978-0-12-818032-7.00021-7

Sinha RK, Agarwal S, Chauhan K et al (2010) Vermiculture technology: reviving the dreams of Sir Charles Darwin for scientific use of earthworms in sustainable development programs. Technol Inv 1(03):155–172. https://doi.org/10.4236/ti.2010.13019

Sogbesan OA, Ugwumba AAA, Madu CT et al (2007) Culture and utilization of earthworm as animal protein supplement in the diet of *Heterobranchus longifilis* fingerlings. J Fish Aquac Sci 2:375–386. https://doi.org/10.3923/jfas.2007.375.386

Sun Z (2015) Earthworm as a biopharmaceutical: from traditional to precise. Eur J Biomed Res 1(2):28–35. https://doi.org/10.18088/ejbmr.1.2.2015.pp28-35

Sun Z, Jiang H (2017) Nutritive evaluation of earthworms as human food. In: Mikkola H (ed) Future foods. IntechOpen, London, pp 127–141. https://doi.org/10.5772/intechopen.70271

Suthar S (2007) Influence of different food sources on growth and reproduction performance of composting epigeics: *Eudrilus euginiae, Perionyx excavates*, and *Perionyx sansibaricus*. Appl Environ Res 5(2):79–92. https://doi.org/10.15666/aeer/0502_079092

Suthar S (2010) Evidence of plant hormone like substances in vermiwash: an ecologically safe option of synthetic chemicals for sustainable farming. Ecol Eng 36(8):1089–1092. https://doi.org/10.1016/j.ecoleng.2010.04.027

United Nations Sustainable Development Goals (UN SDGs) (2022) Goal 2: zero hunger. https://www.un.org/sustainabledevelopment/hunger. Accessed 25 Mar 2022

Venter JM, Reinecke AJ (1988) The life-cycle of the compost worm *Eisenia fetida* (Oligochaeta). S Afr J ZooL 23(3):161–165. https://doi.org/10.1080/02541858.1988.11448096

World Food Programme (2022) Food security highlights. https://fscluster.org/sites/default/files/documents/food_security_highlights_issue_3_january_2022.pdf. Accessed 30 Mar 2022

Zhang BG, Li GT, Shen TS et al (2000) Changes in microbial biomass C, N, and P and enzyme activities in soil incubated with the earthworms *Metaphire guillelmi* or *Eisenia fetida*. Soil BioL Biochem 32:2055–2062. https://doi.org/10.1016/S0038-0717(00)00111-5

Chapter 8
Prospects of Vermicompost and Biochar in Climate Smart Agriculture

P. Nyambo, L. Zhou, T. Chuma, A. Sokombela, M. E. Malobane, and M. Musokwa

Abstract Agricultural food production in Africa is low and cannot sustain its ever-growing population. Climate change-induced high rainfall variability and temperature fluctuations are blamed for the low agricultural productivity amongst African smallholder farmers. Additionally, land degradation is causing shrinkages in the cultivated land and abandonment of crop fields in most African countries. Land degradation is worsened by the anthropogenic activities that lead to the depletion of soil organic carbon, a vital constituent of soil quality. Most farmers in the continent are resource-poor and therefore practise low input agriculture, continuous crop monoculture and use heavy soil tilling methods that enhance the decomposition of the already low soil organic carbon. Researchers devised several interventions like conservation agriculture, but adoption remains poor owing to various associated issues like trade-offs. In addition, the response of soil to these interventions requires several years in most cases, and most smallholder farmers are not patient for that long. Researchers have recently been calling for other supplementary and climate-smart strategies that increase soil organic matter, nutrients and reduce greenhouse gas emissions. This chapter aims to review the potential benefits of biochar-vermicompost on crop and soil productivity with a specific focus on Africa. Despite the limited information on the effect of biochar-vermicompost mixture, adding biochar, especially to sewage sludge, can reduce the amount of hazardous

---

P. Nyambo (✉) · L. Zhou
Risk and Vulnerability Science Centre, Alice Campus, University of Fort Hare, Alice, South Africa
e-mail: pnyambo@ufh.ac.za

T. Chuma · A. Sokombela
Department of Agronomy, University of Fort Hare, Alice, South Africa

M. E. Malobane
Department of Agriculture and Animal Health, University of South Africa, Roodepoort, South Africa

M. Musokwa
Southern African Confederation of Agricultural Unions, Centurion, South Africa

© The Author(s), under exclusive license to Springer Nature Singapore Pte Ltd. 2023
H. A. Mupambwa et al. (eds.), *Vermicomposting for Sustainable Food Systems in Africa*, Sustainable Agriculture and Food Security,
https://doi.org/10.1007/978-981-19-8080-0_8

contaminants like heavy metals, *E. coli* and reduce the moisture content in vermicompost. With the benefits of both biochar and vermicompost in agriculture well recorded, the critical question is if biochar and vermicompost mixture will synergistically benefit soil quality and crop yield.

Keywords Biochar · Soil fertility · Crop yields · Biochar-vermicomposting

8.1 Introduction

Africa is the second-largest and second-most-populous continent, covering about 6% of the land surface (Richardson et al. 2022). The continent is projected to double in population from 1 billion to nearly 2.4 billion inhabitants by 2050, the majority of whom live in the rural areas mostly as smallholder farmers (Hamann and Tuinder 2012). Not enough food is produced to meet the continents' subsistence requirements posing a challenge of increased hunger and poverty. In most countries, vast land is lying idle in many rural areas, and estimates show a substantive percentage remains uncultivated due to low fertility and high erodibility problems that have resulted in a decline in crop and soil productivity (Laker 2004). Soil degradation is further worsened by anthropogenic activities like land-use change and unsustainable farming practices, which result in the loss of soil organic matter, an essential component of soil quality (Nyambo et al. 2020a). Soil organic carbon loss contributes significantly to the ongoing climate change phenomenon (Baveye et al. 2020). Lal (2009) estimates that about 50–100 GT of carbon was lost from the soil into the atmosphere because of the industrial revolution and modern-day farming techniques. IPBES (2018) predicts that by 2050, the combination of land degradation and climate change will reduce global crop yields by an average of 10% and up to 50% in some regions.

Most African smallholder farmers are highly vulnerable to climate change because of their over-reliance on rainfed agriculture and weak adaptation potential (Morton 2007). Overcoming these challenges calls for the adoption of resilient methods of food production. This is especially important since meeting the increased demand for agriculture with existing farming practices will likely lead to more intense competition for natural resources, increased greenhouse gas emissions, and further deforestation and land degradation (FAO 2017). Therefore, there is a need to identify ways to attain a sustainable agricultural system, and coping strategies that mitigate these problems need to be employed.

Of late, attention has been shifting toward the use of various organic amendments as sources of soil organic matter; for example, food waste (Risse and Faucette 2009), leaf litter (Parwada and Van Tol 2018), livestock manure (Gerzabek et al. 2001) and crop residues (Nyambo et al. 2020a). However, one of the significant drawbacks of adding fresh organic amendments is their quick mineralisation in the soil, increased pest, disease, and weed infestation incidences, high electrical conductivity, and their long-term application can increase soil salinity (Naiji and Souri 2018). Furthermore, organic matter such as poultry manure has disadvantages of eutrophication, phytotoxic substances, and releases greenhouse gases (Bolan et al. 2010; Nyambo et al. 2020a).

Consequently, attention is shifting toward sustainable bio-processed products such as biochar and vermicompost. Several studies have reported that amending soil with biochar or vermicompost enhances cation exchange capacity, pH, nutrient, water retention, bulk density, soil microbial communities and absorption of faecal contamination (Mupambwa et al. 2017; Nyambo et al. 2018, Mamera et al. 2021; Ajibade et al. 2022; Nyambo et al. 2022). Still, other researchers argue that the sole application of biochar or vermicompost has its own challenges. Positive synergistic effects of combining biochar with different compost types have been reported on soil properties (Alvarez et al. 2017). Nonetheless, studies focusing on the combined effect are still limited, and information is still scarce. Therefore, this chapter aims to review the potential benefits of adding biochar to vermicomposts on crop and soil productivity with a specific focus on Africa.

8.2 Agricultural Production Challenges in Africa

Soil degradation is a severe challenge that threatens the sustainability of crop production systems globally. Soil degradation in cropping systems is caused by substandard and unsustainable management practices, which reduce soil biological, chemical and physical quality, diminishing the capacity of the soil to sustain production. The effect of soil degradation is critical in sub-Saharan Africa (SSA), where approximately 65% of the land area is degraded (Vlek et al. 2008). It contributes to roughly 350 million ha or 20–25% of the total land area, of which about 100 M ha is approximated to be acutely degraded mainly due to poor agricultural activities. Soil degradation costs SSA around US\$68bn annually and diminishes the regional agricultural gross domestic product by 3% yearly. Soil degradation is regarded as a significant root cause of the low crop productivity and the high incidence of malnutrition in SSA (Sanchez 2002), and impacts the livelihoods of many people who rely straight on agriculture for food, including income. The yield of cereal crops in SSA has remained static at less than 1.5 t/ha for the past 50 years, even though the potential yield of almost all crop varieties surpasses 5 t/ha (FAO 2010). Soil productivity is an important environmental factor that needs to be well managed to keep up with food production for the current growing population.

Soil organic carbon is the basis of soil fertility and is a dominant fraction of soil organic matter (Lal 2016). Soil characteristics such as structure, aeration, pollutant filter, aggregation, water retention, pH, bulk density, biodiversity (rhizosphere processes), nutrient holding capacity, nutrient turnover and stability are dependent on the amount of soil organic carbon (Liu et al. 2006). Therefore, it is essential to maintain soil organic carbon to ensure soil quality, health, fertility and food production. Unsustainable management practices such as excessive irrigation, continuous monoculture, conventional tillage and crop residue removal have caused soil organic carbon loss and massive erosion. Therefore, there is a need to promote practices that enhance build-up of soil carbon.

Organic matter has been used to increase soil organic carbon. Soil organic matter is a primary carbon source, giving energy and nutrients to soil organisms. Furthermore, it supports soil functionality because it improves nutrient retention and turnover, moisture retention and availability, degradation of pollutants, carbon sequestration, and improves aggregate reformation and stability, giving the soil the capacity to withstand erosion. However, as discussed earlier, adding fresh organic matter has its fair share of drawbacks. Anthropogenic practices like the continuous use of tillage accelerate soil degradation through increased decomposition and release of carbon into the atmosphere, thus significantly threatening African sustainable crop production. These adverse effects of fresh organic manure can be countered by using bio-processed products like biochar and vermicompost.

8.3 Biochar

Biochar is a carbon-rich product made from diverse biomass feedstock obtained through heating the biomass in an oxygen-limited environment (Steiner 2016). In most cases, the choice of biomass depends on the availability of materials and associated costs such as of acquisition and transportion the (Gwenzi et al. 2015). Biochar can be produced from any organic biomass, e.g. leaves, crop residues or wood materials. However, the type of biomass determines the quality of biochar; for example, biochar produced from non-woody feedstocks, i.e. manures and plant residue, has been richer in nutrient content and has higher pH values and less stable carbon than that of biochar produced from lignocellulosic feedstocks as wood (Dahlawi et al. 2018). According to the International Biochar Initiative (IBI) (2011), biochar can be produced either by (a) pyrolysis systems or (b) gasification systems at temperatures between 300 °C and 700 °C. In the pyrolysis system, the natural polymeric constituents, lignin, cellulose, fats and starches are thermally broken down into three different fractions: bio-oil (condensed vapours), char (solid fraction) and non-condensable gases called syngas (Mohan et al. 2006). Pyrolysis involves using kilns or drums and vented retorts to contain the baking biomass in the absence of O_2. This pyrolysis system is sub-divided into two types: (i) fast pyrolysis, which tends to produce more oil and liquids and (ii) slow pyrolysis, which produces more pyrolysis gasses called syngas. A large amount of variation in output and characteristics of biochar produced depends on the type of pyrolysis reactor type, process condition and the type of feedstock used (Fig. 8.1).

Fundamental properties such as surface area and micropores presence contribute to the adsorptive properties of biochar and can potentially alter soil's surface area, pore size distribution, bulk density, penetration resistance and water holding capacity (Mukherjee and Lal 2013). The surface area of biochar increases with an increase in the peak temperature of biochar production (Mukherjee et al. (2011).

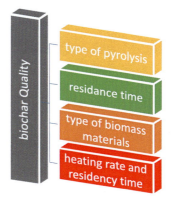

Fig. 8.1 Factors affecting the physio-chemical quality of biochar

8.4 Potential of Biochar in Improving Productivity

Numerous studies have reported the impact of biochar on soil properties (Table 8.1). Biochar can affect soil's physical properties in several ways; however, the degree of that is dependent on factors such as feedstock type, production process, application rate and environmental conditions.

Biochar was identified as a promising soil ameliorant which can enhance soil organic matter and reduce its decomposition (Nyambo et al. 2018). Biochar plays a key role during carbon and nitrogen cycle processes by associating relevant microorganisms that participate during these cycles (Purakayastha et al. 2019). Moreover, biochar addition to agricultural soil increases carbon sequestration, minimises greenhouse gas emissions and halts toxic organic pollutants and elements (Nyambo et al. 2020b). These characteristics of biochar are mainly due to its large surface area, porous structure and a negatively charged surface with high charge density Igalavithana et al. 2017).

The application of biochar has great potential to enhance soil fertility and, thus, crop growth and yields (Tomczyk et al. 2020). A study by Kamau et al. (2019) in Kenya, Nairobi, found that biochar application as soil ameliorant enhanced soil organic carbon by approximately 50% compared to the control treatment. Biochar mineralises slowly in soil, unlike fresh organic matter (Calderon et al. 2015), can improve soil properties and act as a habitat for microorganisms in the soil. It has a carbon content which ranges from <1 to >80% (Cimo 2014). With its higher surface area per unit mass and high charge density compared to organic matter, biochar can give a higher capacity to sorb cations per unit mass (Liang et al. 2006). Biochar can function as a binding agent connecting soil microaggregates (< 0.25 mm) to form macroaggregates (> 0.25 mm) (Jien and Wang 2013), hence increasing aggregate stability.

Biochar amendment enhances soil quality, increasing crop growth by adding nutrients directly to the plants or indirectly by upgrading the soil status, consequently increasing nutrient use efficiency (Nyambo et al. 2020c). Biochar can modify the root morphology of crops by increasing root biomass, specific root length, root

Table 8.1 Summary of research on the effects of biochar on soil properties

Soil type	Biochar	Study type	Rate of biochar application (%)	Observations	References
Residue	Municipal Green waste, 450	Laboratory	0, 2.6 and 5.2 (g g^{-1})	5.2% application significantly increased porosity compared to control. Bulk density low treatments with biochar	Jones et al. (2010)
Sandy soil	Jarrah wood (*Eucalyptus marginata*), 600 °C	Greenhouse	0, 0.45 and 2.27 (g g^{-1})	Porosity significantly increased with an increase in biochar application rate	Dempster et al. (2012)
Loamy sand	Peanut hulls, slow pyrolysis, 500 °C	Laboratory	0, 25, 50, 75 and 100 (g g^{-1})	The higher the application rate of biochar, the higher the porosity.	Githinji (2014)
Hydroagric stagnic anthrosol	Wheat (*Triticum* spp.) straw, 350–550 °C	Field	0, 1.1, 2.2 and 4.4 (g g^{-1})	Bulk density is reduced with an increase in biochar applied	Mankasingh et al. (2011)
Albic Luvisol	Hydrochar, 220 °C	Laboratory Greenhouse	0, 5 and 10 (g g^{-1})	Aggregate stability was highest in treatments with 5% biochar applied and least in control in both the glasshouse and laboratory experiments.	George et al. (2012)
Acidic Hutton soils	Maize residue	Laboratory and glasshouse	0%, 2.5%, 5%, 7.5% and 10% (w/w)	Biochar increased soil organic carbon, stability and water erosion resistance.	Nyambo et al. (2018)
–	Corn straw biochar	Laboratory	20 mg	Dissolved organic matter can increase the adsorption capacity of biochar for metal-ions. The biochar had preferential binding for protein-like substances> humic-like substances and the presence of dissolved organic matter significantly increased the adsorption capacity of biochar for Pb(II).	Wang et al. (2022)
Calcaric Cambisol	Rice husk	Pot experiment	0.4, 0.8, 1.6, 2.4 and 3.3% (w/w)	Biochar application significantly increased soil organic carbon, cation exchange capacity, available potassium and below-ground biomass of lentil	Abrishamkesh et al. (2015)
Slightly Alkaline clay loam	Wheat straw	Laboratory	–	Biochar (IBWS700) significantly decreased the lead bio-availability, affected available phosphorus (AP), cation exchange capacity (CEC), organic matter (OM) and activity of urease, alkaline phosphatase, sucrase and catalase	Zhu et al. (2022)

Table 8.2 Summary of research on the effects of biochar application on crop yields

Country	Soil textural class	Crop	Main findings	Reference
Kenya	Clay	Maize and soybean	Biochar application increased maize seasonal yield by 1.2 mg ha^{-1} and 0.4 mg ha^{-1} for soybean compared to control.	Kätterer et al. (2019)
Ghana	Sandy loam	Okra and cassava	Okra yield increased by 11.4% and cassava by 71.4% in biochar treatment compared to control.	Frimpong et al. (2021)
Ethiopia	Clay	Barely	Biochar application increased barley grain yield by up to 67% compared to control.	Agegnehu et al. (2016)
Ghana	Clay	Rice	71 and 134% compared to the control.	MacCarthy et al. (2020)
Zambia	Sandy loam	Soybean	Soybean grain yield increased by up to 43% compared to control.	Munera-Echeverri et al. (2020)
Zimbabwe	Clay	Maize	Maize biomass was 270% more under biochar application compared to control.	Gwenzi et al. (2016)
Nigeria	Sandy loam	Cocoyam	Biochar increased the yield of cocoyam by up to 8.1% compared to the control.	Adekiya et al. (2020)
Kenya	Clay	Maize	Maize grain yield was 5.6 mg ha^{-1} under biochar, which was about six times higher than that from control at 0.9 mg ha^{-1}.	Kamau et al. (2019)
Benin	Sandy loam	Rice	The rice grain yield was 1.8 tha^{-1} for biochar-amended plots and 1.3 tha^{-1} for control. Thus, biochar application increased rice yield by 38.5% compared to control.	Partey et al. (2016)
Ethiopia	Silty clay	Maize and soybean	Maize yield increment was 1.4 mg ha-1 and 1.9 mg ha-1 for soya bean in biochar treatment compared to control.	Wakweya et al. (2022)

volume, surface area and root tip number (Purakayastha et al. 2019). These improve the roots' ability to secure more nutrients, improving plant growth and yield. Table 8.2 summarises some of the studies that showed the influence of biochar application on crop yields.

8.5 Vermicomposting

Vermicomposting is a process by which biodegradable wastes such as agricultural wastes, kitchen wastes and food wastes are converted while passing through the worm gut to nutrient-rich humus used as a soil conditioner or fertiliser (Adhikary 2012; Lim et al. 2015). Mitchell (1997) reported that vermicompost production from animal manure resulted in decreased electrical conductivity and a change in pH toward a neutral state. Vermicompost involves a symbiotic interaction between earthworms and microorganisms to produce humus used as a soil conditioner or fertiliser.

The incorporation of vermicompost as a soil amendment tool enhances the soil nutrient status but also helps to boost productivity by 40% at 20–60% at reduced inputs (Srivastava et al. 2020). Yatoo et al. (2021) reported that using vermicompost over time resulted in reduced pests and disease occurrence, contributing to reduced herbicides' costs. Moghadam et al. (2012) reported that vermicompost contains similar benefits as conventional composts, such as being a good source of organic matter, increased moisture-holding capacity, improved nutrient uptake and plant hormone-like activity. Furthermore, the literature indicates that vermicompost results enhanced plant growth and development and altered morphology of plants grown in vermicompost amended substrate. Bachman and Metzger (2008) suggested that enhanced growth in plants could be ascribed to plant hormone-like activity linked to microflora related to vermicomposting and metabolites produced as a result of secondary metabolism. Chaoui et al. (2003) reported that vermicompost secretes enzymes such as amylase, lipase, cellulase and chitinase that can break down the organic matter in the soil to release the nutrients and make them available to be used by the roots.

Vermicompost has been reported to contain 40–60% levels of humic compounds compared to conventional composts (Arancon et al. 2004). Humus stimulates plant nutrient uptake and metabolism, thus encouraging protein synthesis and producing hormone-like activity (Nardi et al. 2002). Humic acid in humus acts as a binding structure for the plant nutrients, such as calcium, iron, potassium, sulphur and phosphorus (Adhikary 2012), and these nutrients are stored in a readily available form for plant uptake.

8.6 Opportunities for Mixing Biochar and Vermicompost

Despite the highlighted positive effects of biochar, there are challenges associated with its use as a soil amendment. In some instances, the biochar particles can contain a greater quantity of toxic pollutants compared to the target soil, thus it can potentially introduce or increase their amount in the soil (Mohanty and Boehm 2015). Depending on the transport and application method, this process can result in air pollution and toxic pollutants than the soil in application areas (Ravi et al. 2020). In some cases, yields have been reported to decline because of the sorption of water and nutrients by the biochar, which reduces the availability of these resources for the crops. According to Godlewska et al. (2021), the environmental ageing of biochar can reduce the biochar's affinity for contaminants, releasing them into the environment where they can come into contact with organisms.

Similarly, the use of vermicompost has its fair share of challenges, when used as a soil amendment, vermicompost, especially from sewage waste materials, is problematic primarily due to the high concentration of various contaminants like heavy metals. The contaminants can be hazardous to humans after consuming crops grown using this vermicompost. Furthermore, the heavy metals can inhibit earthworms' activity, resulting in reduced growth, reproduction and mortality, and thus the efficiency of converting sewage sludge into vermicompost. Therefore, amending the soil with either vermicomposting or biochar as standalone applications requires the addition of supplementary materials to ensure optimal soil and crop productivity.

8.7 Potential of Biochar-Vermicomposting in Improving Productivity

Biochar and vermicompost are implemented to manage problems associated with manure application but create more stable products with high nutrient status than in manure form (Najafi-Ghiri et al. 2019). Biochar and vermicompost are applied in agricultural soils because they can enhance soil quality and crop yields (Kheir et al. 2021). With the benefits of both biochar and vermicompost in agriculture well recorded, the critical question is if biochar and vermicompost mixture will synergistically benefit soil quality and crop yield. To answer that question, different research results are summarised in this section. Due to limited research on the co-application of biochar and vermicompost in Africa, research beyond Africa was included to effectively show the potential benefits of applying a combination of biochar and vermicompost. The information can help address land degradation as one of Africa's significant threats to food security (Gomiero 2016; Kibugi 2018; King-Okumu et al. 2021).

Research on the co-application of biochar and vermicompost shows that the combination increased the yield of Knolkhol by between 6% and 11.9% compared to the sole application of compost (Sharma et al. 2021). In a study by Di et al. (2019), biochar-vermicompost significantly increased rice yield by between 26.5% and 35.3% compared to the sole application of biochar. The increase in crop yield after the co-application of biochar and vermicompost might be because their mixture increases nutrient load and stimulates microorganisms (Vandecasteele et al. 2016). Ebrahimi et al. (2021) demonstrated that the full benefit of biochar and vermicompost mixture in increasing crop yield is increased soil water content. Infiltration rate, soil water retention and hydraulic conductivity were found to be at the highest under co-application of biochar and vermicompost compared to their sole application (Ebrahimi et al. 2021; Sharma et al. 2021). Soil aggregate stability was also reported to be higher under biochar and vermicompost mixture than in their sole applications (Sharma et al. 2021). The increase in nutrient loading, aggregate ability and water retention is essential to African agriculture as it receives low and highly variable rainfalls (Kimani et al. 2017). The soils are generally dominated by low soil fertility status. Co-application of biochar and vermicompost decreased soil N_2O emissions compared to their sole application (Di et al. 2019). Thus, the co-application of biochar and vermicompost can serve as a better climate change mitigation strategy than their sole applications.

Biochar vermicomposting, also known as vermi-char, is a product developed from the inclusion of biochar during vermicomposting (Huang et al. 2022). The addition of biochar during vermicomposting can fasten the vermicomposting process and enhance the quality of the final product (Malińska et al. 2017; Huang et al. 2022). The study by Malińska et al. (2017) found that the application of biochar during sewage sludge vermicompost increased the number of juvenile earthworms and cocoons compared to where biochar was not applied. The findings were also supported by the study of Ameen and Al-Homaidan (2022) and Huang et al. (2022), who also found that adding biochar enhances the number of earthworms during vermicomposting. The addition of biochar during vermicomposting also enhanced the number of active bacteria, eukaryotes (Huang et al. 2022), actinobacteria, firmicutes, celluloses, alkaline phosphatase and protease (Cao et al. 2021; Gong et al. 2021). Biochar vermicomposting turns out to have higher active enzymes and microorganisms which are important in nutrient cycling and carbon stabilisation (Huang et al. 2022). According to Malińska et al. (2017), the increase in organisms in vermi-char is the reason why the conversion of waste materials into vermicompost is fast and efficient.

Ameen and Al-Homaidan (2022) reported that biochar application during vermicomposting reduces heavy metals, thus, resting in an eco-friendlier product. In addition to reducing heavy metals, the addition of biochar during vermicomposting was found to reduce pathogens (Paul et al. 2020). The addition of biochar during vermicomposting was found to enhance nitrate by 47.32–122.64% compared to where biochar was not applied (Huang et al. 2022). The study was in line with Gong et al. (2021), who also found that vermi-char had higher total nitrogen, phosphorus, potassium and nitrate compared to where biochar was not added. The addition of biochar during vermicompost was reported to reduce total carbon and nitrogen losses compared to where biochar was not applied (Cao et al. 2021). Thus, the addition of biochar during vermicomposting enhances the nutritional value of the final product (Paul et al. 2020).

Currently, studies on the effect of vermi-char on soil properties and crop yields are limited. However, a study by Gong et al. (2021) found that the application of vermi-char resulted in increased germination and tomato seedlings growth compared to the solo application of biochar and vermicompost. it is important to note that the characteristics and quality of vermi-char depend also on the type of the biochar in addition to the waste stock (Huang et al. 2022).

8.8 Conclusion

The enhancement of soil productivity and increased crop yields is vital in Africa to keep up with food security for the growing population. Applying biochar and vermicompost can enhance soil productivity by improving soil quality indices in

Africa. In addition, biochar and vermicompost application can enhance crop yields, thus aiding in meeting food security in Africa. Studies showed that the benefits of biochar and vermicompost on soils and crop yield is higher when they are co-applied than their solo application. In addition, biochar application during vermicomposting can enhance the vermicomposting process and result in a climate-smart product. Currently, studies on the application of vermi-char in agricultural lands are limited. Thus, local studies are recommended under different agro-climatic conditions, soil types and crops.

Acknowledgement This work is based on the research supported by the National Research Foundation (NRF) of South Africa and the Dutch Research Council (de Nederlandse Organisatie voor Wetenschappelijk Onderzoek - NWO) Project UID 129352. Any opinion, finding, conclusion or recommendation expressed in this material is that of the author(s) and not that of NRF or NWO.

References

Abrishamkesh S, Gorji M, Asadi H, Bagheri-Marandi GH, Pourbabaee AA (2015) Effects of rice husk biochar application on the properties of alkaline soil and lentil growth. Plant Soil Environ 61:475–482

Adekiya AO, Agbede TM, Olayanju A, Ejue WS, Adekanye TA, Adenusi TT, Ayeni JF (2020) Effect of biochar on soil properties, soil loss, and cocoyam yield on a tropical sandy loam Alfisol. Sci World J 2020. https://doi.org/10.1155/2020/9391630

Adhikary S (2012) Vermicompost, the story of organic gold: a review. Agricultural Sciences 03 ((07):905–917

Agegnehu G, Nelson PN, Bird MI (2016) Crop yield, plant nutrient uptake and soil physicochemical properties under organic soil amendments and nitrogen on Nitisols. Soil Tillage Res 160: 1–13

Ajibade S, Mupambwa HA, Manyevere A, Mnkeni PN (2022) Vermicompost amended with rock phosphate as a climate smart technology for production of organic Swiss Chard (Beta vulgaris subsp. vulgaris). Front Susain Food Syst 6

Alvarez JM, Pasian C, Lal R, Lopez R, Fernández M (2017) Vermicompost and biochar as growing media replacement for ornamental plant production. J Appl Hortic 19:205–214

Ameen F, Al-Homaidan AA (2022) Improving the efficiency of vermicomposting of polluted organic food wastes by adding biochar and mangrove fungi. Chemosphere 286:131945

Arancon NQ, Edwards CA, Atiyeh R, Metzger JD (2004) Effects of vermicomposts produced from food waste on the growth and yields of greenhouse peppers. Bioresour Technol 93(2):139–144

Bachman GR, Metzger JD (2008) Growth of bedding plants in commercial potting substrate amended with vermicompost. Bioresour Technol 99(8):3155–3161

Baveye PC, Schnee LS, Boivin P, Laba M, Radulovich R (2020) Soil organic matter research and climate change: merely re-storing carbon versus restoring soil functions. Front Environ Sci 8: 579904

Bolan NS, Szogi AA, Chuasavathi T, Seshadri B, Rothrock MJ, Panneerselvam P (2010) Uses and management of poultry litter. Worlds Poult Sci J 66(4):673–698

Calderon FJ, Benjamin J, Vigil MF (2015) A comparison of Corn (Zea mays L.) residue and its biochar on soil C and plant growth. PLoS One 10(4):e0121006

Cao Y, Tian Y, Wu Q, Li J, Zhu H (2021) Vermicomposting of livestock manure as affected by carbon-rich additives (straw, biochar and nanocarbon): a comprehensive evaluation of earthworm performance, microbial activities, metabolic functions and vermicompost quality. Bioresour Technol 320:124404

Chaoui HI, Zibilske LM, Ohno T (2003) Effects of earthworms casts and compost on soil microbial activity and plant nutrient availability. Soil Biol Biochem 35:295–302. https://doi.org/10.1016/S0038-0717(02)00279-1

Cimo G (2014) Characterization of chemical and physical properties of biochar for energy purposes and environmental restoration. https://www.researchgate.net/publication/259901709_characterization_of_chemical_and_physical_properties_of_biochar_for_energy_purposes_and_environmental_restoration. Accessed 29 Aug 2018

Dahlawi S, Naeem A, Rengel Z, Naidu R (2018) Biochar application for the remediation of salt-affected soils: challenges and opportunities. Sci Total Environ 625:320–335

Dempster DN, Gleeson DB, Solaiman ZI, Jones DL, Murphy DV (2012) Decreased soil microbial biomass and nitrogen mineralisation with Eucalyptus biochar addition to a coarse textured soil. Plant Soil 354(1):311–324

Di WU, Yanfang FENG, Lihong XUE, Manqiang LIU, Bei YANG, Feng HU, Linzhang YANG (2019) Biochar combined with vermicompost increases crop production while reducing ammonia and nitrous oxide emissions from a paddy soil. Pedosphere 29(1):82–94

Ebrahimi M, Souri MK, Mousavi A, Sahebani N (2021) Biochar and vermicompost improve growth and physiological traits of eggplant (Solanum melongena L.) under deficit irrigation. Chem Biol Technol Agric 8(1):1–14

FAO F (2017) The future of food and agriculture–trends and challenges. Annu Rep 296:1–180

FAO (2010) Agriculture data, agricultural production. http://faostat.fao.org/site/567

Frimpong KA, Phares CA, Boateng I, Abban-Baidoo E, Apuri L (2021) One-time application of biochar influenced crop yield across three cropping cycles on tropical sandy loam soil in Ghana. Heliyon 7(2):e06267

George C, Wagner M, Kucke M, Rillig MC (2012) Divergent consequences of hydrochar in the plant-soil system: arbuscular mycorrhiza, nodulation, plant growth and soil aggregation effects. Appl Soil Ecol 59:68–72

Gerzabek MH, Haberhauer G, Kirchmann H (2001) Soil organic matter pools and carbon-13 natural abundances in particle-size fractions of a long-term agricultural field experiment receiving organic amendments. Soil Sci Soc Am J 65(2):352–358

Githinji L (2014) Effect of biochar application rate on soil physical and hydraulic properties of a sandy loam. Arch Agron Soil Sci 60:457–470

Godlewska P, Ok YS, Oleszczuk P (2021) The dark side of black gold: ecotoxicological aspects of biochar and biochar-amended soils. J Hazard Mater 403:123833. https://doi.org/10.1016/j.jhazmat.2020.123833

Gomiero T (2016) Soil degradation, land scarcity and food security: reviewing a complex challenge. Sustainability 8(3):281. https://doi.org/10.3390/su8030281

Gong X, Zhang Z, Wang H (2021) Effects of Gleditsia sinensis pod powder, coconut shell biochar and rice husk biochar as additives on bacterial communities and compost quality during vermicomposting of pig manure and wheat straw. J Environ Manag 295:113136

Gwenzi W, Chaukura N, Mukome FN, Machado S, Nyamasoka B (2015) Biochar production and applications in sub-Saharan Africa: opportunities, constraints, risks and uncertainties. J Environ Manag 150:250–261

Gwenzi W, Muzava M, Mapanda F, Tauro TP (2016) Comparative short-term effects of sewage sludge and its biochar on soil properties, maize growth and uptake of nutrients on a tropical clay soil in Zimbabwe. J Integr Agric 15(6):1395–1406

Hamann M, Tuinder V (2012) Introducing the Eastern Cape: A quick guide to its history, diversity and future challenges. Stockholm Resilience Centre, Stockholm University, Stockholm

Huang K, Guan M, Chen J, Xu J, Xia H, Li Y (2022) Biochars modify the degradation pathways of dewatered sludge by regulating active microorganisms during gut digestion of earthworms. Sci Total Environ 828:154496

Igalavithana AD, Ok YS, Niazi NK, Rizwan M, Al-Wabel MI, Usman AR, Moon DH, Lee SS (2017) Effect of corn residue biochar on the hydraulic properties of sandy loam soil. Sustainability 9(2):266

International Biochar Initiative (IBI) (2011) Biochar production units. https://biochar-international. org/biochar-production-technologies/. Accessed 03 Oct 2018

IPBES (2018) Media release: worsening worldwide land degradation now 'critical', undermining well-being of 3.2 billion people. El-Afry

Jien SH, Wang CS (2013) Effects of biochar on soil properties and erosion potential in a highly weathered soil. CATENA 110:225–233

Jones BE, Haynes RJ, Phillips IR (2010) Effect of amendment of bauxite processing sand with organic materials on its chemical, physical and microbial properties. J Environ Manag 91(11): 2281–2288

Kamau S, Karanja NK, Ayuke FO, Lehmann J (2019) Short-term influence of biochar and fertilizer-biochar blends on soil nutrients, fauna and maize growth. Biol Fertil Soils 55(7):661–673

Kätterer T, Roobroeck D, Andrén O, Kimutai G, Karltun E, Kirchmann H, Nyberg G, Vanlauwe B, de Nowina KR (2019) Biochar addition persistently increased soil fertility and yields in maize-soybean rotations over 10 years in sub-humid regions of Kenya. Field Crop Res 235:18–26

Kheir AM, Ali EF, Ahmed M, Eissa MA, Majrashi A, Ali OA (2021) Biochar blended humate and vermicompost enhanced immobilization of heavy metals, improved wheat productivity, and minimized human health risks in different contaminated environments. J Environ Chem Eng 9(4):105700. https://doi.org/10.1016/j.jece.2021.105700

Kibugi R (2018) Soil health, sustainable land management and land degradation in Africa: legal options on the need for a specific African soil convention or protocol. In: International yearbook of soil law and policy 2017. Springer, Cham, pp 387–411

Kimani MW, Hoedjes JC, Su Z (2017) An assessment of satellite-derived rainfall products relative to ground observations over East Africa. Remote Sens 9(5):430. https://doi.org/10.3390/rs9050430

King-Okumu C, Tsegai D, Sanogo D, Kiprop J, Cheboiwo J, Sarr MS, da Cunha MI, Salman M (2021) How can we stop the slow-burning systemic fuse of loss and damage due to land degradation and drought in Africa? Curr Opin Environ Sustain 50:289–302

Laker MC (2004) Advances in soil erosion, soil conservation, land suitability evaluation and land use planning research in South Africa, 1978–2003. S Afr J Plant Soil 21:345–368

Lal R (2009) Sequestering atmospheric carbon dioxide. Crit Rev Plant Sci 28(3):90–96

Lal R (2016) Soil health and carbon management. Food Energy Secur 5(4):212–222

Liang B, Lehmann J, Solomon D, Kinyangi J, Grossman J, O'Neill B, Skjemstad JO, Thies J, Luizão FJ, Petersen J, Neves EG (2006) Black carbon increases cation exchange capacity in soils. Soil Sci Soc Am J 70(5):1719–1730

Lim SL, Wu TY, Lim PN, Shak KPY (2015) The use of vermicompost in organic farming: overview, effects on soil and economics. J Sci Food Agric 95(6):1143–1156

Liu X, Herbert SJ, Hasheni AM, Zhang X, Ding G (2006) Effects of agricultural management on soil organic matter and carbon transfornmation-a review. Plant Soil Environ 52:531–543

MacCarthy DS, Darko E, Nartey EK, Adiku SG, Tettey A (2020) Integrating biochar and inorganic fertilizer improves productivity and profitability of irrigated rice in Ghana, West Africa. Agronomy 10(6):904

Malińska K, Golańska M, Caceres R, Rorat A, Weisser P, Ślęzak E (2017) Biochar amendment for integrated composting and vermicomposting of sewage sludge–the effect of biochar on the activity of Eisenia fetida and the obtained vermicompost. Bioresour Technol 225:206–214

Mamera M, van Tol JJ, Aghoghovwia MP, Nhantumbo AB, Chabala LM, Cambule A, Chalwe H, Mufume JC, Rafael R (2021) Potential use of biochar in pit latrines as a faecal sludge management strategy to reduce water resource contaminations: a review. Appl Sci 11(24): 11772. https://doi.org/10.3390/app112411772

Mankasingh U, Choi PC, Ragnarsdottir V (2011) Biochar application in a tropical, agricultural region: a plot scale study in Tamil Nadu, India. Appl Geochem 26:218–221

Mitchell A (1997) Production of Eisenia foetida and vermicompost from feedlot manure. Soil Biol Biochem 29(3–4):763–766

Moghadam ARL, Ardebili ZO, Saidi F (2012) Vermicompost induced changes in growth and development of Lilium Asiatic hybrid var. Navona. Afr J Agric Res 7(17):2609–2621

Mohan D, Pittman CU, Steele PH (2006) Pyrolysis of wood/biomass for bio-oil: a critical review. Energy Fuel 20(3):848–889

Mohanty SK, Boehm AB (2015) Effect of weathering on mobilization of biochar particles and bacterial removal in a stormwater biofilter. Water Res 85:208–215

Morton JF (2007) The impact of climate change on smallholder and subsistence agriculture. Proc Natl Acad Sci 104(50):19680–19685

Mukherjee A, Lal R (2013) Biochar impacts on soil physical properties and greenhouse gas emissions. Agronomy 3(2):313–339

Mukherjee A, Zimmerman AR, Harris W (2011) Surface chemistry variations among a series of laboratory-produced biochars. Geoderma 163(3–4):247–255

Munera-Echeverri JL, Martinsen V, Strand LT, Cornelissen G, Mulder J (2020) Effect of conservation farming and biochar addition on soil organic carbon quality, nitrogen mineralization, and crop productivity in a light textured Acrisol in the sub-humid tropics. PLoS One 15(2): e0228717

Mupambwa HA, Lukashe NS, Mnkeni PNS (2017) Suitability of fly ash vermicompost as a component of pine bark growing media: effects on media physicochemical properties and ornamental marigold (Tagetes spp.) growth and flowering. Compost Sci Util 25(1):48–61

Najafi-Ghiri M, Razeghizadeh T, Taghizadeh MS, Boostani HR (2019) Effect of sheep manure and its produced vermicompost and biochar on the properties of a calcareous soil after barley harvest. Commun Soil Sci Plant Anal 50(20):2610–2625

Naiji M, Souri MK (2018) Nutritional value and mineral concentrations of sweet basil under organic compared to chemical fertilisation. Acta Sci Pol Hortorum Cultus 17(2):167–175

Nardi S, Pizzeghello D, Muscolo A, Vianello A (2002) Physiological effects of humic substances on higher plants. Soil Biol Biochem 34(11):1527–1536

Nyambo P, Taeni T, Chiduza C, Araya T (2018) Effects of maize residue biochar amendments on soil properties and soil loss on acidic Hutton soil. Agronomy 8(11):256

Nyambo P, Chiduza C, Araya T (2020a) Carbon input and maize productivity as influenced by tillage, crop rotation, residue management and biochar in a semiarid region in South Africa. Agronomy 10(5):705

Nyambo P, Chiduza C, Araya T (2020b) Carbon dioxide fluxes and carbon stocks under conservation agricultural practices in South Africa. Agriculture 10(9):374. https://doi.org/10.3390/agriculture10090374

Nyambo P, Mupambwa HA, Nciizah AD (2020c) Biochar enhances the capacity of climate-smart agriculture to mitigate climate change. Handbook of climate change management: research, leadership, transformation, pp 1–8

Nyambo P, Chiduza C, Araya T (2022) Effect of conservation agriculture on selected soil physical properties on a haplic cambisol in Alice, Eastern Cape, South Africa. Arch Agron Soil Sci 68(2): 195–208. https://doi.org/10.1080/03650340.2020.1828578

Partey ST, Saito K, Preziosi RF, Robson GD (2016) Biochar use in a legume–rice rotation system: effects on soil fertility and crop performance. Arch Agron Soil Sci 62(2):199–215

Parwada C, Van Tol J (2018) Effects of litter source on the dynamics of particulate organic matter fractions and rates of macroaggregate turnover in different soil horizons. Eur J Soil Sci 69(6): 1126–1136

Paul S, Kauser H, Jain MS, Khwairakpam M, Kalamdhad AS (2020) Biogenic stabilization and heavy metal immobilization during vermicomposting of vegetable waste with biochar amendment. J Hazard Mater 390:121366

Purakayastha TJ, Bera T, Bhaduri D, Sarkar B, Mandal S, Wade P, Kumari S, Biswas S, Menon M, Pathak H, Tsang DC (2019) A review on biochar modulated soil condition improvements and nutrient dynamics concerning crop yields: pathways to climate change mitigation and global food security. Chemosphere 227:345–365

Ravi S, Li J, Meng Z, Zhang J, Mohanty S (2020) Generation, resuspension, and transport of particulate matter from biochar-amended soils: a potential health risk. GeoHealth 4(11): e2020GH000311

Richardson DM, Witt AB, Pergl J, Dawson W, Essl F, Kreft H, van Kleunen M, Weigelt P, Winter M, Pyšek P (2022) Plant invasions in Africa. In: Global plant invasions. Springer, Cham, pp 225–252

Risse LM, Faucette B (2009) Food waste composting: institutional and industrial applications. University of Georgia

Sanchez P (2002) Ecology. Soil fertility and hunger in Africa. Science 295:2019–2020. Smaling EMA, Nandwa SM, Janssen BH (1997). In: Buresh RJ, Sanchez PA, Calhoun F (eds) Replenishing soil fertility in Africa. SSSA, Madison, USA

Sharma P, Abrol V, Sharma V, Chaddha S, Rao CS, Ganie AQ, Hefft DI, El-Sheikh MA, Mansoor S (2021) Effectiveness of biochar and compost on improving soil hydro-physical properties, crop yield and monetary returns in inceptisol subtropics. Saudi J Biol Sci 28(12):7539–7549. https://doi.org/10.1016/j.sjbs.2021.09.043

Srivastava V, Vaish B, Singh RP, Singh P (2020) An insight to municipal solid waste management of Varanasi city, India, and appraisal of vermicomposting as its efficient management approach. Environ Monit Assess 192(3):1–23

Steiner C (2016) Considerations in biochar characterization. In: Agricultural and environmental applications of biochar: advances and barriers, SSSA Special Publication 63. SSSA, Madison, pp 87–100

Tomczyk A, Sokołowska Z, Boguta P (2020) Biochar physicochemical properties: pyrolysis temperature and feedstock kind effects. Rev Environ Sci Biotechnol 19(1):191–215

Vandecasteele B, Sinicco T, D'Hose T, Nest TV, Mondini C (2016) Biochar amendment before or after composting affects compost quality and N losses, but not P plant uptake. J Environ Manage 168:200–209

Vlek PLG, Le QB, Tamene L (2008) CGIAR science council secretariat. Italy, Rome

Wakweya T, Nigussie A, Worku G, Biresaw A, Aticho A, Hirko O, Ambaw G, Mamuye M, Dume B, Ahmed M (2022) Long-term effects of bone char and lignocellulosic biochar-based soil amendments on phosphorus adsorption–desorption and crop yield in low-input acidic soils. Soil Use Manag 38(1):703–713

Wang Y, van Zwieten L, Wang H, Wang L, Li R, Qu J, Zhang Y (2022) Sorption of Pb (II) onto biochar is enhanced through co-sorption of dissolved organic matter. Sci Total Environ 825: 153686

Yatoo AM, Ali M, Baba ZA, Hassan B (2021) Sustainable management of diseases and pests in crops by vermicompost and vermicompost tea. A review. Agron Sustain Dev 41(1):1–26

Zhu X, Li X, Shen B, Zhang Z, Wang J, Shang X (2022) Bioremediation of lead-contaminated soil by inorganic phosphate-solubilising bacteria immobilised on biochar. Ecotoxicol Environ Saf 237:113524

Chapter 9
Wild Birds Animal Manure Vermicomposting: Experiences from Namibia

Brendan M. Matomola, Simon T. Angombe, Rhoda Birech, and Hupenyu A. Mupambwa

Abstract Wild animal manures, especially birds that tend to roost at defined places each day, present a potentially valuable source of organic matter and nutrients that can be used in agriculture. However, due to the limited quantities of the availabilities of these manures, very few researchers have looked at using processes like vermicomposting to process these manures. This chapter presents insights on on-farm research being done on the processing of animal manures particularly guinea fowl manure, using vermicomposting. The results and experiences of this research have shown that the vermicomposting of guinea fowl manure without any pretreatment is not possible as it resulted in the death of the earthworm specie *Eisenia fetida*. However, pretreatment over a 2-week period allowed for survival and growth of these earthworms, while the vermicompost that came from the guinea fowl manure proved more superior as a seedling planting media. The insights presented in this chapter seeks to further invoke research in this unexplored organic nutrient source.

Keywords Guinea fowl · Dove manure · Sheep and goat manure mixture · Eisenia fetida · Windrow vermicomposting

---

B. M. Matomola (✉)
Agronomy & Horticulture Section, Neudamm Farming, Neudamm Campus, University of Namibia, Windhoek, Namibia
e-mail: bmatomola@unam.na

S. T. Angombe · R. Birech
Faculty of Agriculture, Engineering & Natural Science; University of Namibia, Windhoek, Namibia

H. A. Mupambwa
Sam Nujoma Marine and Coastal Resources Research Center, University of Namibia, Henties Bay, Namibia

© The Author(s), under exclusive license to Springer Nature Singapore Pte Ltd. 2023
H. A. Mupambwa et al. (eds.), *Vermicomposting for Sustainable Food Systems in Africa*, Sustainable Agriculture and Food Security,
https://doi.org/10.1007/978-981-19-8080-0_9

9.1 Introduction

Vermicomposting is a process that involves the biodegradation of organic materials with the use of earthworms mainly the epigeic species of earthworms that feed on fresh but dead organic matter (Mupambwa and Mnkeni 2018). With the increase in global population, there has been an increase in the generation of organic wastes from farming and municipal activities, with potential to cause environmental challenges if not properly managed. Therefore, vermicomposting has been proposed as one such technology that is environmentally friendly that can be used to manage these organic wastes. However, much of the research has thus focused on organic wastes that are generated through human activities such as cow, pig, goat, chicken and sheep manures; municipal wastes, among others, with very limited research focusing on manures or organic wastes generated in the natural environment. However, manures that are generated by wild animals like elephants, bats, seabirds, among others, have been shown to have potential as nutrient sources in agriculture.

In a recent review, De La Peña-Lastra (2021) indicated that seabird fecal material also called guano when fresh may contain up to 7.3% nitrogen, 1.5% phosphorus and 60% water. However, much of this nitrogen within the fresh guano is available as uric acid and protein with 80% and 10%, respectively. Similarly, African elephants (*Loxodonta africana*) can deposit up to 150 kg of wet dung per day, presenting an important resource that can be harnessed for agricultural purposes (Stanbrook 2018). Deposition of dung per elephant was reported to result in the accumulation of 0.01 kg N/ha, 0.26 kg carbon/ha and 0.01 kg P/ha in soil (Stanbrook 2018). Free ranging animal manures present a potentially important source of manure, though there is general lack of research information available on the nutrient status of these manures. In countries like Namibia, there are large areas that are designated as conservancies where wild animals exist and have specific areas where they normally sleep and breed. Large quantities of wild animal manures that include birds like Guinea fowl and doves tend to accumulate at these resting areas, and our chapter presents experiences on the processing on such wild animal manures using earthworms into valuable organic fertilizers.

9.2 Collection and Preliminary Characterization of Wild Animal Manure in Namibia

Poultry manure obtained from domesticated chickens, ducks, etc. is one of the most common potential known sources of manure in improving soil fertility to crop farmers, more so with chicken manure used as fertilizer for many horticultural crops. Little is known about wild animal manures such as those from wild guinea fowls, doves, etc. as sources of manure except sea bird guano which is commercially known. Although there is extensive research on guinea fowls, most of the research is based on meat not manure. Organic crop farming requires the use of sustainable crop

Fig. 9.1 Sequence of guinea fowl manure cleaning from original bush collected sample

production practices that do no harm to the environment and consumers of products produced through organic farming. We have observed that some wild animals may possess the potential to produce manures which can be processed into compost or vermicompost and applied to crops as soil fertilizer, a source of organic matter and carbon.

After identifying the resting sites for the wild birds, wild guinea fowl manures were collected along riverbanks under trees, Acacia species. Collection was done by raking, filling carry bags and transporting manures to the site for cleaning and screening. The typical screening process for guinea fowl manure is shown in Fig. 9.1. Screening is a process of cleaning the manure to remove foreign materials such as pieces of twigs, twigs, pebbles, other unwanted animal manures, plastics, pieces of wire, seed pods, acacia seeds, grasses etc. before use as shown in Figs 9.2 and 9.3.

After thorough cleaning, guinea manure samples were collected and sent to the Namibian Ministry of Agriculture, Water and Land Reform soil laboratory in 2018 and 2020. Dove manures was also collected on the farm around old buildings, during the dry season though samples are yet to be fully analyzed for chemical properties. Chemical laboratory analysis showed that raw wild guinea fowl manures (not processed) contain sufficient quantities of mineral nutrients comparable to poultry manure with a near neutral pH in all samples. Samples of raw wild GFM manures and domesticated animal manures were sent for lab analysis and the following results (Table 9.1) were obtained in comparison to other manures at different periods.

From the laboratory results, it stands to support that wild guinea fowl manures contain sufficient amounts of macro- and micronutrients with a potential to produce manure as soil fertilizer and the slightly acidic to neutral pH ranging from 6.21 to 8.86. The pH range is important for nutrient availability to plants. However, lab chemical analysis on the manures of wild guinea fowl indicates that there is a variation of the mineral composition of the same manures. This depends on the age of the manure, the season and screening methods applied (Figs. 9.2, 9.3, and 9.4).

Table 9.1 Preliminary lab results of guinea fowl manure (GFM) and cattle dung (CD) chicken manure (CM), goat sheep manure (SGM), collected in 2018 and 2020

Parameter	Units	2018		2020						
		GFM	CD	GFM01A	GFM01B	GFM02	GFM	CM	SGM	CD
Organic matter	%	24.49	27.18	68.6	62.04	58.94	67.89	57.84	65.67	73.87
Nitrogen	%	–	–	1.57	1.16	1.42	1.55	3.38	1.60	1.07
Phosphorus	%	0.53	0.49	0.123	0.05	0.21	0.120	0.35	0.20	0.38
Potassium	%	0.40	1.19	1.12	0.73	0.69	0.51	2.66	3.22	1.53
Magnesium	%	0.30	0.84	0.3	0.20	0.27	0.06	0.47	0.62	0.39
Calcium	%	0.68	2.00	1.62	0.80	0.98	0.25	4.61	1.82	1.20
Sodium	%	0.01	0.24	0.05	0.01	0.00	0.00	0.31	0.12	0.18
Iron	ppm	15603.3	16000.2	402.3	3516.1	6518.5	889.9	72.0	2816.3	2427.3
Copper	ppm	25.9	37.6	0.0	0.0	0.0	24.2	26.7	9.9	25.7
Zinc	ppm	34.4	155.7	30.5	10.4	17.2	13.3	207.7	41.1	84.0
Manganese	ppm	329.1	532.8	155.7	85.8	149.7	29.5	463.2	238.1	212.8
Salt content	%	0.31	1.04	1.45	1.07	0.81	1.78	2.98	2.88	1.28
pH		6.27	8.17	6.57	6.22	6.21	6.28	7.65	8.69	8.26

GFM01A = 4–5 months old manure fully cleaned.
GFM01B = 2 weeks old manure fully cleaned.
GFM02 = 4–5 months partly cleaned manures.
GFM = 5–6 months old manure fully cleaned.

9 Wild Birds Animal Manure Vermicomposting: Experiences from Namibia

Fig. 9.2 Guinea fowl manure weighed after cleaning in 2020

Fig. 9.3 Vermicomposting cattle dung (CD) two vermibeds. On the left (Plot 15) without guinea fowl manure amendment and the other on the right (Plot 16) amended with 5% guinea fowl manure

Fig. 9.4 Vermicomposting sheep-goat manure (SGM) in two vermibeds. On the left (Plot 13) without guinea fowl manure amendment and the other on the right (Plot 14) amended with 5% guinea fowl manure

9.3 Vermicomposting of Wild Animal Manures

Vermicompost is one of safest and sustainable methods of biodegrading organic material into a finished product suitable for soil application as a fertilizer, with sufficient nutrients and microorganisms essential for plant growth while suppressing pathogens. However, very little research has been done on these wild animal manures indicating not only chemical changes but also earthworm growth and development as indicated in Figs 9.3 and 9.4. In an attempt to use guinea fowl manure as soil fertilizer, the first approach was to decompose it just like cattle dung (CD) and sheep goat manure (SGM).

9.4 Container Vermicomposting Experience

9.4.1 Vermicomposting of Sole Animal Manures

The earthworm specie *Eisenia fetida* was used on different feedstocks, i.e., cattle dung (CD), guinea fowl manure (GFM), sheep goat manure mixture (SGM), and chicken manure (CM). For each feedstock type, four plastic containers of 20 L capacity with holes at the bottom were used and filled with manures (about 5 kg) after water at ratio of 1:1 (a kg of feedstock with 1 liter of water) and allowed to stand for 1 day (curing). After a day of soaking in water 200 juvenile (non-clitellated) earthworms of *E. fetida* were inoculated in each container for all four feedstock types.

9.4.2 Preliminary Results

After five hours of observation, the earthworms introduced in the chicken and guinea fowl manures were trying to crawl out of both types of manures except for the cattle and sheep goat manure mixture. The guinea fowl manure and chicken manure had very strong odors and produced more heat for several days compared to CD and SGM manures. After 8 hours all earthworms in chicken, and guinea fowl manure showed signs of dehydration as they could not enter into the manures and eventually died. Earthworms in the CD and SGM manures did not show any signs of stress. As a result, the chicken manure (CM) was discarded after a two-week attempt to cure and stabilize it failed. However, it was possible to stabilize the guinea fowl manure within the same period of two weeks through a process of soaking in water, removing the water after every two days. New juveniles of earthworm species *E. fetida* were re-introduced after two weeks of curing and stabilization for the second time. After few hours of observation, it was possible for the earthworms to adapt to the conditions of the guinea fowl manures and did not show any signs of attempting to leave the manures. Continuous observation indicated that earthworms tolerated the guinea fowl manure in a similar manner compared to CD or SGM manures after stabilization. All containers were watered by hand twice a week for months. However, after 45–60 days the guinea fowl manure feedstock produced numerous cocoons and a few weeks later numerous hatchlings much faster compared to earthworms in CD and SGM manures, although the size of worms did not differ in all feedstock types. After five months, the vermicomposts of GFM, CD, and SGM were sieved and screened to get the final product. The vermicompost (VC) produced from guinea fowl appeared smooth and very fine with no odor or foul smell with a brownish color while CD and SGM vermicompost appeared black and dark black. The next stage was to test the VCs as growth media to produce vegetable seedlings. The GFM and other vermicompost were sent to the lab for chemical analysis and the results are yet to be obtained. The indications of this test trial are that guinea fowl manures (GFM) can only be vermicomposted if stabilized for certain period of time. In addition, it is possible to use containers as vermi-reactors for small-scale operations. The next stage was to test the GFM vermicompost on vegetable crops.

9.4.3 Vermicomposting of Animal Manures Mixed with Other Organic materials

Different manures such CD, GFM, and SGM were mixed with dried "maize chopped stalks" (MCS) at 1:1 each with six plastic containers. Guinea Fowl Manure, Chicken Pellet Starter Feed (CPSF), and Dove manure (DM) were added at 10% of the total feedstock to the mix of either CD or SGM manures each with a control where no additions were made. After several days it was observed that the additions of 10% of GFM, DM, and CPSF to cattle manure and sheep goat manure do not seem to affect the reproductive capacity and the decomposition process of the earthworm species *E. fetida*. It was observed that where CPSF was added, the earthworm size increased significantly two to three times than in control containers compared to those where GFM or Dove Manure was applied. The addition of MCS helped to reduce the tendency of the manure to pack and condense, especially the goat sheep manure mixture. Preliminary results of this trail are that *E. fetida* can successfully adapt to different feedstocks for vermicomposting provided favorable conditions are created especially through the addition of bulking materials like maize stalks.

9.5 Windrow Vermicomposting Experience

Surface vermibeds (VB) were used as a method vermicomposting due to its simplicity. In this test trial, CD and SGM manures were selected as main feed stocks and then amended using guinea fowl manure. Vermibeds for each feedstock type were prepared in a shaded tunnel in dimensions of 5 m long × 1.2 m wide × 0.30-meter high. The vermibeds were based on the following treatments, i.e., SGM alone (Control), SGM plus 5% GFM, CD alone (Control) and CD plus 5% GFM added. After two months of observations, vermibeds that received 5% addition of GFM progressed faster than those that did not receive additions of GFM (control VBs) regardless of the feedstock type. This indicated that GFM contains unknown elements worth to consider for accelerating the vermicomposting process either by itself or an additional ingredient to other manures. The question is what is it that GFM manures and chicken pellet starter (CPSF) feed contain to influence the vermicomposting process of different manures? Is it the same element(s) found in GFM and CPSF that influences the vermicomposting process or are they different unknown elements?

9.6 Crop Growth Experiments with Wild Animal Manure Vermicomposts

9.6.1 Vegetable Seedlings Germination and Growth

Vermicomposts from GFM, SGM, CD and commercial vermicompost as control (C. VC) were tested on the germination, growth and vigor of tomato vegetable seedlings using commercial raising trays for 35 days. Treatments were: **T1** = Control (commercial VC), **T2** = Cattle Dung vermicompost, **T3** = Sheep Goat Manure mixture vermicompost and **T4** = Guinea Fowl Manure vermicompost. During observation GFM (VC.GFM) and cattle dung vermicompost (VC.CD) had the first seed germination, reaching 50% germination after four days compared to the control and SGM vermicomposts. Most of the parameters measured from 15 DAS (Days After Sowing in trays) such as plant height, plant canopy, and stem girth were reasonably higher in GFM vermicompost compared to the commercial vermicompost (control) and that of CD or SGM. Seedlings growing in GFM vermicompost had bigger leaves, thick stems, were taller than other vermicomposts and had strong vigor after 35 days as shown in Fig. 9.5.

However, it was noted that after 20–25 DAS symptoms of nitrogen deficiency was observed in all treatments. At this stage seedlings are actively growing and the demand for nutrition is higher, with limited space in cell cavities, deficiencies are likely to happen. The options can be either to transplant as soon as possible or additional foliar spray to keep seedlings in good condition for longer. This test trial showed that vermicomposted guinea fowl manure has a potential for use as a

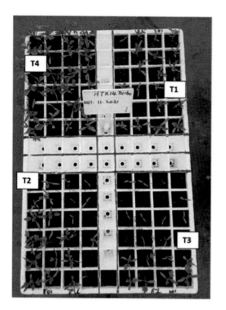 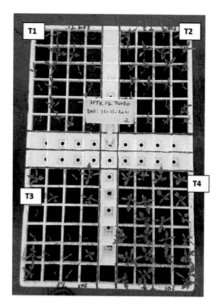

Fig. 9.5 Seedling trays showing germination and growth of tomato seedlings under different treatments, i.e., T1 (commercial vermicompost), T2 (Vermicomposted cow manure), T3 (Vermicomposted sheep and goat manure mixture) and T4 (Vermicomposted guinea fowl manure)

substrate and source mineral nutrients for raising quality seedlings better than commercial vermicompost and comparable to vermicomposts obtained through CD and SGM manures.

9.7 Conclusions

The use of wild animal manures is a new concept in pursuit of sustainable alternative sources of organic fertilizers in light of climate change and the reduction in the use of chemical fertilizers in crops. The test trials showed that wild animal manures from guinea fowls as with poultry can be a good source of plant available nutrients, a soil conditioner and carbon source for microorganisms. More research is required into wild animal manures especially guinea fowl and dove manures as potential sources of soil amendments to improve crop growth and yields. It is important to note that the test trials conducted so far are meant to pave a way forward for future research into the potential use of wild animal manures as a source of fertilizer for crops through vermicomposting methods. The second aspect is to conduct field and greenhouse experiments on different important crops to improve research in the subject.

References

De La Peña-Lastra S (2021) Seabird droppings: effects on a global and local level. Sci Total Environ 754:142148

Roisin Aoife Stanbrook (2018) Assessing the nutrient status of elephant dung in the Aberdare National Park, Kenya. Pachyderm No. 59

Mupambwa HA, Mnkeni PNS (2018) Optimizing the vermicomposting of organic wastes amended with inorganic materials for production of nutrient-rich organic fertilizers: a review. Environ Sci Pollut Res 25:10577. https://doi.org/10.1007/s11356-018-1328-4

Chapter 10
Rock Phosphate Vermicompost with Microbial Inoculation Potential in Organic Soil Fertility

Sinazo Ajibade, Hupenyu Allan Mupambwa, Barbara Simon, and Miklos Gulyas

Abstract Phosphorus (P), an essential macronutrient and the 11th most abundant element in the earth's crust, represents a small fraction of the soil. To enhance P content in the soil, the commercial P fertilizer is produced from high-grade rock phosphate (RP), which will be exhausted in the future, explaining why low-grade RP is being promoted as P fertilizer to be used during RP vermicomposting, with or without microbial inoculation. As the most effective P solubilizers, phosphorus solubilizing microorganisms have increased P from RP during vermicomposting. The cost-effectiveness of rock phosphate, phosphorus solubilizing microorganisms, and its P solubilizing capability make it an interesting and valuable organic fertilizer that farmers can use and exploit to improve soil fertility and plant growth. This chapter highlights that RP vermicompost inoculated with phosphorus solubilizing microorganisms can be used as a cost-effective and environmentally sustainable alternative for sustainable agricultural development. With few studies on the use of this effective fertilizer on plant growth and soil fertility, more research in the field will help save the production cost and the soil ecosystem function.

Keywords Rock phosphate · Vermicompost · Phosphorus · Microbial inoculation · Soil fertility

---

S. Ajibade (✉) · B. Simon · M. Gulyas
Department of Soil Science, Institute of Environmental Sciences, Hungarian University of Agriculture and Life Sciences, Gödöllő, Hungary

H. A. Mupambwa
Sam Nujoma Marine and Coastal Resources Research Center, University of Namibia, Henties Bay, Namibia

© The Author(s), under exclusive license to Springer Nature Singapore Pte Ltd. 2023
H. A. Mupambwa et al. (eds.), *Vermicomposting for Sustainable Food Systems in Africa*, Sustainable Agriculture and Food Security,
https://doi.org/10.1007/978-981-19-8080-0_10

10.1 Introduction

In the soil, P is absorbed by plants in the form of phosphate, which forms low solubility complexes with soil materials (Abd-Alla 1994), precipitated as poorly soluble salts, and bound within organic matter. Additionally, the most crucial potential loss of P from the ecosystem is the soil loss due to erosion of water (Carpenter and Bennett 2011), which might leach as dissolved soluble phosphate leading to very small P fractions available for plant uptake in bedrocks and soils (Riskin et al. 2013; Helfenstein et al. 2018). Hence, P has low mobility in the soil and low availability to plants even at high concentrations. Moreover, when P in the soil is available to plants in adequate quantities, it improves plant growth and increases stalk strength, root growth, and earlier crop maturity (Meyer et al. 2011). Compared to other nutrients, the concentration of P in soil solution ranges from 0.001 to 1 mg L (Brady and Weil 2002). To increase the soil P content, vermicompost has been adopted since the earthworms; during the vermicomposting process can transform insoluble P into soluble P. According to Sharma and Garg (2018) and Yuvaraj et al. (2020), the vermicompost has a considerable amount of P compared to the initial feedstock. Even with the increased P content compared to the initial feedstock, the adoption of vermicompost in conservation agriculture is limited by the low concentration of macronutrients (Mupambwa and Mnkeni 2018; Baversiha et al. 2018).

Most P fertilizers are currently derived from high-grade rock phosphate (RP) with more than 13.1% P (Liu et al. 2020). The global reserve of this high-grade RP is approximately 60,000 million tonnes (Kauwenbergh 2010), which according to Roy et al. (2018), may be exhausted within 75 to 175 years. Hence, it is crucial to efficiently utilize the low-grade P, which contains less than 8.73% P, to reduce the dependency of the agricultural sector on high-grade RP. For instance, the low bioavailability of low-grade RP type found in South Africa, due to low reactivity and solubility (Mineral Information Institute 2003), could be improved through the co-application with vermicompost and P solubilizing microorganisms (PSMs). Based on Dinesh et al. (2018), low solubility P compounds can be converted by various species of fungi and bacteria into plant-available forms. For example, in the 1990s, Karmakar and Barthakur (1995) investigated the use of an indigenous Mussoorie RP applied at different levels (40, 80, 120, and 160 kg P_2O_5 ha^{-1}) in acid soil belonging to Aeric Haplaquept, with or without inoculation of phosphate solubilizing fungus (*Aspergillus niger Van Tiegh*). The authors reported higher Bray 2 extractable P of post-harvest soil, with the contribution of about 97% from three P fractions, i.e., Fe-P, Al-P, and organic P. In the same study, the three P fractions contributed to 83 and 94% of grain yield and P uptake of Kharif rice (*Monohar sali*) grown in acid soil. It is worth noting that the suitability of RP in acid soils was also reported by Khasawneh and Doll (1978), who observed the availability of P from RP applied to acid soils.

In a recent study by Busato et al. (2021), two Trichoderma fungi strains (i.e., *T. virens* (F1d5c1) and *T. asperellum* (Tr266B)) were inoculated alone or combined with RP vermicompost, 15 and 30 days after the beginning of the vermicomposting process. After 120 days of vermicomposting, the authors observed increased citric acid-soluble P content, which was 62.2% higher than the control treatment without *T. virens* (F1d5c1) and *T. asperellum* (Tr266B), including increased humic acids content of 38.20%, and a stability index based on the results of humic/fulvic acids (HA/FA) ratio. However, compared to fungi, various species of bacteria solubilize P more effectively and are abundant in most soils (Alam et al. 2002), and these include *Bacillus, Pseudomonas, Rhizobium, Achromobacter, Flavobacterium, Agrobacterium, Burkholderia,* and *Burkholderia* (Alireza Fallah 2006). Regarding the solubility of insoluble mineral phosphate, the species of *Pseudomonas,* such as *P. aeruginosa, P. fluorescens,* and *P. putida,* have the high solubilizing capacity (Sharma et al. 2013), and these have been used as biofertilizers since the 1990s (Kudashev 1956; Krasilinikov 1957). In a recent study, Ajibade et al. (2020) investigated the effect of inoculating animal manure RP vermicompost with *P. fluorescens* on nutrient release and vermidegradation. The authors reported the capacity of *P. fluorescence* to solubilize the low-grade RP during vermicomposting, as the results showed high Olsen extractable P content of 3342.24 and 3052. 20 mg kg^{-1} in the cow and pig manure RP vermicompost treatment inoculated with *P. fluorescens* after week 6 of vermicomposting. In a more recent study, Ajibade et al. (2022) reported the effect of supplementing an Oxisol soil with animal manure RP vermicompost that has been inoculated with *P. fluorescence* on the growth of Swiss chard (*Beta vulgaris subsp. Vulgaris*) and its heavy metals uptake. The authors recorded the highest total P tissue content, leaf area, and fresh weight of 326.91 mg kg^{-1}, 240.41 cm^2, and 39.78 g, and the tissue Zn, Cr, Cu, and Pb content was below the permissible levels in the treatment with the highest application rate of 50 mg kg^{-1} P as pig manure RP vermicompost. In this chapter, the research on the use of RP vermicompost that has been inoculated with biofertilizers as a sustainable waste management strategy will be highlighted.

10.2 The Need for a Sustainable Waste Management Strategy

The global nutrient budget of the entire agricultural sector is dominated by the fast-growing livestock farming sector, with, however, its low recovery of nutrients. In 2007, WHO (2012) reported that the worldwide production of livestock manure is estimated at 22.5 billion tons. As the world's fastest-growing population, Sub-Saharan Africa accounts for more than 950 million people, which is approximately 13% of the global population, with about 250 to 300 million people in Africa depending on livestock for their livelihood and income, with livestock representing 10% of the total GDP, and an average of 30% of the agricultural gross domestic product (GDP) (Enahoro et al. 2019). Moreover, as early as 2007, Petersen et al. reported that the environmental impact of livestock raised great concern due to the substantial amount of manure produced that is characterized by a high mineral and organic load of phosphorus (P), nitrogen (N), and potassium (K), and the environmental impact of inadequate manure disposal on soil acidification (Giola et al. 2012), reduced soil fertility, biodiversity loss, water eutrophication, and groundwater contamination. Due to increased manure production, the annual emission of greenhouse gases (GHG) from livestock farming is estimated at 7100 million tons of CO_2e (Kumari et al. 2019). Also, apart from GHG emissions, soil water contamination with heavy metals has been reported from the improper disposal of cattle solid waste (Ba et al. 2020; Qi et al. 2020). This is because manure has a significant amount of heavy metals (e.g., zinc, cadmium, arsenic, and copper) and antibiotics (e.g., fluoroquinolones, sulfonamides, ionophores, pleuromutilins, and tetracyclines) (Larson 2015).

Based on Bhat et al. (2018) and UNEP (2021), the global production of solid waste is estimated at 11.2 billion tonnes per year, with 1.2 billion tonnes as non-hazardous industrial waste. In developing countries, the recycling and treatment of solid waste are more efficient due to the implementation of advanced technologies, waste segregation, and better collection system (Lim et al. 2015). For effective management and treatment of solid and livestock waste vermicompost technology (Ganguly and Chakraborty 2020; Falco et al. 2021) has been widely adopted as a cost-effective (Joshi et al. 2015) and environmentally friendly approach to promote sustainable agriculture. For instance, in India, due to the increase in per capita solid waste generated (CPCB 2018), and the challenges of waste management in rural and urban areas, the use of vermicompost in nutrient and waste management is embraced across India (Gupta et al. 2019; Balachandar et al. 2021). Compared to traditional compost, vermicompost has a high percentage of N, P, and Na, the lowest C: N ratio, and electrical conductivity (Bhat et al. 2013). In addition, during organic waste composting, nitrogen loss which occurs through nitrogen oxides and ammonia promotes greenhouse gas emissions (GHG), while the controlled vermicompost process reduces GHG emissions (Lv et al. 2018; Rini et al. 2020). Vermicompost involves the addition of earthworms, and several groups of microorganisms (Domínguez et al. 2019) for increased degradation and enrichment of the final vermicompost product (Busato et al. 2017), as the final vermicompost is rich in nutrients, valuable soil microorganisms, and stabilized organic materials (de Souzaa et al. 2013). The earthworm gut and associated microorganisms contain beneficial degradative enzymes, cellulose, lipase, amylase, phosphatase, urease, protease, etc. that stabilize various components of organic waste material (Ravindran et al. 2015; Gusain and Suthar 2020), with most of these enzymes, including xylanase, and invertase being involved in N-mineralization (Fu et al. 2015). After N, P is an essential macronutrient that plays a major role in many plants' physiological processes, which include photosynthesis, energy transfer, cell division, respiration, and the synthesis of carbohydrates, proteins, and nucleic acids (Elser 2012).

10.3 Rock Phosphate: An Acceptable Phosphorus Source in Organic Farming

Even with no current imminent global shortages of rock phosphate (RP), with the global RP resources of more than 300 billion tons, principally RP resources occur as sedimentary marine phosphorite with the largest deposits in Northern Africa, China, the USA, and the Middle East (US geological survey, mineral commodity summaries 2022), and the current reserves are presented in Fig. 10.1. The future supply security of both N and K is practically a given, and both macronutrients are available unlimitedly in either air (N) or seawater (K), while the supply of primary P is limited to mining of infinite deposits of RP (Wellmer and Scholz 2015). About 87% of phosphate resources in the world are produced from sedimentary phosphate rock deposits of marine origin, and only 13% are from igneous and weathered deposits (Geissler et al. 2018). The guano rock type which was mainly found on the Islands is of negligible significance in modern RP production, as noted by Udert (2018). The sedimentary deposits are, however, negatively affected by chemical impurities. For instance, the igneous deposits found in South Africa, Russia, and Brazil contain less than 2 mg Cd/kg P_2O_5, while the sedimentary deposit contains more than 2 mg Cd/kg P_2O_5 (Geissler et al. 2018). Based on Sattari et al. (2012), RP reserves will be exhausted in 100 to 200 years. Apart from the anticipated depletion of RP, Niu et al. (2012) reported that 5.7 billion hectares of land worldwide are P deficient, which explains the relationship between agricultural production and the 85% global demand for RP (IFA 2011). It is worth mentioning that the growing demand for fertilizer has resulted in an increase in the cost of RP from 80 dollars per US ton in 1961 to 700 dollars per ton in 2015, with relatively large year-to-year fluctuations (Amundson et al. 2015; USDA 2019).

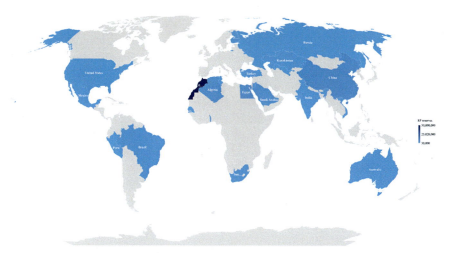

Fig. 10.1 Global rock phosphate reserves. **Source:** Data extracted from Mineral Commodity Summaries (*US geological survey, mineral commodity summaries* 2022)

Therefore, as stated by Ditta et al. (2015), what is currently needed in agricultural production is to discover different approaches to exploit the resources of indigenous RP in bioavailable forms without compromising crop yield. For example, the low-grade igneous phosphate ore with P_2O_5 of less than 5% "can be upgraded to high-grade products (from 35% to over 40% P_2O_5)" (Van Kauwenbergh 2010). As an effective approach, RP has been applied during the vermicomposting process, using the igneous deposit carbonatite-peralkaline complexes, like the one found in the Phalaborwa complex, South Africa, with 40.3% P_2O_5 (Van der linde 2004). Also, the valuable minerals found in the igneous RP in South Africa such as magnesium oxide—0.26%, calcium oxide—54.6%, zinc—13.22 mg/kg, chromium-18.05 mg/kg, cadmium—2.2 mg/kg, Pb-6.05 mg/kg, copper—5.85 mg/kg, and the mineralization, and vermiremediation potential of earthworms and microorganisms found in vermicompost makes its application during the vermicomposting process an effective approach. During vermicomposting, the rate of nutrient mineralization is increased by earthworms, which increases P availability and uptake of nutrients by plants (Alikhani et al. 2017). Mainly, the earthworm gut contains P solubilizing microorganisms involved in P's mineralization process (Sharma and Garg 2019).

Also, adding RP powders during vermicompost can increase the quality of vermicompost, as the solubility of minerals is accelerated and nutrients are released from the rock powder (de Souza et al. 2013). The vermicomposting of RP-enriched organic materials has also been found to enhance P release from RP (Unuofin and Mnkeni 2014; Mupondi et al. 2018). In the intestinal organs of the earthworms, the suitability of RP can be increased during vermicomposting through the effects of digestive enzymes (Carpenter et al. 2007). Mupondi (2010) reported the successful conversion of cow manure enriched RP vermicompost into organic-rich fertilizer. The author reported that including South African igneous RP to supply between 2% and 8% elemental P during cow dung wastepaper vermicomposting increased plant-available P from 434 mg kg^{-1} to 1312 mg kg^{-1} and also accelerated the biodegradation of the cow dung wastepaper mixture. In another study, Unuofin and Mnkeni (2014), using igneous RP-enriched vermicomposting, concluded that the treatment with the highest stocking density of 22.5 g-worm/kg of E. fetida resulted in the highest P release within 42 days. However, these two studies on the igneous RP were done over a short period between 42 and 56 days, beyond which there was still the potential for more P release from the RP. Busato et al. (2012), in another study, also used igneous RP incorporated into cow dung vermicompost with or without microbial inoculation at an earthworm stocking density of 50 worms/kg of the substrate and vermicomposted for 120 days. Their study reported positive results on water-soluble P being 106% more for the vermicompost compared to the vermicompost with no inoculation (Jacobs 2020).

10.4 Factors Affecting the Efficiency of RP Vermicomposting

The successful conversion of RP vermicomposting into a nutrient-rich fertilizer can be archived when all the parameters involved during vermicompost are at their optimum. During RP vermicomposting, the C: N is used as one of the maturity parameters, and when it declines during vermicomposting, this reflects the mineralization and decomposition of waste. It also shows the active role played by earthworms during vermicomposting. The RP content is also important in determining the best treatment combinations for effective nutrient mineralization and P release. For example, researchers observed that the incorporation of 2% P as RP resulted in improved biodegradation of compost in the treatment with decreased C: N (Mupondi 2010; Unuofin and Mnkeni 2014; Mupondi et al. 2018). The diagram below shows the factors affecting RP vermicompost efficiency (Fig. 10.2). The earthworms, during vermicompost, have the potential to transform insoluble P into soluble P due to the P-mineralization process by P solubilizing microorganisms found in the earthworm gut (Sharma and Garg 2019). The E. fetida is the most effective earthworm to use during RP vermicomposting as they can break down wastes and release phosphatase enzymes from their gut that facilitate the release of P, making it available for plant uptake (Mupondi 2010; Busato et al. 2012; Unuofin and Mnkeni 2014). However, vermicomposting efficiency, like composting, is influenced by several factors such as initial C: N ratio, pH, feeding stock, earthworm stocking density, moisture, and temperature. Singh et al. (2017) noted that nutrient or physiochemical characteristics of the waste might be linked to the earthworm growth with moisture content and temperature. The performance and growth of E. fetida

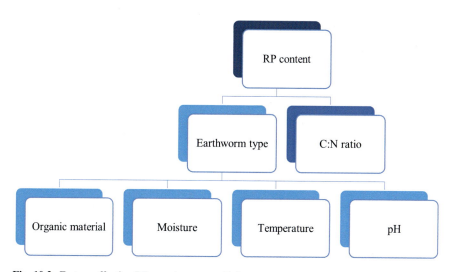

Fig. 10.2 Factors affecting RP vermicompost efficiency

during RP vermicomposting are, in turn, affected by the organic matter, moisture content, temperature, and the pH of the mixture, as listed below (Jacobs 2020).

Organic material: The nature and suitability of the feedstock affect the final vermicompost in that their quality determines the speed of the breakdown by earthworms and microorganisms as well as the nutrient content of the resultant vermicomposts (Lim et al. 2015, 2016; Alikhani et al. 2017).

Moisture: The moisture content for most of the earthworm species ranges from 60% to 85%, which ensures that, during vermicomposting, the earthworms absorb a lot of moisture since it may be lost through evaporation (Amaravathi and Reddy 2014). Earthworms' moisture requirement differs depending on their growth stages. The moisture required by clitellate cocoon producing or the adult *E. fetida* in separate cow dung ranges between 50% and 80% for adults, but the immature or juvenile needs moisture between 65% and 70%.

Temperature: One of the determining factors of earthworm growth, such as cacoon production, is the temperature. E. fetida thrives very well at temperatures ranging from 25 °C–30 °C, and according to Edwards (1988); the E. fetida can tolerate a temperature range of 0 °C to 35 °C. This might be why Munroe (2007) suggested that the temperature is maintained within the range of 15 °C–30 °C as temperatures above 35 °C kill the earthworms. However, under laboratory conditions, the maturation and, therefore, the growth of earthworms is best at 20 °C (Edwards et al. 2011).

pH: The process of vermicomposting can be affected by the pH of the feedstock. Earthworms in the vermicompost mixture can only survive at pH values of between 5 and 9, and the most desirable pH for its effective performance is between 7.5 and 8 (Garg and Gupta 2011).

10.5 Acid and Alkaline Soil Reactivity with RP for Improved P Release

As reported in the study of Jacobs (2020), soil properties such as pH, organic matter content, Ca concentration, P concentration, and P sorption influence the dissolution of apatite materials in rock phosphate (RP). Low soil pH levels (i.e., <5.5) enhance the dissolution of RP Eq. (10.1) (Rafael et al. 2018) since the dissolution of RP requires an H^+ proton that can be gained through chemical and microbial intervention (Roy et al. 2018). The dissolution of apatite (Ca_{10} (PO_4)6F_2) is driven by a neutralization reaction between the proton (H^+) ion concentration and apatite in the RP.

$$Ca_{10} (PO_4)_6 F_2 \rightarrow 10\, Ca_2 + 6H_2PO_4{}^- + 2F^- \qquad (10.1)$$

The chemical reaction is driven from left to right mainly by increasing H^+ ion concentration or soil acidification. Hence, the application of RP to acid soils will be more beneficial compared to its application to alkaline soils. In acid soils, RP quickly reacts with Fe and Al oxides, making P available for plant uptake, as shown in Eq. (10.2) below.

$$Al^{3+} \qquad Fe^{3+} \qquad \Longrightarrow \qquad \begin{cases} AlPO_4.2H_2O \\ \\ FePO_4.2H_2O \end{cases}$$

$$(10.2)$$

In alkaline soils, Ca and Mg are the most abundant cations, and these react with phosphate compounds, making P available for plant uptake Eq. (10.3). It is worth noting that the dissolution of RP by soil organic matter is related to organic matter and Ca complexes. When the formation of Ca in the soil is reduced, the dissolution of RP increases. Therefore, the more the organic matter in the soil, the better the dissolution of RP with 50% or more of the organic total P.

$$Ca^{2+} + HPO_4^{2-} \rightarrow Ca_3 (PO_4)_2 \qquad (10.3)$$

10.6 Impact of Microbial Inoculated RP Vermicompost on Soil Fertility, Health, and Crop Productivity

10.6.1 Role of Phosphorus Solubilizing Microorganisms on P Release

Phosphorus solubilizing microorganisms (PSMs) play an essential role in maintaining the availability of P in soil, more especially for P deficient and weathered soils (Meena and Biswas. 2015), and maintain soil nutrient status of agricultural soils, soil structure, as well as sustaining the production base (Rathi and Gaur 2016; Satya prakash et al. 2017). PSM includes microorganisms like bacteria, fungi, actinomycetes, and algae (Antoun 2012). Khan et al. (2009a) found out that the PSB is more effective in solubilizing P than fungi, as it constitutes 1–50%, while fungi are only 0.1–0.5% in phosphate solubilizing potential (Chen et al. 2006). In many studies, the *PSMs* have been used as inoculants, and the potential mechanism underlying the role of PSMs in making P available includes the microbial synthesis of a range of exudate-like metabolites and enzymes, including phosphatase, low molecular weight organic acids like gluconic acid, releasing H^+ and chelating compounds (Behera et al. 2014; Khan et al. 2014), including production of carbonic acid from the reaction between H_2O and the carbon dioxide (CO_2) produced by microbial respiration (Khan et al. 2009b). Jilani et al. (2007) also reported that the PSB and plant growth-promoting rhizobacteria (PGPR) can reduce the application of P fertilizer by 50% without reduction in crop yield. Pseudomonas sp. improves plant growth in salt stress conditions and shows tolerance toward drought stress (Ali et al. 2014). For example, in a recent study by Biswas et al. 2022, a low-grade RP was mixed with Pseudomonas striata and sub-tropical inceptisol for wheat production. The authors reported the production of organic acids, phosphatase enzymes, and sidospores by the PSB prevented the fixation of P in soil, solubilized soil P, and increased inherent microbial activities by 10 to 24%, and concluded that the soil application of PSB inoculated low-grade RP could be recommended for crop production to save approximately 50% of the chemical P fertilizer.

10.6.2   Role of Phosphorus Solubilizing Microorganisms During RP Vermicomposting Process

Kumar et al. (2017) note that earthworms and microorganisms are essential organisms that assist nature in maintaining the flow of nutrients from one system to another and reducing the degradation of the environment. Microorganisms both in the intestinal organ of the worms and organic waste, in combination with earthworms, produce a significant amount of enzyme and organic acids, which are mainly responsible for P-mineralization in the vermicomposting mixtures" (Fu et al. 2015; Das et al. 2016). Through hydrolysis, alkaline and acid phosphatase enzymes mineralize organic P and split inorganic P from organic residues (Tarafdar and Claassen 1988; Hilda and Fraga 1999). During RP vermicomposting, these enzymes and organic acids are mainly responsible for nutrient release (Sharma et al. 2013). The commonly secreted enzymes are phosphatase, described as a broad group of enzymes responsible for catalyzing the hydrolysis of phosphoric acid (H_3PO_4) monoesters and anhydride into P ion and a molecule with a free hydroxyl group (Othman and Panhwar 2014). The greater capacity of organic acids for releasing P from RP minerals, and the release of P from the pore water of the soils, is a result of their chelating abilities with Fe, Al, and Ca compared to mineral acids with comparable strength and different soil pH. Another significant factor in P release from RP is the participation of OH groups in organic acids. For example, citric acid with three carboxyl groups (COOH) and OH group has been shown to release P from RP compared to cis-aconitic acid, which is of the same carboxyl group but with no OH group (Kpomblekou and Tabatabai 1994). It is also noted that organic acids like oxalic, malic, acetic, and citric acids could reduce Al toxicity and increase nutrient availability of P (Van Hees and Lundstrom 2000). Bhattacharyya (2012) states that Pseudomonas, Serratia, Rhizobium, and Mycobacterium are the most significant PSB. The inoculation of RP-enriched vermicompost with PSB increases P (Ajibade et al. 2020), while the application of RP in the soil does not directly increase P availability to plants (Kaushik and Garg 2004; Jacobs 2020).

10.7 Microbe Inoculated RP Vermicompost Applied Under Different Soil Conditions on Plant Growth

Recent investigations have established the effectiveness of microbial inoculated RP vermicompost on soil fertility and plant growth (Jahil and Kamal 2021; Ajibade et al. 2022). In one study, a silt clay loamy soil with an initial pH of 7.6 was applied with RP vermicompost that had been inoculated with *Bacillus megaterium* to investigate its effect on the growth of sunflower (*Helianthus annuus L*) (Jahil and Kamal 2021). Using different rates of RP (0, 100, and 200 kg ha^{-1}), the authors reported that the inoculation of RP vermicompost with *Bacillus* at an RP application rate of 100 kg ha^{-1} resulted in a significant increase in leaf area, chlorophyll content, grain yield, and plant height with the highest average values of 7250 cm^2, 48. 70 mg kg^{-1}, 2808 kg ha^{-1}, and 134.3 cm plant^{-1}. In a more recent study, Ajibade et al. (2022) investigated the potential impact of cow and pig manure RP-enriched vermicompost inoculated with *Pseudomonas fluorescence* applied to an Oxisol on the growth of Swiss chard (*Beta vulgaris subsp. Vulgaris*). From the three RP vermicompost applications rates (i.e., 0, 25, and 50 mg P kg^{-1} soil), the results revealed that the use of 50 mg P kg^{-1} soil as VP gave the highest fresh weight of 39.78 g, leaf area of 240.41 cm^2, and a total P in tissues of 326.91 mg kg^{-1} at 8 weeks after transplanting with a slight increase in soil pH, with a significant increase in Bray 1 extractable P, respectively.

10.8 Conclusion

The mechanisms of P solubilization by phosphorus solubilizing microorganisms and earthworms in the literature are well documented with however, few studies on the effect of microbial inoculated RP vermicompost on soil fertility and soil remediation. Due to the properties of RP vermicompost that has been inoculated with phosphorus solubilizing, its performance under different soil conditions needs more attention, to reduce the dependency on commercial P fertilizers. This chapter presents information on improved organic phosphorus fertilizers with rock phosphate amendment, processed through vermicomposting. The fertilizer processing using vermicomposting presents an opportunity to improve the fertilizer value of low grade igneous rock phosphate, which has an opportunity to create sustainable P fertilizers.

References

Abd-Alla MH (1994) Phosphatases and the utilization of organic phosphorus by rhizobium leguminosarum biovar viceae. Lett Appl Microbiol 18:294–296

Ajibade S, Mupambwa HA, Manyevere A, Mnkeni PNS (2022) Vermicompost amended with rock phosphate as a climate smart Technology for Production of organic Swiss chard (Beta vulgaris subsp. vulgaris). Frontiers in Sustainable Food Systems 6

Ajibade S, Mupambwa HA, Manyevere A, Mnkeni PNS (2020) Influence of microbial inoculation of igneous rock phosphate-amended cow and pig manures on vermidegradation and nutrient release. Agronomy 10(10):1587

Alam S, Khalil S, Ayub N, Rashid M (2002) In vitro solubilization of inorganic phosphate by phosphate solubilizing microorganisms (PSM) from maize rhizosphere. Int J Agric Biol 4(4): 454–458

Ali S, Charles TC, Glick BR (2014) Amelioration of high salinity stress damage by plant growth-promoting bacterial endophytes that contain ACC deaminase. Plant Physiol Biochem 80:160–167

Alikhani HA, Hemati A, Rashtbari M, Tiegs SD, Etesami H (2017) Enriching vermicompost using P-solubilizing and N-fixing bacteria under different temperature conditions. Commun Soil Sci Plant Anal 48(2):139–147

Alireza Fallah SWRI (2006, July) Abundance and Distribution of Phosphate Solubilizing Bacteria and Fungi in Some Soil Samples from North of Iran. In: *The 18th World Congress of Soil Science*

Amaravathi G, Reddy RM (2014) Effect of substrate composition on the nutrients of vermicompost prepared by different types of earthworms. Am Int J Contemp Sci Res 1(3):85–95

Amundson R, Berhe AA, Hopmans JW, Olson C, Sztein AE, Sparks DL (2015) Soil and human security in the 21st century. Science 348(6235):1261071

Antoun H (2012) Beneficial microorganisms for the sustainable use of phosphates in agriculture. Procedia Eng 46:62–67

Ba S, Qu Q, Zhang K, Groot JCJ (2020) Meta-analysis of greenhouse gas and ammonia emissions from dairy manure composting. Biosyst Eng 193:126–137

Balachandar R, Biruntha M, Yuvaraj A, Thangaraj R, Subbaiya R, Govarthanan M, Kumar P, Karmegam N (2021) Earthworm intervened nutrient recovery and greener production of vermicompost from ipomoea staphylina–an invasive weed with emerging environmental challenges. Chemosphere 263:128080

Baversiha N, Parvanak K, Nasrabadi M (2018) Reduction of adverse environmental impacts caused by urban sewage: application of green soil fertilizers. Ukr J Ecol 8:437–440

Behera BC, Singdevsachan SK, Mishra RR, Dutta SK, Thatoi HN (2014) Diversity, mechanism and biotechnology of phosphate solubilising microorganism in mangrove—a review. Biocatal Agric Biotechnol 3(2):97–110

Bhat SA, Singh J, Vig AP (2013) Vermiremediation of dyeing sludge from textile mill with the help of exotic earthworm Eisenia fetida Savigny. Environ Sci Pollut Res 20(9):5975–5982

Bhat SA, Singh S, Singh J, Kumar S, Vig AP (2018) Bioremediation and detoxification of industrial wastes by earthworms: vermicompost as powerful crop nutrient in sustainable agriculture. Bioresour Technol 252:172–179

Bhattacharyya PN, Jha DK (2012) Plant growth-promoting rhizobacteria (PGPR): emergence in agriculture. World J Microbiol Biotechnol 28(4):1327–1350

Biswas SS, Biswas DR, Ghosh A, Sarkar A, Das A, Roy T (2022) Phosphate solubilizing bacteria inoculated low-grade rock phosphate can supplement P fertilizer to grow wheat in sub-tropical inceptisol. Rhizosphere 23:100556

Brady NC, Weil RR (2002) The nature and properties of soils, 13th edn. Prentice Hall of India, New Delhi, p 960

Busato JG, Lima LS, Aguiar NO, Canellas LP, Olivares FL (2012) Changes in labile phosphorus forms during maturation of vermicompost enriched with phosphorus-solubilizing and diazotrophic bacteria. Bioresour Technol 110:390–395

Busato JG, Ferrari LH, Chagas Junior AF, da Silva DB, dos Santos Pereira T, de Paula AM (2021) Trichoderma strains accelerate maturation and increase available phosphorus during vermicomposting enriched with rock phosphate. J Appl Microbiol 130(4):1208–1216

Busato JG, Zandonadi DB, Mól AR, Souza RS, Aguiar KP, Júnior FBR, Olivares FL (2017) Compost biofortification with diazotrophic and P-solubilizing bacteria improves maturation process and P availability. J Sci Food Agric 97(3):949–955

Carpenter D, Hodson ME, Eggleton P, Kirk C (2007) Earthworm induced mineral weathering: preliminary results. Eur J Soil Biol 43:S176–S183

Carpenter SR, Bennett EM (2011) Reconsideration of the planetary boundary for phosphorus. Environ Res Lett 6(1):014009

Central pollution control board (CPCB) India (2018) MSW-generation and composition. https://cpcb. nic. in/ uploa ds/ MSW/ Waste_ gener ation_Composition.pdf

Chen YP, Rekha PD, Arun AB, Shen FT, Lai WA, Young CC (2006) Phosphate solubilizing bacteria from subtropical soil and their tricalcium phosphate solubilizing abilities. Appl Soil Ecol 34(1):33–41

Das D, Bhattacharyya P, Ghosh BC, Banik P (2016) Bioconversion and biodynamics of Eisenia foetida in different organic wastes through microbially enriched vermiconversion technologies. Ecol Eng 86:154–161

de Souzaa MEP, de Carvalhoa AMX, Deliberalia DDC, Juckscha I, Brown GG, Mendonca ES, Cardoso IM (2013) Vermicomposting with rock powder increases plant growth. Appl Soil Ecol 69:56–60

Dinesh R, Srinivasan V, Hamza S, Sarathambal C, Gowda SA, Ganeshamurthy AN, Gupta SB, Nair VA, Subila KP, Lijina A, Divya VC (2018) Isolation and characterization of potential Zn solubilizing bacteria from soil and its effects on soil Zn release rates, soil available Zn and plant Zn content. Geoderma 321:173–186

Ditta A, Arshad M, Zahir ZA, Jamil A (2015) Comparative efficacy of rock phosphate enriched organic fertilizer vs. mineral phosphatic fertilizer for nodulation, growth and yield of lentil. Int J Agric Biol 17(3):589–595

Domínguez J, Aira M, Kolbe AR, Gómez-Brandón M, Pérez-Losada M (2019) Changes in the composition and function of bacterial communities during vermicomposting may explain beneficial properties of vermicompost. Sci Rep 9(1):1–11

Edwards CA (1988) Breakdown of animal, vegetable and industrial organic wastes by earthworms. In: Edwards CA, Neuhauser EF (eds) Earthworms in waste and environmental management

Edwards CA, Subler S, Arancon NQ (2011) Quality criteria for vermicomposts. Vermiculture technology. Earthworms, organic wastes and environmental management, pp 287–301

Elser JJ (2012) Phosphorus: a limiting nutrient for humanity. Curr Opin Biotechnol 23:833–838

Enahoro D, Mason-D'Croz D, Mul M, Rich KM, Robinson TP, Thornton P, Staal SS (2019) Supporting sustainable expansion of livestock production in South Asia and sub-Saharan Africa: scenario analysis of investment options. Glob Food Sec 20:114–121

Falco ED, Celano G, Vittal A (2021) Effect of growing media added with vermicompost on different horticultural species in a controlled environment. JSSPP 3(1):131

Fu X, Huang K, Cui G, Chen X, Li F, Zhang X, Li F (2015) Dynamics of bacterial and eukaryotic community associated with stability during vermicomposting of pelletized dewatered sludge. Int Biodeterior Biodegrad 104:452–459

Ganguly RK, Chakraborty SK (2020) Eco-management of industrial organic wastes through the modified innovative vermicomposting process: a sustainable approach in tropical countries. In: *Earthworm assisted remediation of effluents and wastes*. Springer, Singapore, pp 161–177

Garg VK, Gupta R (2011) Optimization of cow dung spiked pre-consumer processing vegetable waste for vermicomposting using Eisenia fetida. Ecotoxicol Environ Saf 74(1):19–24

Geissler B, Mew MC, Matschullat J, Steiner G (2018) The phosphate rock mining-innovation nexus. Gaia, forthcoming

Giola P, Basso B, Pruneddu G, Giunta F, Jones JW (2012) Impact of manure and slurry applications on soil nitrate in a maize–triticale rotation: field study and long term simulation analysis. Eur J Agron 38:43–53

Gupta C, Prakash D, Gupta S, Nazareno MA (2019) Role of vermicomposting in agricultural waste management. In: Sustainable green Technologies for Environmental Management. Springer, Singapore, pp 283–295

Gusain R, Suthar S (2020) Vermicomposting of duckweed (Spirodela polyrhiza) by employing Eisenia fetida: changes in nutrient contents, microbial enzyme activities and earthworm biodynamics. Bioresour Technol. 311:123585

Helfenstein J, Tamburini F, von Sperber C, Massey MS, Pistocchi C, Chadwick OA, Vitousek PM, Kretzschmar R, Frossard E (2018) Combining spectroscopic and isotopic techniques gives a dynamic view of phosphorus cycling in soil. Nat Commun 9(1):1–9

Hilda R, Fraga R (1999) Phosphate solubilizing bacteria and their role in plant growth promotion. Biotechnol Adv 17:319–339

IFA (2011) Production and trade statistic

Jacobs S (2020) Improving the vermicomposting of igneous rock phosphate enriched animal manures through inoculation with phosphate solubilizing bacteria for spinach (Spinacia oleracea L.) growth. MSc. Thesis, University of Fort Hare, Alice, South Africa

Jahil HM, Kamal JAK (2021, April) Effect of bacterial inoculation, bacillus megaterium, vermicompost, and phosphate pock on growth and yield of sunflower (Helianthus annuus L.). In: IOP conference series: earth and environmental science, vol 735, No. 1. IOP Publishing, p 012087

Jilani G, Akram A, Ali RM, Hafeez FY, Shamsi IH, Chaudhry AN, Chaudhry AG (2007) Enhancing crop growth, nutrients availability, economics and beneficial rhizosphere microflora through organic and biofertilizers. Ann Microbiol 57(2):177–184

Joshi R, Singh J, Vig AP (2015) Vermicompost as an effective organic fertilizer and biocontrol agent: effect on growth, yield and quality of plants. Rev Environ Sci Bio 14:137–159

Karmakar RM, Barthakur HP (1995) Phosphorus availability from Mussoorie rock phosphate in an Aeric Haplaquept in a Rice-Rice sequence. Agropedology 5:37–42

Kauwenbergh, S., 2010. International fertilizer development centre. Washington, DC, USA

Kaushik P, Garg VK (2004) Dynamics of biological and chemical parameters during vermicomposting of solid textile mill sludge mixed with cow dung and agricultural residues. Bioresour Technol 94(2):203–209

Khan MS, Zaidi A, Wani PA (2009a) Role of phosphate solubilizing microorganisms in sustainable agriculture-a review. Sustainable agriculture, pp 551–570

Khan AA, Jilani G, Akhtar MS, Naqvi SMS, Rasheed M (2009b) Phosphorus solubilizing bacteria: occurrence, mechanisms and their role in crop production. J Agric Biol sci 1(1):48–58

Khan MD, Zaidi A, Ahmad E (2014) Mechanism of phosphate solubilization and physiological functions of phosphate-solubilizing microorganisms. In: Phosphate solubilizing microorganisms. Springer, Cham, pp 31–62

Khasawneh FE, Doll EC (1978) The use of phosphate rock for direct application to soils. Adv Agron 30:159–206

Kpomblekou-a K, Tabatabai MA (1994) Effect of organic acids on release of phosphorus from phosphate rocks1. Soil Sci 158(6):442–453

Krasilinikov NA (1957) On the role of soil microorganism in plant nutrition. Microbiologiya 26: 659–672

Kudashev IS (1956) The effect of phosphobacterin on the yield and protein content in grains of Autumm wheat, maize and soybean. Doki Akad Skh Nauk 8:20–23

Kumar A, Gupta RK, Kumar S, Kumar S (2017) Nutrient variations in vermicompost prepared from different types of straw wastes. Forage Res 42:267–270

Kumari S, Hiloidhari M, Narayan Naik S, Pal Dahiya R (2019) Methane emission assessment from Indian livestock and its role in climate change using climate metrics. In: Saddam Hussain (ed) Climate change and agriculture. IntechOpen, pp 1–16

Larson C (2015) China's lakes of pig manure spawn antibiotic resistance. Science 80:347–704

Lim PN, Wu TY, Clarke C, Nik Daud NN (2015) A potential bioconversion of empty fruit bunches into organic fertilizer using Eudrilus eugeniae. Int J Environ Sci Technol 12(8):2533–2544

Lim SL, Lee LH, Wu TY (2016) Sustainability of using composting and vermicomposting technologies for organic solid waste biotransformation: recent overview, greenhouse gases emissions and economic analysis. J Clean Prod 111:262–278

Liu J, Qi W, Li Q, Wang SG, Song C, Yuan XZ (2020) Exogenous phosphorus-solubilizing bacteria changed the rhizosphere microbial community indirectly. 3 Biotech 10(4):1–11

Lv B, Zhang D, Cui Y, Yin F (2018) Effects of C/N ratio and earthworms on greenhouse gas emissions during vermicomposting of sewage sludge. Bioresour Technol 268:408–414

Meyer J, Rein P, Turner P, Mathias K, McGregor C (2011) Good management practices for the cane sugar industry (Final).

Mineral Information Institute, M. I. F. (2003). Rocks are made of minerals. Available online at: http://www.mii.org/pdfs/every/rocksare.pdf~[Online]

Munroe G (2007) Manual of on-farm vermicomposting and vermiculture, vol 39. Organic Agriculture Centre of Canada, p 40

Mupambwa HA, Mnkeni PNS (2018) Optimizing the vermicomposting of organic wastes amended with inorganic materials for production of nutrient-rich organic fertilizers: a review. Environ Sci Pollut Res 25(11):10577–10595

Mupondi LT, Mnkeni PNS, Muchaonyerwa P (2010) Effectiveness of combined thermophilic composting and vermicomposting on biodegradation and sanitization of mixtures of dairy manure and waste paper. Afr J Biotechnol 9(30):4754–4763

Mupondi LT, Mnkeni PNS, Muchaonyerwa P, Mupambwa HA (2018) Vermicomposting manure-paper mixture with igneous rock phosphate enhances biodegradation, phosphorus bioavailability and reduces heavy metal concentrations. Heliyon 4(8):e00749

Niu N, Wang D, Huang S, Li C, He F, Gai S et al (2012) Controlled synthesis of luminescent F-substituted strontium hydroxyapatite with hierarchical structures for drug delivery. CrystEngComm 14(5):1744–1752

Othman R, Panhwar QA (2014) Phosphate-solubilizing bacteria improves nutrient uptake in aerobic rice. In: Phosphate solubilizing microorganisms. Springer, Cham, pp 207–224

Qi Z, Gao X, Qi Y, Li J (2020) Spatial distribution of heavy metal contamination in mollisol dairy farm. Environ Pollut 263:114621

Rafael RBA, Fernandez-Marcos ML, Cocco S, Ruello ML, Weindorf DC, Cardelli V, Corti G (2018) Assessment of potential nutrient release from phosphate rock and dolostone for application in acid soils. Pedosphere 28(1):44–58

Rathi M, Gaur N (2016) Phosphate solubilizing bacteria as biofertilizer and its applications. J Pharm Res 10(3):146–148

Ravindran B, Contreras-Ramos SMM, Sekaran G (2015) Changes in earthworm gut associated enzymes and microbial diversity on the treatment of fermented tannery waste using epigeic earthworm Eudrilus eugeniae. Ecol Eng 74:394–401

Rini J, Deepthi MP, Saminathan K, Narendhirakannan RT, Karmegam N, Kathireswari P (2020) Nutrient recovery and vermicompost production from livestock solid wastes with epigeic earthworms. Bioresour Technol 313:123690

Riskin SH, Porder S, Neill C, Figueira AMES, Tubbesing C, Mahowald N (2013) The fate of phosphorus fertilizer in Amazon soya bean fields. Philosophical Transactions of the Royal Society B: Biological Sciences 368(1619):20120154

Roy T, Biswas DR, Datta SC, Sarkar A, Biswas SS (2018) Citric acid loaded nano clay polymer composite for solubilization of Indian rock phosphates: a step towards sustainable and phosphorus secure future. Arch Agron Soil Sci 64(11):1564–1581

Sattari SZ, Bouwman AF, Giller KE, van Ittersum MK (2012) Residual soil phosphorus as the missing piece in the global phosphorus crisis puzzle. Proc Natl Acad Sci 109(16):6348–6353

Sharma K, Garg VK (2019) Recycling of lignocellulosic waste as vermicompost using earthworm Eisenia fetida. Environ Sci Pollut Res 26(14):14024–14035

Sharma K, Garg VK (2018) Comparative analysis of vermicompost quality produced from rice straw and paper waste employing earthworm Eisenia fetida (Sav.). Bioresour Technol 250:708–715

Sharma SB, Sayyed RZ, Trivedi MH, Gobi TA (2013) Phosphate solubilizing microbes: sustainable approach for managing phosphorus deficiency in agricultural soils. Springerplus 2(1):1–14

Singh P, Mitra S, Majumdar D, Bhattacharyya P, Prakash A, Borah P et al (2017) Nutrient and enzyme mobilization in earthworm casts: a comparative study with addition of selective amendments in undisturbed and agricultural soils of a mountain ecosystem. Int Biodeterior Biodegradation 119:437–447

Tarafdar JC, Claassen N (1988) Organic phosphorus compounds as a phosphorus source for higher plants through the activity of phosphatases produced by plant roots and microorganisms. Biol Fertil Soils 5:308–312

Udert KM (2018) Phosphorus as a resource. Phosphorus: polluter and resource of the future: motivations, technologies and assessment of the elimination and recovery of phosphorus from wastewater, p 57

UNEP (2021) Solid waste management. UN Environment Programme https:// www unep org/ explo re- topics/ resou rce- effic iency/ what- we-do/cites/solid-waste-management

Unuofin FO, Mnkeni PNS (2014) Optimization of Eisenia fetida stocking density for the bioconversion of rock phosphate enriched cow dung–waste paper mixtures. Waste Manag 34 (11):2000–2006

US Geological Survey (2022, January) Mineral Commodity Summaries. Government Printing Office

USDA. USDA Economic Research Service (2019)

Van der linde G (2004) All rock phosphate not the same; a caution. http://www.fssa.org.za/pebble. asp

Van Hees PAW, Lundström US (2000) Equilibrium models of aluminium and iron complexation with different organic acids in soil solution. Geoderma 94(2–4):201–221

Van Kauwenbergh SJ (2010) World phosphate rock reserves and resources. Ifdc, Muscle Shoals, p 48

Wellmer FW, Scholz RW (2015) The right to know the geopotential of minerals for ensuring food supply security: the case of phosphorus. J Ind Ecol 19(1):3–6

WHO (2012) In: Dufour A, Bartram J, Bos R, Gannon V (eds) *Animal Waste,* Water quality, and human health. IWA Publishing, London

Yuvaraj A, Karmegam N, Tripathi S, Kannan S, Thangaraj R (2020) Environment-friendly management of textile mill wastewater sludge using epigeic earthworms: bioaccumulation of heavy metals and metallothionein production. J Environ Manag 254:109813

Part III
Vermicomposts on Soil Quality and Crop Growth

Chapter 11
A Farmers' Synthesis on the Effects of Vermicomposts on Soil Properties

Adornis D. Nciizah, Hupenyu A. Mupambwa, Patrick Nyambo, and Binganidzo Muchara

Abstract Soil degradation is one of the major challenges affecting crop productivity across the world. However, the effects on productivity are more pronounced in sub-Saharan Africa (SSA) where the soils are highly susceptible to various forms degradation. Moreover, climate change exacerbates soil degradation, further exposing the SSA population to food insecurity. Whilst practices such as conservation agriculture (CA) can be employed to improve productivity, it is important to complement CA with technologies that organically contribute to soil fertility, such as vermicomposting. The objectives of this chapter were to examine the contribution of vermicompost to the improvement of soil chemical, biological and physical properties of degraded soils. Various studies carried out across the world showed that vermicompost can enhance soil biological, chemical and physical properties due to high levels of soil organic matter, nutrients and enzymes. Therefore, vermicompost can be a good approach to complement CA in places where soil organic matter and soil fertility are too low to support meaningful biomass production to enable residue retention in order to fulfil the three principles of CA. However, more research work needs to be done on the effects of vermicomposts on soil physical properties in SSA.

A. D. Nciizah (✉)
Agricultural Research Council – Natural Resources & Engineering, Pretoria, South Africa

Department of Agriculture and Animal Health, University of South Africa, Roodepoort, South Africa
e-mail: NciizahA@arc.agric.za

H. A. Mupambwa
Sam Nujoma Marine and Coastal Resources Research Center, University of Namibia, Henties Bay, Namibia

P. Nyambo
Risk and Vulnerability Science Centre, Alice Campus, University of Fort Hare, Alice, South Africa

B. Muchara
Graduate School of Business Leadership, University of South Africa, Midrand, South Africa

© The Author(s), under exclusive license to Springer Nature Singapore Pte Ltd. 2023
H. A. Mupambwa et al. (eds.), *Vermicomposting for Sustainable Food Systems in Africa*, Sustainable Agriculture and Food Security,
https://doi.org/10.1007/978-981-19-8080-0_11

Keywords Biomass production · Climate change · Soil degradation · Soil fertility

11.1 Introduction

One of the most daunting tasks facing humanity is feeding an estimated 9 billion people by 2050 (FAO 2009). Consequently, food production needs to increase by up to 70% in order to meet the global food requirements. Achieving this formidable task requires overcoming several challenges posed by climate change, competition for land resources, changing diets due to urbanization and improved incomes, etc. These challenges place the ever-growing global population especially in semi-arid to arid regions of sub-Saharan (SSA) where soils are highly degraded creating a risk of food insufficiency. Whilst several researchers and international research organizations have over the years recommended several interventions and approaches as sure means of ensuring food sufficiency not much success has been achieved especially in SSA. For instance, conservation agriculture has been widely recommended for smallholder farmers across SSA and yet uptake remains very low due to a number of barriers such as limited access to appropriate farm equipment and tools, inadequate farm inputs and materials, low volumes of biomass, etc. (Barnard et al. 2015). Low volumes of biomass is particularly an important barrier because it hinders famers from retaining the requisite amount of soil cover required to fully practise conservation agriculture. For CA to be effective, approximately 4 t/ha of organic residues should be retained on the soil surface (Mupangwa et al. 2007). Such high biomass production is not easily attained in semi-arid parts of SSA and instead requires the use of commercial fertilizers, which are unfortunately beyond the reach of many resource poor smallholder farmers. Moreover, the excessive use of inorganic fertilizer such as urea is associated with increased greenhouse gas (GHG) emission, which exacerbates climate change. To avert this challenge, some researchers have recommended the use of organic nutrient sources, which increase soil organic carbon (SOC) without increasing GHG emission (Wakindiki et al. 2019). The incorporation of organic fertilizer sources (biofertilizer) into CA has been suggested as a possible solution because CA physically protects SOM, which delays decomposition thus giving rhizosphere microbes ample time to mine nutrients and thus increase biomass production (Wakindiki and Njeru 2017; Wakindiki et al. 2019).

It is therefore vital to recommend appropriate nutrient organic sources that can be incorporated into CA such as vermicomposts. Vermicomposting is an innovative and low-cost biotechnological approach for the valorization of agro-industrial wastes (Garg and Gupta 2009). The vermicomposting process produces a humus like finely granulated and friable material high in organic matter, which can improve the physical and chemical properties of soils (Mupambwa et al. 2022). Vermicompost is rich in both macro- and micronutrients and enzymes which improve soil properties and aid the decomposition of organic matter. This chapter synthesizes available information on the application of vermicompost in marginal soils, which are predominant in sub-Saharan Africa. Soils in this region are usually highly prone to various forms of soil degradation.

11.2 Sub-Saharan Soils

Sub-Saharan Africa is dominated by highly weathered soils with low nutrient and water holding capacity and are hence of poor intrinsic fertility (Rattan 1987). These soils are highly prone to crusting compaction and erosion due to their weak structure. As a result, up to 65% of SSA agricultural land is degraded due to soil erosion, soil acidification, and low nutrient application (Zingore et al. 2015). Some of the characteristics of these soils include weak structure and high susceptibility to both crusting and compaction, which consequently increases susceptibility to soil erosion. The high susceptibility of SSA soils to degradation is worsened by climate change, which has now become a profound challenge to agricultural production. Climate change directly impacts both desertification and soil erosion through its influence on rainfall amount, duration, intensity, interval between rainfall events, increasing evaporation due to high temperatures, high runoff rate and decrease in runoff (Lal 2012). High frequency of drought and high rates of temperature induced evaporation result in poor plant biomass production. This in most cases leaves soil surfaces bare which increases the effects of raindrop impacts on mostly poorly aggregated soil, which break-up leading to poor infiltration, runoff and soil erosion. The low production of biomass also results in low litter inputs into the soil leading to low organic carbon in these soils. Soils low in organic carbon are poorly aggregated and have low water holding capacity hence the low water storage in these soils, which has deleterious effects on crop production.

Although SSA soils are inherently poor and highly susceptible to degradation, several socio-economic and political drivers have historically played a significant role in driving soil degradation. For instance, during the pre-independence era across much of SSA, people were displaced into highly unproductive marginal lands. These overpopulated areas received very little infrastructural and technological support, which exacerbated degradation. Since most of the people in these areas were resource poor, there was very little use of agricultural inputs, leading to highly depleted soils. Moreover, farmers in these areas use unsustainable agricultural practices due to poor technical support, which significantly degrades the soils (Tully et al. 2015). Whilst soil cannot be restored to their original state, there are various technologies and practices that can be employed to improve the condition and productivity of the soil. It is therefore important to adopt technologies that can improve the soil health and hence improve agricultural productivity.

11.3 What Is Vermicompost?

The term vermicompost emanates from two words, i.e. the Latin word vermis, which means worm and compost, which describes a process of controlled microbial degradation of organic matter (Mupambwa et al. 2022). Unlike traditional composting, which is as old as agriculture itself, vermicomposting, which is the process of deliberately using organic matter decomposing earthworms to degrade organic and inorganic matter, has been scientifically defined by Gómez-Brandón and Domínguez (2014) as a *"bio-oxidative process in which detrivorous earthworms interact with microorganisms and other fauna within the decomposer community, thus accelerating the stabilization of organic matter (OM) and greatly modifying its physical and biochemical properties."* The process of vermicomposting thus does not only involve the earthworms alone but various trophic levels, i.e. primary consumers (naturally occurring microbes), secondary consumers (involves that earthworms and their gut associated activities) and tertiary consumers (which involves microbes and bio-chemicals) (Mupambwa et al. (2022).

Vermicomposts have gained momentum in organic soil fertility management as they have proved to be more nutrient rich and also contain important functional molecules relative to traditional composts. During traditional composting, it has been indicated that nitrogen is lost mainly due to volatilization during the thermophilic stages of composting. As indicated by Szántó (2009), in composts, nitrogen is present in organic matter as mainly proteins or urea, and during hydrolysis of these organic compounds, ammonia (NH_3) is produced, which then requires microbes to convert it into ionic forms. This conversion process is controlled by pH and temperature and due to heating and acidification happening in the early stages of composting, most of the N (up to 50% of the nitrogen) is lost through volatilization of NH_3. However, in vermicomposts, the entire composting process involves mesophilic temperatures which promote microbes (nitrifiers), thus significantly reducing the loss of nitrogen. The characteristics of vermicomposts relative to traditional composts are summarized in Table 11.1, indicating that in terms of chemistry, vermicomposts are superior to composts.

11.4   Degraded Soils Amended with Vermicompost

11.4.1   Chemical Properties

Vermicompost use is a sustainable agricultural practice that can potentially reduce the deterioration of soil fertility and groundwater caused by intensive farming practices. Vermicompost use for soil fertility improvements has gained considerable momentum due to its contribution to agro-ecological sustainability (Xu and Mou 2016). It can be applied to improve soil fertility by increasing soil organic matter, CEC, nutrient content and structure (Srivastava et al. 2011). Furthermore, vermicompost gradually releases nutrients into the soil, thus ameliorating issues like salinity, nutrient toxicity and loss; however, the rate of nutrient release into the soil is influenced by weather conditions, soil type, presence and type of plants (Ilker et al. 2016). For example, a study by Arancon et al. (2005) reported that paper waste vermicompost applied at a rate of 10 t ha^{-1} contained more extractable N under DoA—Doles silt loam soils, whilst food waste vermicomposts applied at the same rate had significantly higher extractable N in Hoytville silty clay loam soil. A study by Hoque et al. (2022) reported higher phosphorus release in terrace soil than in floodplain.

Table 11.1 A summary of selected characteristics of composts and vermicomposts based on results of different organic materials

Organic material	Compost type		Reference
	Compost	Vermicompost	
Cow manure	C/N ratio (17.5); Electrical Conductivity (2.13 dSm^{-1}) pH (8.86) Dissolved organic carbon (9338 mg kg^{-1}) Ammonium (1235 mg kg^{-1}) Nitrate (721 mg kg^{-1})	C/N ratio 11.3 Electrical Conductivity (0.78 dSm^{-1}) pH (7.73) Dissolved organic carbon (5249 mg kg^{-1}) Ammonium (276 mg kg^{-1}) Nitrate (917 mg kg^{-1})	Lazcano et al. (2008)
Municipal waste	pH (9.2) Organic matter (20%) NH_4 (7.3 mg kg^{-1}) NO3 (90 mg kg^{-1}) P (127 mg kg^{-1}) C/N (11.7)	pH (8.7) Organic matter (24%) NH_4 (25.2 mg kg^{-1}) NO3 (203 mg kg^{-1}) P (207 mg kg^{-1}) C/N (11.1)	Tognetti et al. (2005)
Cow manure with fly ash	Humification index (1) Increase in Olsen P (39.3%) Nitrate (8 mg kg^{-1}) Humification ratio (7) Polymerization index (0.8)	Humification index (2.5) Increase in Olsen P (87.5%) Nitrate (18 mg kg^{-1}) Humification ratio (9) Polymerization index (6)	Mupambwa and Mnkeni (2015)
Tomato crop residues and almond shells	pH (8.9) Electrical conductivity (252 mS/m) Organic matter (73%) Fulvic acid carbon (1.4%) Humic acid carbon (10.2%)	pH (8.7) Electrical conductivity (61 mS/m) Organic matter (67.1%) Fulvic acid carbon (0.7%) Humic acid carbon (8.6%)	Fornes et al. (2012)

Various studies have reported the effects of vermicomposts on soil chemical properties (Table 11.2). According to Oo et al. (2013), the salt concentration is maintained in the soils with vermicompost in the form of leachate. After vermicompost application, improvements in plant-available microelements such as Ca, Mg, Na, N, P, K, Zn, Fe, Cu, and Mn have also been reported (Esringü et al. 2022). According to Gutiérrez-Miceli et al. (2007), the ion concentration and raw materials used for vermicompost affect the electrical conductivity. A reduced electrical conductivity appears because of stabilized raw material in the vermicompost (Lim et al. 2011). The number of exchangeable Na^+ and Ca^+ ions increases with vermicompost use and subsequently lowers the electrical conductivity of the soil (Oo et al. 2013). Vermicompost application differs considerably according to plant species and genotype (Lazcano et al. 2011).

Table 11.2 A summary of studies on the effects of vermicomposts on soil chemical properties

Country	Type of experiment	Crops used	Soil type	Main findings	References
Bangladesh	Incubation and field experiment	Wetland rice	Highly weathered terrace soil and very young floodplain soil.	Significant higher nitrogen and phosphorous was released in floodplain soil compared to terrace soil	Hoque et al. (2022)
Turkey	Field experiment	Celery (*Apium graveolens L. var. dulce mill.*)	Alkaline soil with high lime content	Vermicomposting was more effective in increasing organic matter, N, P, and Ca relative to farmyard manure	Ilker et al. (2016)
Bangladesh	Incubation study/pot experiment	Kalmi (*Ipomoea Aquatica Forsk*)	Mixed soil 1: 1 (i.e. acid: Calcareous)	Application of 12 t/ha vermicomposting significantly improved soil pH; electrical conductivity, organic carbon, available N, P, K, S, Ca, Mg, Na, Fe and Zn compared to the control, 4 t/ha vermicomposting, 8 t/ha vermicomposting, 4 t/ha compost, 8 t/ha compost and 12 t/ha compost.	Nasrin et al. (2019)
Mexico	Greenhouse	Beans (*Phaseolus vulgaris L.*)	Clayey	Vermicomposting application decreased NH_4^+, whilst NO_3^- increased. Electrical conductivity was significantly higher in vermicomposting-amended soil compared to the control.	Valdez-Perez et al. (2011)
South Africa	Greenhouse	Organic Swiss Chard (Beta	Oxisol soil and pine bark growing media	Amending pine bark with vermicomposting increased both pH	Ajibade et al. (2022)

(continued)

11 A Farmers' Synthesis on the Effects of Vermicomposts on Soil Properties 195

Table 11.2 (continued)

Country	Type of experiment	Crops used	Soil type	Main findings	References
		vulgaris subsp. vulgaris)		and EC. The higher concentration of vermicomposting applied to pine bark the higher the pH and EC.	
Namibia	Incubation experiment	–	Abandoned gold mine site soil, coal mine waste contaminated soil and soil from a mine tailing	Nitrogen mineralization was enhanced with addition of the vermicompost. The Mn, Zn and Pb solubility was reduced with addition of the vermicompost.	Lukashe et al. 2019
USA	Greenhouse	Pea sorghum bi-Colour	Mineral soil (Crosby-silt-loam and sphagnum peat moss)	Addition of vermicompost increased available N, P and K in the media	Cavender et al. (2003)

11.4.2 Physical Properties

Several studies have shown that vermicomposts improve soil physical properties such as soil structure, soil water holding capacity, penetration resistance, bulk density, soil organic carbon, aggregation and nutrient content. Improvement in these physical soil properties provides an improved medium for root and overall plant growth. The ability of vermicomposting to improve soil structural conditions is a result of high amounts of organic matter and polysaccharides. The benefits of organic matter on soil structural formation and stability are well documented (Tisdall and Oades 1982; Six et al. 1998; Blanco-Canqui and Lal 2004). Organic matter not only binds both primary and secondary soil particles, which contributes to soil aggregation but also reduces aggregate breakdown by reducing clay wettability due to hydrophobicity (Blanco-Canqui and Lal 2004). Soils with good aggregate stability have good water infiltration, porosity and low bulk density, which promote good root growth. Aksakal et al. (2016) applied vermicompost at three rates (0·5%, 1%, 2% and 4% w/w) on sandy loam, loam and clay textured soils and observed increased wet aggregate stability and decreased dispersion ratio in all the experimental soils in all aggregate size fractions. In addition, a concomitant increase in soil organic matter was observed, which explained the increase in soil aggregation. Additionally, the application of vermicompost resulted in lower bulk density than the control, which reduced penetration resistance and increased porosity. Such conditions provide optimum conditions for root growth, which ensures good crop growth and yield. Similarly, Demir (2019) reported improved bulk density, available water capacity, hydraulic conductivity, and field capacity with application of vermicompost. An increase in total porosity with increase in vermicompost rate was also reported in this study. This was attributed to improved soil aggregation, which was improved by an elevated concentration of polysaccharides, which act as binding agents (Lim et al. 2014). In a study on short-term effects of compost amendments to soil on soil structure, hydraulic properties, and water regime, Rivier et al. (2022) reported reduced irrigation demand and evaporation losses and increased water use efficiency of the plant. This was attributed to the positive effect of vermicompost on soil structure and the pore-size distribution. Using X-Ray tomography, the authors confirmed that vermicomposting resulted in a more complex and diverse porous system, which increased soil macro-porosity. Moreover, vermicomposting resulted in the formation and stabilization of soil structure as evidenced by increased fractal dimensions. This finding on soil structure is in line with numerous other studies that reported improved soil structural properties with application of vermicompost. Azarmi et al. (2018) reported a decrease in bulk density and an increase in porosity with an increase in vermicompost application rate. The increase in porosity was attributed to an increased number of pores in the 30–50μm and 50–500 size ranges and a decrease in number of pores greater than those 500μm in diameter. The increase in porosity increases pore space or total pore volume, which explains the reduction of bulk density with application of vermicompost.

11.4.3 Biological Properties

The recent promotion of organic fertilizers has mainly not been due to their superiority in providing nutrients to the plant but rather due to their ability to also feed the soil, which is a living ecosystem of various organisms at different trophic levels. Soil organisms require organic matter, which is a source of energy and other nutrients, which cannot be supplied by inorganic fertilizers. Through feeding on this organic matter, microbes also drive several bio-geo-chemical processes in the soil that are critical in improving soil quality. Soil biology is quite a critical parameter that can be influenced through the application of organic matter sources such as vermicomposts. Biological properties such as microbial activity, diversity and biomass are crucial soil management indicators that are sensitive to soil management especially in degraded soils (Mupambwa et al. 2020). As a result, any changes in these properties may precede detectable changes in other soil properties (chemical and physical properties), thus providing early signal of improvement or warning of degradation. Apart from triggering high microbial activity, vermicomposts also contain large quantities of different microbes, and the summary of research that has looked on the effects on soil biological properties is presented in Table 11.3. What is important

Table 11.3 Summary of research done on application of vermicompost in soils and the changes in soil biological properties

Vermicompost type	Influence on soil biological properties	Reference
Cow dung and green forage consisting of grasses, the green leaves of vegetables, herbs, and other plant materials and vermicomposting done with *Eisenia fetida* earthworms.	• The addition of vermicompost at the highest level increases soil respiration measured as CO_2-C. • Across the 3 years of the study, the microbial biomass and the activities of the enzymes dehydrogenase; urease; β-glucosidase, phosphatase, and arylsulfatase all increased significantly whilst the control remained almost unchanged. • Soil microbial biomass in soils amended with cow dung vermicompost was 21.8% higher than in soils amended with green forage vermicompost. • The soil biological properties improved in soils amended with the cow dung vermicompost, rich in fulvic-like acids, possibly due to the greater labile fraction of organic matter than in the green forage vermicompost, making it more degradable and susceptible to rapid mineralization.	Tejada et al. (2010)
Vermicompost was produced mainly from farmyard manure obtained from the dairy farm.	• In soils receiving organic materials, dehydrogenase enzyme activity started to increase from the first week to the 4th–7th weeks of the experiment and then decreased to control (no amendment) levels at around 10th–13th weeks though the control treatment showed gradual decrease from the beginning of the experiment. • From the beginning of the incubation period, soils with organic materials showed higher β-glucosidase, alkaline phosphatase and urease activity compared to the control. • Total aerobic mesophilic bacterial numbers immediately increased at the beginning of the experiment in soils treated with organic materials and followed a steady trend after the fourth week with no significant increase being recorded in the control treatment.	Uz and Tavali (2014)
Vermicompost from source separated household solid waste (HSW), horse and rabbit manure (HRM) and chicken manure (CM).	• An increase in microbial respiration in all the amended plots was observed with respect to the controls.	Ferreras et al. (2006)

(continued)

Table 11.3 (continued)

Vermicompost type	Influence on soil biological properties	Reference
Rice straw and cattle manure were used to prepare vermicompost using E. fetida earthworms.	• Organic amendments increased microbial carbon biomass (24–45%). • The effectiveness of organic amendments on enzyme activity in rhizosphere of inoculated plants increased compared to non-inoculated plants.	Ghadimi et al. (2021)

to note is that the addition of vermicomposts in soil significantly improves the microbial activity, enzyme activity and soil respiration, which indicates the increased soil ecological activity, which drives soil nutrient cycling.

11.5 Conclusions

This chapter has highlighted the benefits of using vermicompost on improving soil chemical, biological and physical properties, e.g. microbial activity, diversity and biomass, soil structure, hydraulic properties and water regime. However, detailed studies on the benefits of vermicomposts in SSA especially effects of soil physical properties are still very few. It is thus important that more work be carried out on the effects of vermicomposting various organic material on soil physical properties under various environmental and soil conditions across SSA. This is particularly important since physical properties influence other soil properties, which influence crop growth. Moreover, vermicompost may play a critical role in moderating the effects of climate change by promoting biomass accumulation, which will improve soil cover.

References

Ajibade S, Mupambwa HA, Manyevere A, Mnkeni PNS (2022) Vermicompost amended with rock phosphate as a climate smart Technology for Production of organic Swiss chard (Beta vulgaris subsp. vulgaris). Front Sustain Food Syst 6. https://doi.org/10.3389/fsufs.2022.75779

Aksakal EL, Sari S, Angin I (2016) Effects of vermicompost application on soil aggregation and certain physical properties. Land Degrad Develop 27:983–995

Arancon NQ, Edwards CA, Bierman P, Metzger JD, Lucht C (2005) Effects of vermicomposts produced from cattle manure, food waste and paper waste on the growth and yield of peppers in the field. Pedobiologia 49(4):297–306

Azarmi R, Giglou MT, Taleshmikail RD (2018) Influence of vermicompost on soil chemical and physical properties in tomato (Lycopersicum esculentum) field. Afr J Biotechnol 7(14): 2397–2401

Barnard J, Manyire H, Tambi E, Bangali S (2015) Barriers to scaling up/out climate smart agriculture and strategies to enhance adoption in Africa forum for agricultural research in Africa, Accra, Ghana. FARA

Blanco-Canqui H, Lal R (2004) Mechanisms of carbon sequestration in soil aggregates. Crit Rev Plant Sci 23:481–504

Cavender ND, Atiyeh RM, Knee M (2003) Vermicompost stimulates mycorrhizal colonization of roots of Sorghum bicolor at the expense of plant growth. Pedobiologia 47(1):85–89

Demir Z (2019) Effects of vermicompost on soil physicochemical properties and lettuce (Lactucasativa Var. Crispa) yield in greenhouse under different soil water regimes. Commun Soil Sci Plant Anal 50:2151. https://doi.org/10.1080/00103624.2019.1654508

Esringü A, Turan M, Sushkova S, Minkina T, Rajput VD, Glinushkin A, Kalinitchenko V (2022) Influence of vermicompost application on the growth of Vinca rosea valiant, Pelargonium peltatum L. and Pegasus patio rose. Horticulturae 8:534. https://doi.org/10.3390/horticulturae8060534

FAO (2009) 2050: A third more mouths to feed. Food and Agricultural Organisation. Available at: https://www.fao.org/news/story/en/item/35571/icode/. Accessed on 20 May 2022.

Ferreras L, Gomez E, Toresani S, Ine's Firpo, Rossana Rotondo. (2006) Effect of organic amendments on some physical, chemical and biological properties in a horticultural soil. Bioresour Technol 97(2006):635–640

Fornes F, Mendoza-Hernández D, García-de-la-Fuente R, Abad M, Belda RM (2012) Composting versus vermicomposting: a comparative study of organic matter evolution through straight and combined processes. Bioresour Technol 118(2012):296–305

Garg V, Gupta R (2009) Vermicomposting of agro-industrial processing waste. In: Singh nee' Nigam P, Pandey A (eds) Biotechnology for agro-industrial residues utilisation. Springer, Dordrecht. https://doi.org/10.1007/978-1-4020-9942-7_24

Ghadimi M, Sirousmehr A, Ansari MH, Ghanbari A (2021) Organic soil amendments using vermicomposts under inoculation of N2-fixing bacteria for sustainable rice production. PeerJ 9:e10833. https://doi.org/10.7717/peerj.10833

Gómez-Brandón M, Domínguez J (2014) Recycling of solid organic wastes through vermicomposting: microbial community changes throughout the process and use of vermicompost as a soil amendment. Crit Rev Environ Sci Technol 44(12):1289–1312. https://doi.org/10.1080/10643389.2013.763588

Gutiérrez-Miceli FA, Santiago-Borraz J, Molina JAM, Nafate CC, Abud-Archila M, Llaven MAO et al (2007) Vermicompost as a soil supplement to improve growth, yield and fruit quality of tomato (Lycopersicum esculentum). Bioresour Technol 98(15):2781–2786

Hoque TS, Hasan AK, Hasan MA, Nahar N, Dey DK, Mia S, Solaiman ZM, Kader MA (2022) Nutrient release from Vermicompost under anaerobic conditions in two contrasting soils of Bangladesh and its effect on wetland rice crop. Agriculture 12(3):376

Ilker UZ, Sonmez S, Tavali IE, Citak S, Uras DS, Citak S (2016) Effect of vermicompost on chemical and biological properties of an alkaline soil with high lime content during celery (Apium graveolens L. var dulce Mill) production. Notulae Botanicae Horti Agrobotanici Cluj-Napoca 44(1):280–290

Lal R (2012) Climate change and soil degradation mitigation by sustainable management of soils and other natural resources. Agric Res 1:199–212. https://doi.org/10.1007/s40003-012-0031-9

Lazcano C, Revilla P, Malvar RA, Dominguez J (2011) Yield and fruit quality of four sweet corn hybrids (Zea mays) under conventional and integrated fertilization with vermicompost. J Sci Food Agric 91:1244–1253

Lazcano C, Gómez-Brandón M, Domínguez J (2008) Comparison of the effectiveness of composting and vermicomposting for the biological stabilization of cattle manure. Chemosphere 72(2008):1013–1019

Lim PN, Wu TY, Sim EYS, Lim SL (2011) The potential reuse of soybean husk as feedstock of Eudrilus eugeniae in vermicomposting. J Sci Food Agric 91:2637–2642

Lim SL, Wu TY, Lim PN, Shak KPY (2014) The use of vermicompost in organic farming: overview, effects on soil and economics. J Sci Food Agric 95(6):1143–1156. https://doi.org/10.1002/jsfa.6849. (wileyonlinelibrary.com)

Lukashe NS, Mupambwa HA, Mnkeni PNS (2019) Changes in nutrients and bioavailability of potentially toxic metals in mine waste contaminated soils amended with fly ash enriched vermicompost. Water Air Soil Pollut 230(12):1–17

Mupangwa W, Twomlow S, Walker S, Hove L (2007) Effect of minimum tillage and mulching on maize (Zea mays L.) yield and water content of clayey and sandy soils. Phys Chem Earth 32: 1127–1134

Mupambwa HA, Ravindran B, Dube E, Lukashe NS, Katakula AAN, Mnkeni PNS (2020) Some perspectives on Vermicompost utilization in organic agriculture. In: Bhat S, Vig A, Li F, Ravindran B (eds) Earthworm assisted remediation of effluents and wastes. Springer, Singapore. https://doi.org/10.1007/978-981-15-4522-1_18

Mupambwa HA, Haulofu M, Nciizah AD, Mnkeni PNS (2022) Vermicomposting technology: A sustainable option for waste beneficiation. In: Jacob-Lopes E, Zepka LQ, Deprá MC (eds) Handbook of waste biorefinery - circular economy of renewable energy, Book Series: biofuel and biorefinery technologies. Springer-Nature, Cham

Mupambwa HA, Mnkeni PNS (2015) Optimization of fly ash incorporation into cow dung–waste paper mixtures for enhanced vermidegradation and nutrient release. J Environ Qual 44:972–981. https://doi.org/10.2134/jeq2014.10.0446

Nasrin A, Khanom S, Hossain SA (2019) Effects of vermicompost and compost on soil properties and growth and yield of Kalmi (Ipomoea aquatica Forsk.) in mixed soil. Dhaka university. J Biol Sci 28(1):121–129

Oo AN, Iwai CB, Saenjan P (2013) Soil properties and maize growth in saline and non-saline soils using cassava– industrial waste compost and vermicompost with or without earthworms. Land Degrad Dev 26(3):300–310

Rattan L (1987) Managing the soils of sub-Saharan Africa. Science 236(4805):1069–1076. https://doi.org/10.1126/science.236.4805.1069

Rivier P-A, Jamniczky D, Nemes A, Makó A, Barna G, Uzinger N, Rékási M, Farkas C (2022) Short-term effects of compost amendments to soil on soil structure, hydraulic properties, and water regime. J Hydrol and Hydromech 70(1):74–88. https://doi.org/10.2478/johh-2022-0004

Six J, Elliott ET, Paustian K, Doran JW (1998) Aggregation and soil organic matter accumulation in cultivated and native grassland soils. Soil Sci Soc Am J 62:1367–1377. https://doi.org/10.2136/sssaj1998.03615995006200050032x

Srivastava PK, Singh PC, Gupta M, Sinha A, Vaish A, Shukla A, Singh N, Tewari SK (2011) Influence of earthworm culture on fertilization potential and biological activities of vermicomposts prepared from different plant wastes. J Plant Nutr Soil Sci 174:420–429

Szántó G (2009) NH3 dynamics in composting – assessment of the integration of composting in manure management chains PhD-thesis Wageningen University, Wageningen, the Netherlands – with summaries in English and Dutch ISBN: 978-90-8585-369-5.

Tejada M, Gómez I, Hernández T, García C (2010) Utilization of Vermicomposts in soil restoration: effects on soil biological properties. Soil Sci Soc Am J 74:525–532. https://doi.org/10.2136/sssaj2009.0260

Tisdall JM, Oades JM (1982) Organic matter and water-stable aggregates in soils. J Soil Sci 33: 141–163

Tognetti C, Laos F, Mazzarino MJ, Hernández MT (2005) Composting vs. vermicomposting: A comparison of end product quality. Compost Sci Utiliz 13(1):6–13. https://doi.org/10.1080/1065657X.2005.10702212

Tully K, Sullivan C, Weil R, Sanchez P (2015) The state of soil degradation in sub-Saharan Africa: baselines, trajectories, and solutions. Sustainability 7(6):6523–6552. https://doi.org/10.3390/su7066523

Uz I, Tavali IE (2014) Short-term effect of vermicompost application on biological properties of an alkaline soil with high lime content from mediterranean region of Turkey. Sci World J 2014: 395282, 11 pages. https://doi.org/10.1155/2014/395282

Valdez-Perez MA, Fernandez-Luqueno F, Franco-Hernandez O, Cotera LF, Dendooven L (2011) Cultivation of beans (Phaseolus vulgaris L.) in limed or unlimed wastewater sludge, vermicompost or inorganic amended soil. Sci Hortic 128(4):380–387

Wakindiki IIC, Njeru SK (2017) Organic carbon associated with tillage-induced aggregates of some quartzdominated loamy soils in a semi-arid region of South Africa. S Afr J Plant Soil 34(3): 239–242. https://doi.org/10.1080/02571862.2017.1281446

Wakindiki IIC, Malobane ME, Nciizah AD (2019) Integrating biofertilizers with conservation agriculture can enhance its capacity to mitigate climate change: examples from southern Africa. In: Leal FW, Leal-Arcas R (eds) University initiatives in climate change mitigation and adaptation. Springer, Cham, pp 277–289

Xu C, Mou B (2016) Vermicompost affects soil properties and spinach growth, physiology, and nutritional value. HortScience 51(7):847–855

Zingore S, Mutegi J, Agesa B, Desta LT, Kihara J (2015) Soil degradation in sub-Saharan Africa and crop production options for soil rehabilitation. Better Crops Plant Food 99(1):24–26

Chapter 12
Vermicompost as a Possible Solution to Soil Fertility Problems and Enrichment in the Semiarid Zones of Namibia

S. N. Nghituwamhata, L. N. Horn, and S. N. Ashipala

Abstract Poor soil fertility is a major constraint to agricultural production in Namibia, a semiarid country with low rainfall and extreme temperatures. The poor soil fertility has made farmers resort to chemical fertilizer use to increase production. However, chemical fertilizers are implicated with heavy metals in the product, making the foods unhealthy compared to organic fertilizers. Chemical fertilizers also kill beneficial soil organic components, destroy the soil's natural fertility, and weaken the crops' ability of natural resistance, increasing their vulnerability to pests and infections. Scientists are discovering vermicompost worldwide as an economically feasible, socially safe, and environmentally sustainable alternative to agrochemicals. Vermicomposting produces organic manure using earthworms mixed with biological and plant residues. The technology is "economically viable," "environmentally sustainable," and "socially acceptable." The practice of vermicomposting in Namibia is at an initial stage, with few businesses venturing into the vermicompost market commercially and unpublished research works. As a result, many farmers do not have access to this limited market and have not been exposed to this incredible technology. This calls for more research and public awareness on vermicompost to increase production and preserve the environment. This chapter will highlight the benefits of using vermicompost to farmers, its possible applications to small- and large-scale farmers, and the current work on this mechanism in Namibia. This will enlighten farmers, provide the gaps in knowledge related to vermicompost, and instigate researchers to focus on these research gaps.

S. N. Nghituwamhata (✉)
Department of Agricultural Production, Makerere University, Kampala, Uganda

Regional Universities Forum for Capacity Building in Agriculture (RUFORUM), Kampala, Uganda

L. N. Horn
Zero Emissions Research Initiatives, University of Namibia (UNAM), Windhoek, Namibia

S. N. Ashipala
Ministry of Agriculture, Water and Land Reform (MAWLR), Windhoek, Namibia

© The Author(s), under exclusive license to Springer Nature Singapore Pte Ltd. 2023
H. A. Mupambwa et al. (eds.), *Vermicomposting for Sustainable Food Systems in Africa*, Sustainable Agriculture and Food Security,
https://doi.org/10.1007/978-981-19-8080-0_12

203

Keywords Chemical fertilizers · Namibia · Organic fertilizers · Soil fertility · Vermicompost · Vermiculture

12.1 Introduction

Due to Namibia's climatic conditions and soil structure, soil degradation is a major threat to agricultural production. The semiarid climatic condition, low rainfall, and poor water retention capacity result in low soil fertility and organic matter status, contributing to low food productivity (Mendelsohn et al. 2002). According to the Namibian Agronomic Board report (NAB) (2018/2019), the production of agronomic crops, mostly white maize, wheat and pearl millet, occurs in commercial and communal areas under irrigation rain-fed. Pearl millet, locally known as "Mahangu," is predominately produced in communal areas under rain-fed production. In contrast, wheat production occurs under irrigation at the commercial level, and horticultural production occurs in the commercial and communal areas, under irrigation only NAB (2018/2019).

The Namibian government, through its fifth National Development Plan (NDP5) (2017/2018–2021/2022), aimed to achieve food security through the increase of agricultural production for cereals and horticulture (small-scale and backyard gardening), including urban and peri-urban communities. However, given the challenge of poor soil fertility and low rainfall, this is only achievable under heavy chemical fertilizers and irrigation. Although these can help increase production, they are associated with high costs, environmental impacts, health concerns and destruction of the soil structure. Therefore, an alternative biological solution to poor soil fertility can be using *Epigeic* earthworms such as *Eisenia fetida* (that have no permanent burrows and feed on decaying organic matter to produce vermicomposts) to improve the soil fertility in Namibia.

Vermicomposting is one of the sustainable approaches for managing organic waste streams using worms in processing organic waste. The practice has been around since the 1970s, with minimal adoption in most parts of the world (Furlong et al. 2017). There is a very slow adoption and limited information regarding vermicomposting in Namibia.

12.2 General Description of Namibia: Climatic Condition and Soil Structure

12.2.1 Climatic Conditions

With an average annual temperature of 20.6 °C and average monthly temperatures ranging from 24 °C (November to March) to 16 °C (April to October), Namibia is

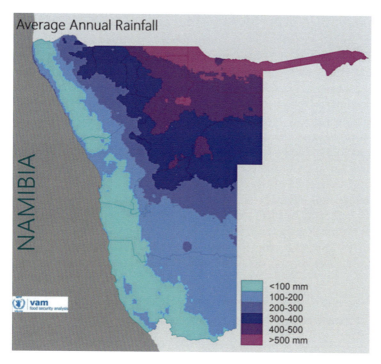

Fig. 12.1 Average rainfall for Namibia, at 5 km resolution, for period 1981–2018 Source: (MEFT 2021)

one of the world's hottest and driest countries (June, July) (Goudie and Viles 2015; Garrard et al. 2017).

Being an arid to semiarid country (Bertram 1999), the Namibia rainfall season is unimodal, lasting from October to April. However, in the southern part of the country, rains fall outside this period with a short peak around August (Garrard et al. 2017). The wettest period within the season is December–March, with January and February being the wettest months across the country. The rainfall distribution is highly erratic and variable (Mendelsohn et al. 2002). This yearly average ranges from 650 mm in the northeast to less than 50 mm in the southwest, including coastal areas (Garrard et al. 2017). The country is arid and water-scarce in terms of hydrology, with high solar radiation, low humidity, and high temperatures, resulting in extremely high evaporation rates (Garrard et al. 2017; Bertram 1999). The evaporation rate varies from 3800 mm per year in the south to 2600 mm per year in the north (Mendelsohn et al. 2002). A summary of the average rainfall is shown in Fig. 12.1 below.

12.3 Soil Structure and Fertility

According to MAWLR (2021), the major soils that are found in Namibia are arenosols (sandy soils), calcisols (calcareous soils), fluvisols (fluvial soils), leptosols (shallow), regosols (skeletic), cambisols (textural change), gypsisols (gypsum rich soils), and luvisols (clay accumulation). Most of these soils are characterized by poor water holding capacity, very low agricultural potential, very low sodic agricultural potential, low to moderate chemical fertility and organic carbon (FAO 2015; MAWLR 2021). The soils that are most fertile in the Namibian context and are found in small parts of the country do not rate highly on the world scale of soil potential (Mendelsohn et al. 2002).

Therefore, soil infertility is a major contributing factor to low productivity in Namibia, especially in small-scale farming, where the soil improvement and management practices are not well considered. Promoting soil organic sequestration on small-scale communal farms in North-Central Namibia will significantly improve soil quality, increase crop yields and reduce agricultural greenhouse gas emissions.

12.4 Crop Production in Namibia

Namibia comprises commercial farming, communal or subsistence farming, and small-scale yard gardening in townships and subsistence communities. Communal farmers are primarily confined to areas predominantly by arenosols or sandy soils and depend on rainfall for production (MAWLR 2021; Mendelsohn et al. 2002), while commercial farming largely depends on irrigation and the use of fertilizers. The communal farmers hardly use fertilizers due to low erratic rainfalls and possible leaching of fertilizers due to poor soil organic matter, which is very low and does not benefit the crop (Rigourd and Sape 1999). Additionally, fertilizers during drought may negatively affect soil (Rigourd and Sape 1999).

Occasionally, communal farmers depend on inherent soil fertility, farm manure (from the cow, goats, donkeys, sheep, chickens, etc.), compost (rarely practiced), and the use of inorganic or chemical fertilizers to improve soil fertility. However, direct use of animal manures for plant nutrition is not recommended since they are organically bound and need to be converted from organic to inorganic nutrients through mineralization (Mupambwa et al. 2020). The small stock manure, also commonly used, is associated with nematodes and thus should be discouraged from being applied directly as a soil improvement strategy. Additionally, inorganic fertilizers are not readily available to the poor resource communal farmers and are associated with destroying the soil structure as they contain trace elements. With the current increase in small-scale yard gardening in townships and subsistence communities, vermicomposting can thus be the best option to convert these farm manure and other wastes into inorganic nutrients readily available to plants.

12.4.1 Vermicomposting

Vermicomposting is a technology for managing organic wastes by converting them into inorganic compounds using earthworms (Sinha et al. 2010). It is a simple biotechnological practice of composting, whereby certain species of earthworms are used to convert wastes into a better product (Adhikary 2012). The process of rearing these earthworms is called vermiculture. These earthworms (Fig. 12.2) can safely convert and manage municipal and industrial organic wastes, including sewage sludge and divert them from landfills to beneficial manures. During the process, the earthworm body acts as a biofilter, purifying, disinfecting and detoxifying municipal and several industrial wastewaters (Rani and Shukla 2014). Vermicomposting takes a short time than composting as the waste materials are digested through the earthworm gut, which later execrates manure enriched microbial activity, plant growth regulators, and fortified with attributes that enable the plant to repel pests, as shown in Fig. 12.3 (Adhikary 2012).

However, not all the wastes are converted directly as some, such as effluents from palm oil mill or olive oil mill industries, require modification before vermicomposting. The feedstocks also require optimization for proper use in vermicomposting. Several studies on optimizing the wastes, such as the mixture of fly ash, cow dung and waste papers (Mupambwa and Mnkeni 2015), fly ash alone (Mupambwa and Mnkeni 2016), chicken manure (Ravindran and Mnkeni 2016), and goat manure (Katakula et al. 2021), proved to be effective in vermicomposting.

Fig. 12.2 Earthworms (*Eisenia fetida*) on crop residues. Source: MAWLR (2012)

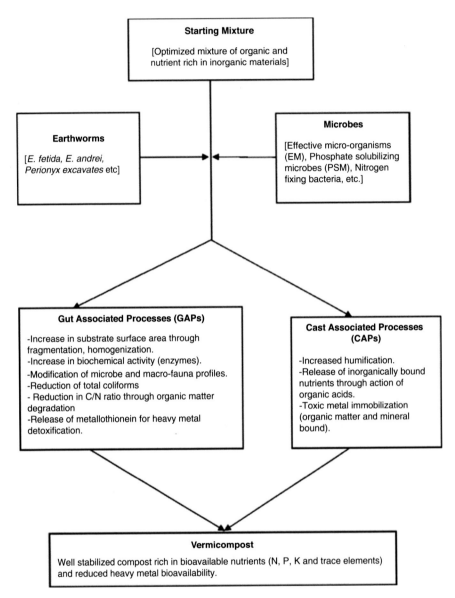

Fig. 12.3 Schematic representation of the different roles of earthworm and microbe in vermicomposting of organic materials. (Source: Mupambwa and Mnkeni 2018)

The representation above in Fig. 12.3 is supported by an experiment to remove heavy metals from urban sewage sludge amended with spent mushroom compost using worms. Heavy metals such as Cr, Cd, and Pb contained in vermicompost were lower than the initial concentrations, with 90–98.7% removal on the tenth week (Azizi et al. 2013). Based on that evidence, it is proven that the earthworms are

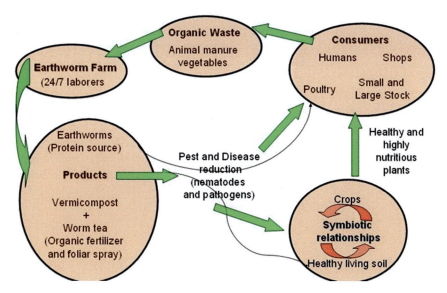

Fig. 12.4 A schematic diagram of the importance of earthworms farming and their contribution to agricultural production

capable to absorb heavy metals from the sewage sludge and discharging them into their surroundings after some time.

Figure 12.4 further highlights the importance of earthworms farming and its significance to agriculture.

12.5 Use of Vermicomposting in Namibia

The Ministry of Agriculture Water and Forestry (MAWF) launched a project to support urban and peri-urban horticulture (UPH). The project was funded by the Kingdom of Belgium and technical expertise from FAO (Fosso 2014). The vermicompost production was one of the project's activities. After its successful implementation, the project was awarded a gold medal as the Best Exhibition Indoor at the Windhoek Agricultural Society Show in October 2007.

Another project on vermicomposting and vermiculture was introduced in Namibia by the Tunweni project, funded by the UNDP in collaboration with the University of Namibia (UNAM) (Okeyo 2000). Tunweni project concept consisted of a vermiculture house, an animal house, settling/filtering/sump tanks, a bio-digester, shallow algae ponds, deep fish ponds, and irrigated land adjacent to a 600 ha irrigated area with orange and mango trees (Okeyo 2000). The Tunweni project in Namibia ceased its operation due to a lack of financial support.

The use of earthworms in developing compost cannot be overemphasized as the technology does not only contribute to soil condition improvement but is also used as an income-generating through the sale of both compost and the worms.

Some private companies and individuals not well-known in Namibia produce and sell earthworms. Their advertisements are usually seen online via Facebook. Furlong et al. (2017) reported the market analysis of composting worms in South Africa and India that the Indian market supplied over 70,000 kg of worms per month, while the South African market supplied 3000 kg of worms per month. According to an online report by Furlong (2014), the price of earthworms in South Africa was between US $15–57 per kg, and worms were exported to other countries through two companies Vermi Trade and Full Cycle. Vermi Trade company had exported worms to Zambia, Tanzania, and Nambia, whereas Full Cycle exported worms to Tanzania, Namibia, Botswana, Lesotho, and Uganda.

12.6 Locally Available Wastes that Can Be Converted into Vermicompost

Vermicomposting can be made from a wide range of organic wastes such as municipal (domestic and commercial) and industrial (livestock, food processing, and paper industries) sources (Sinha et al. 2010).

12.6.1 Municipal Organic Wastes

Vermicomposting feedstocks include food waste from homes and restaurants, garden waste (leaves and grass clippings), and sewage sludge (biosolids). However, grass clippings (a high carbon waste) must be correctly combined with nitrogenous wastes. The worms decompose the sludge and convert a significant amount of it into vermicompost (Sinha et al. 2010).

12.6.2 Agriculture and Animal Husbandry Wastes

Farm wastes like crop residues, dry leaves, grasses, and livestock waste like bovine, pig, and poultry make ideal earthworm fodder (Sinha et al. 2010).

12.6.3 Industrial Organic Wastes

These are solid waste (wastewater sludge from the paper pulp and cardboard industries), food processing sectors (breweries and distilleries, vegetable oil manufacturers, potato and maize chip manufacturing, sugarcane, and aromatic oil extraction). Sericulture, forestry, and woodworking waste all provide good feed for earthworm vermicomposting.

12.7 Conclusions

Namibian communal farmers have picked interest in small-scale gardening recently, with horticultural products as the main crops. Direct application of farm manure is not suitable for crops, and inorganic and foliar fertilizer application may be costly. Vermicompost is an excellent technology for waste and land management, improving soil fertility to promote crop productivity in Namibia. It is easy to use, and the feedstocks are readily available. However, there is minimal knowledge about the technology among farmers resulting in low utilization. There is a need to train farmers on the proper use of the technology and increase awareness of the benefits.

References

Adhikary S (2012) Vermicompost, the story of organic gold: a review. Agric Sci 3(7):905–917

Azizi AB, Lim MPM, Noor ZM, Abdullah N (2013) Vermiremoval of heavy metal in sewage sludge by utilizing *Lumbricus rubellus*. Ecotoxicol Environ Saf 90:13–20

Bertram S (1999) Assessment of soils and geomorphology in Central Namibia, issue 71 minor field studies / Swedish University of Agricultural Sciences, International Office, ISSN 1402-3237

Food and Agriculture Organisation of the United Nations (FAO) (2015) World Reference Base for Soil Resources, International soil classification system for naming soils and creating legends for soil maps

Fosso A (2014) Integrated initiatives in support of urban and Peri-urban horticulture in Namibia: project achievements. In: *International Symposium on Urban and Peri-Urban Horticulture in the Century of Cities: Lessons, Challenges, Opportunities 1021*, pp 239–242

Furlong C (2014) Worm farming in South Africa. Assessing the potential to supply composting worms for the humanitarian sector in Africa. https://www.researchgate.net/profile/ClaireFurlong/publication/303702502_Worm_Farming_in_South_Africa/links/574eb8210 8aefc38ba1122db/Worm-Farming-in-South-Africa.pdf

Furlong C, Rajapaksha NS, Butt KR, Gibson WT (2017) Is composting worm availability the main barrier to large-scale adoption of worm-based organic waste processing technologies? J Clean Prod 164:1026–1033

Garrard S, Heyns P, Pfaffenthaler M, Schneider G (2017) Environmental awareness for sustainable development. In: A resource book for Namibia. Solitaire, Republic of Namibia

Goudie A, Viles H (2015) Landscape and landform of Namibia. University of Oxford, Oxford

Katakula AAN, Handura B, Gawanab W, Itanna F, Mupambwa HA (2021) Optimized vermicomposting of a goat manure-vegetable food waste mixture for enhanced nutrient release. Sci Afr 12:e00727

Mendelsohn, J., Jarvis A, Carole Robert C., Robertson T (2002) Atlas of Namibia-A portrait of the Land and its People, New Africa Books (Pty) Ltd, Cape Town.

Ministry of Agriculture Water and Forestry (MAWLR) (2012) Farmer's training presentation on vermiculture and vermicomposting by Sarafia Ashipala. Power point presentation

Ministry of Agriculture Water and Land Reform (2021) Agro-Ecological Zones and Carrying Capacity Maps

Ministry of Environment, Forestry and Tourism (MEFT) (2021) Namibia's climate change adaptation communication to the united nations framework convention on climate change (UNFCCC). First adaptation communication. Republic of Namibia. https://unfccc.int/sites/default/files/resource/namibia-adaptation-communication-to-the-unfccc.pdf

Mupambwa HA, Mnkeni PN (2015) Optimization of fly ash incorporation into cow dung–waste paper mixtures for enhanced vermidegradation and nutrient release. J Environ Qual 44(3):972–981

Mupambwa HA, Mnkeni PN (2016) Eisenia fetida stocking density optimization for enhanced bioconversion of fly ash enriched vermicompost. J Environ Qual 45(3):1087–1095

Mupambwa HA, Mnkeni PNS (2018) Optimizing the vermicomposting of organic wastes amended with inorganic materials for production of nutrient-rich organic fertilizers: a review. Environ Sci Pollut Res 25(11):10577–10595

Mupambwa HA, Ravindran B, Dube E, Lukashe NS, Katakula AA, Mnkeni PN (2020) Some perspectives on Vermicompost utilization in organic agriculture. In: Earthworm assisted remediation of effluents and wastes. Springer, Singapore, pp 299–331

Namibia Agronomic Board (NAB) annual report 2018/2019. https://www.nab.com.na

Namibia's 5th National Development Plan (NDP5), 2017/2018–2021/2022

Okeyo DO (2000) The application of the zero emissions research and initiatives (ZERI) concept in an integrated industry-polyculture-farm system in Namibia: the case of Tunweni sorghum brewery. South Afr J Aquat Sci 25(1):71–75

Rani S, Shukla DV (2014) Potential of earthworms in various fields of life. Journal of International Academic Research for Multidisciplinary 2(4):616–621

Ravindran B, Mnkeni PNS (2016) Bio-optimization of the carbon-to-nitrogen ratio for efficient vermicomposting of chicken manure and waste paper using *Eisenia fetida*. Environ Sci Pollut Res 23(17):16965–16976

Rigourd C, Sape T (1999) Investigating into soil fertility in the North Central regions, Northern Namibia Rural Development Project, Davies, J., 2017. Climate change impacts and adaptation in north-central Namibia. https://idlbncidrc.dspacedirect.org/bitstream/handle/10625/58611/IDL-58611.pdf?sequence=2

Sinha RK, Agarwal S, Chauhan K, Chandran V, Soni BK (2010) Vermiculture technology: reviving the dreams of sir charles Darwin for scientific use of earthworms in sustainable development programs. Technol Invest 01(03):155–172

Chapter 13
Role of Vermicompost in Organic Vegetable Production Under Resource-Constrained Famers in Zimbabwe

Cosmas Parwada and Justin Chipomho

Abstract Organic farming has emerged as a sustainable alternative strategy for achieving food security in Zimbabwe because the system creates a sustainable high farm output with minimal reliance on off-farm inputs. Contrary to industrialized agriculture, organic farming does not have negative impacts on people's health, ecosystem stability and immensely reduces climate change. Moreover, there is a significant increase in demand for organically produced vegetables because they contain zero chemical residues as no synthetic chemicals are used. Generally, crop yield under organic management is relatively lower per unit area compared to the non-organic production system. However, the yield under organic farming can be improved by enhancing the efficacy of the organic manure though a challenge. Organic manures from composts, leaf litter, cattle, goat, and chicken are usually used for soil fertility management in communal areas of Zimbabwe. Unfortunately, most of the available livestock manures are of poor quality (C:N > 35) caused by poor quality fodder in the communal rangelands. Poor quality manure has low efficacy, thereby reducing soil and crop productivity. The low crop productivity can scare-out farmers from organic farming regardless of associated socio-economic and environmental benefits. It is imperative to devise technologies to enhance the efficacy of organic manures, e.g., vermicomposting. Vermicomposting turns the organic wastes into high-quality (C:N < 35) nutrient-rich organic fertilizer called vermicompost. Vermicompost is an acceptable source of fertilizer in organic farming as it is cheap and readily available hence a sustainable fertilizer among the resource-constrained farmers. Unfortunately, the role of vermicompost as a vegetable fertilizer is poorly researched in Zimbabwean horticulture so numerous gray areas on its effects on soil and plant quality and application rates in different soils exist.

C. Parwada (✉)
Faculty of Agricultural Sciences, Department of Agricultural Management, Zimbabwe Open University, Hwange, Zimbabwe

J. Chipomho
Faculty of Agricultural Sciences and Technology, Crop Science Department, Marondera University of Agricultural Sciences and Technology, Marondera, Zimbabwe

© The Author(s), under exclusive license to Springer Nature Singapore Pte Ltd. 2023
H. A. Mupambwa et al. (eds.), *Vermicomposting for Sustainable Food Systems in Africa*, Sustainable Agriculture and Food Security,
https://doi.org/10.1007/978-981-19-8080-0_13

Therefore, this chapter assessed use of vermicompost among the resource-constrained organic vegetable producers. The aim is to increase vegetable productivity, thereby achieving food security among resource-constrained farmers in Zimbabwe.

Keywords Crop productivity · Nutrient concentration · Organic agriculture · Soil fertility · Waste management

13.1 Introduction

Zimbabwe's agriculture is highly industrialized and characterized by heavy use of off-farm inputs like inorganic fertilizers and synthetic pesticides and herbicides as well as poor cropping systems such as monocropping (Svotwa et al. 2008). This has negative impacts on people's health and the environment. They also accelerate the changing of climate which is a menace to many agricultural activities. Besides, these off-farm inputs are expensive and beyond reach to most smallholder communal farmers and this reduces productivity causing acute food shortages. The smallholder farmers account for up to >70% of the farmers in Zimbabwe so their productivity is very critical to national food security. However, productivity among these farmers has always been low even with the introduction of high input-based production systems, the inputs are expensive to purchase. The communal farmers are resource-constrained hence depend on multinationals for seed, fertilizers, and chemicals inputs. These production systems have cost the country's food security and led to loss of livelihoods (Parwada et al. 2022). Sustainable production systems that rely on readily available non-inorganic fertilizer sources are needed to mitigate the impacts of the inorganic fertilizers. Organic farming systems alternatively reduce dependency on harmful chemicals, promotes healthy people and enhances heath of the ecosystem. Therefore, sustaining livelihood even to the resource-constrained farmers who cannot afford the chemicals. The organic farming is advocated by environmentally sensitive people both in the developed and developing nations as it is ecologically, economically, and socially beneficial. Organic farming encourages the build-up of soil structure, thereby mitigating soil degradation caused by the heavy use of inorganic fertilizer (Svotwa et al. 2008).

High frequency and non-judicious use of inorganic fertilizers degrade the environment, cause food pollution, reduce soil quality and overall result in loss of agricultural biodiversity (Savci 2012). Overreliance on inorganic fertilizers has reduced soil value that ultimately lowers the horticultural crop yield among the communal farmers in Zimbabwe. A large proportion (70%) of soils in the communal areas are characteristically sandy soils of granitic origin and naturally infertile, low pH (<4.5), low nutrient holding capacity because of low clay and organic matter content (< 2% soil organic carbon) (Parwada et al. 2022). Vegetable productivity in such sandy soils is generally low without fertilizer or manure, hence proper fertilization can increase productivity (Abdulraheem et al. 2021).

Application of inorganic fertilizers is a common practice to increase crop yield in Zimbabwe (Svotwa et al. 2008). Nevertheless, there are growing concerns around this crop production system because it feeds the crop rather than feeding the soil and this degrades the soil (Savci 2012). Paradoxically, many farmers are resource-constrained so cannot afford to purchase off-farm inorganic fertilizers. Resultantly they grow crops without or underapply the inorganic fertilizers leading to soil degradation through nutrient mining. However, there are many cheap and readily available sources of fertilizers at the farmers' disposal. On-farm soil organic amendments include among others cattle, poultry, and goat manures and these are affordable sources of plant nutrients among the resource-constrained farmers. However, the chemical composition of livestock manure is generally of poor quality (C: N > 35). This is mainly caused by degraded, poor quality rangelands which affects the feed quality and quantity (Fudzagbo and Abdulraheem 2020) and the cattle breed (Sørensen et al. 2003). The amount of nitrogen taken through feed influences the ratio at which the N is excreted via either dung or urine (metabolized N, mostly urea and NH_4+) (Chadwick et al. 2018). If the animals are fed on a low-nitrogen content diet, then a larger proportion of N is excreted via dung and the proportion of N excreted via urine will decrease. Grazing areas in the communal lands of Zimbabwe are characterized by poor quality forage grass and trees resulting in low efficacy manure (Svotwa et al. 2008). The mean fecal-N: urinary-N ratio of cattle is 66:34 in poor rangelands (Fudzagbo and Abdulraheem 2020), compared to ratios ranging from 50:50 to 25:75 in high-quality rangelands (Chadwick et al. 2018). Efficacy of the manure should therefore be enhanced in order to attain high crop production when applied to the soil (Abdulraheem et al. 2021). Use of technologies such as vermicomposting that lowers the C:N < 35 in the low efficacy manure is necessary among farmers. Nevertheless, the vermicomposting technology is still uncommon and given less attention by both farmers and researchers in Zimbabwe.

Vermicomposting is a process of making compost by earthworms that transform low-quality organic waste like agro-wastes, animal manure, and domestic refuse into high-quality (C:N < 35) compost (Habetamu et al. 2021). The vermicomposting of organic wastes is an alternatively a cheaper source of plant nutrients for sustainable vegetable production both at small- and large-scale levels. The worm cast (vermicompost) is a finely divided peat-like material which is excellently structured, highly porous, well aerated and draining and high in moisture-holding capacity (Gopal et al. 2009). Besides, the vermicompost is an effective organic fertilizer rich in nitrogen (N), phosphorus (P) and potassium (K) (NPK), micronutrients and beneficial soil microbes (nitrogen fixing and phosphate solubilizing bacteria and actinomycetes), humic substances and plant growth hormones (Fig. 13.1) (Mupambwa et al. 2017). This makes it a perfect sustainable alternative to inorganic fertilizers in crop production.

Recently, the vermicompost is fast becoming a cornerstone source of fertilizer in organic farming. It is easy to prepare, has excellent inherent properties that improves soil health and is harmless to plants. Some studies showed that application of vermicompost improves nutrient status as well as biological characters of soil (Gopal et al. 2009; Mupambwa et al. 2017). The application of vermicompost to

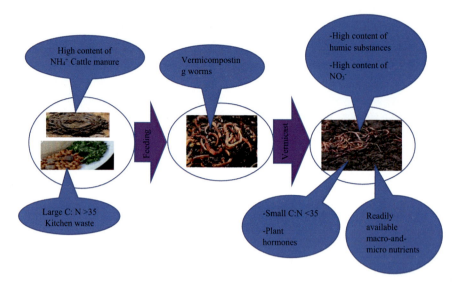

Fig. 13.1 The diagramatic illustration of vermicomposting process

the soil promotes many biological interactions and processes in the soil, thereby creating a complex environment that ensures nutrient availability to the crops. It is agreed that the vermicompost can positively influence the physicochemical and biological properties of the soil as well as raising the soil organic matter content (Lim et al. 2015; Habetamu et al. 2021). However, the suitability of vermicompost as vegetable fertilizer, its influence on soil nutrient holding capacity and plant growth, and mechanisms by which it affects plant yield have received little attention in Zimbabwe. Hence, the objective of this chapter was to review the role of vermicompost in organic vegetable production under the resource-constrained famers. The thrust was to achieve sustainable soil fertility management and constantly high crop productivity for improving livelihoods and achieving food security among the resource-constrained farmers in Zimbabwe.

13.2 Production of Vermicompost

As explained above, vermicompost technology is a biotechnological process of converting organic wastes into compost using specialized earthworms (Fudzagbo and Abdulraheem 2020). Vermicompost is made by vermicomposting any organic waste that is at the farmer's disposal such as food wastes, e.g., fruit and vegetables or livestock manure. Therefore, this type of fertilizer is easily affordable even to the resource-constrained farmers. Cultivated earthworms, e.g., red wigglers like *Eisenia foetida* and white worms are usually kept inside a vermicompost bin and where they are fed with organic wastes during vermicomposting. Inside the earthworm gut, the wastes pass through metabolic activity and are excreted as odorless cast rich in available plant nutrients (macro-and-micronutrients) blended with some plant growth hormones (Fig. 13.1).

The vermicomposting is affected by the environmental conditions (e.g., temperature and pH), worm stocking densities, and type of earthworm species apart from the characteristics of waste. These factors should be monitored throughout the process in order to enhance both quality and quantity of the vermicompost. An ideal worm stocking density was observed to range from 2.0 to 4.0 kg m^{-2} and an optimum of 3.0 kg m^{-2} as far as the feed processing is concerned (Rostami et al. 2009). The vermicast production was observed to vary from 0.20–0.29 g-cast/g-worm/day with the *E. foetida* during the vermicomposting of solid waste, this translated to approximately 0.870 kg-cast/3 kg/m^2-worm/day when a production rate of 0.29-cast/g-worm/day is used. These production rates are low considering that farmers may need large amounts of the vermicompost to apply on their large production areas. The vermicompost production rates could be a limiting factor in using vermicompost as a fertilizer. Vermicompost production areas larger than 1 m^2 are therefore required to increase the vermicompost production per day.

13.3 Properties of Vermicompost

Quality of the vermicompost is highly dependent on the origin so the vermicompost differs chemically (Mupambwa et al. 2017). However, vermicompost from the same waste origin was observed to have the same intrinsic chemical characteristics (Habetamu et al. 2021). Vermicomposting can reduce pH towards neutral which is an important factor in influencing nitrogen retention in the vermicast. The low pH associated with vermicomposting could be due to the production of CO_2 and organic acids by microbial activity during the bioconversion of the different substrates in the beds (Lim et al. 2015). Plus, the high solubility of nutrients in earthworm casts can increase pH of cast. Rostami et al. (2009) reported high 7.37% N and 19.58% P_2O_5 in worm's vermicast and lower values of the mineral nutrients in the conventional cattle manure. Sinha et al. (2010) asserts that vermicompost is at least 4 times more nutritious than ordinary cattle manure.

Electrical conductivity (EC) which indicates the concentrations of soluble salts content was also higher in the vermicompost than in the conventional manure (Table 13.1.). The increase in EC with vermicomposting might be due to the presence of exchangeable calcium, magnesium, and potassium in worm cast unlike in the conventional manure (Anoop and Garg 2011). However, the level of organic carbon (OC) content in the vermicompost was reduced compared to the original conventional manure (Table 13.1.). Microorganisms utilize carbon as their energy source and nitrogen for growth during the decomposition of the organic wastes in which carbon is lost as CO_2. The CO_2 is also lost during the respiration of earthworms combustion of carbon into CO_2 and further reduces the OC content (Table 13.1.).

Table 13.1 Comparative of nutrient composition of vermicompost and conventional cattle manure

Nutrient	Vermicompost	Conventional manure
Total N (%)	1.6 ± 0.2	0.5 ± 0.2
Total C (%)	28 ± 3.3	36 ± 3.3
C:N ratio	$1:18 \pm 0.5$	$1:72 \pm 0.5$
Total P (%)	0.7 ± 0.1	0.2 ± 0.1
K (%)	0.8 ± 0.2	0.5 ± 0.2
Ca (%)	0.9 ± 0.3	0.5 ± 0.3
Mg (%)	0.2 ± 0.01	0.1 ± 0.01
Fe (ppm)	175.0 ± 10.2	146.5 ± 10.2
Cu (ppm)	5.0 ± 1.1	2.8 ± 1.1
Zn (ppm)	24.5 ± 0.2	14.5 ± 0.2
Mn (ppm)	96.6 ± 7.5	69.0 ± 7.5
pH (H_2O)	6.79 ± 1.4	5.43 ± 1.4
EC (dSm^{-1})	1.2 ± 0.3	0.7 ± 0.3

EC, electrical conductivity (2) Data are means $\pm$ standard error of the means for three replicates
Source: Sinha et al. 2010

The excreta of earthworms have more nitrogen compared to the conventional organic manure. The raised nitrogen content in the vermicompost could be due to the addition of muco-proteins secreted from the body wall of the earthworms (Mupambwa et al. 2017). Abdulraheem et al. (2021) also confirmed the presence of the muco-proteins secreted from the body surface in the drilosphere. In a different study, Rostami et al. (2010) noted that earthworms can increase nitrogen availability by reducing microbial immobilization, thereby lowering the C:N ratio (Table 13.1.). The C:N ratio is significant in composting because microorganisms need a good balance of carbon and nitrogen (ranging from 25 to 35) in order to remain active. The lower the C:N ratio, the more rapidly nitrogen will be released into the soil for immediate crop use (Lim et al. 2015). A C:N ratio > 35 results in microbial immobilization and a ratio of 20–30 is required to achieve an equilibrium state between mineralization and immobilization (Rostami et al. 2010).

Phosphorus compounds in vermicast are higher than in the conventional manure. Sinha et al. (2010) reported that 15–30% more phosphorus was made available to plants by vermicomposting. Hanc and Pliva (2013) found that water-soluble potash of convectional manure increased by >8.4 times when vermicomposted compared to conventional cattle manure. Vermicomposting increased the calcium content because of the excretion of calcium from the calciferous glands in the earthworm gut (Mupambwa et al. 2017). Anoop and Garg (2011) suggested that the chemical changes that occur in the gut of earthworms enabled various metals more readily available to plants and mineralization of dead earthworms would release accumulated heavy metals into the soil. Hanc and Pliva, (2013) observed availability of metals such as Pb, Cu, Cd, Zn, Mn, and Ca in the tissues of earthworm which are embedded into the vermicompost, thereby increasing their concentrations in the excreta (Gopal et al. 2009). The ionic regulatory mechanism in earthworms involves uptake of Fe, Mn, and Na from ingesta and its excretion via calciferous glands

13.4 Effects of Vermicompost on Soil Properties and Plant Nutrients

Just like the conventional manure, the vermicompost is an important source of organic matter. Soil organic matter (SOM) is an indicator of soil health hence a sustainable form of fertilizer for sustainable agricultural production. Physically, vermicompost supplemented soils have better aeration, porosity, lower bulk density, and higher water retention capacity (Gopal et al. 2009). Vermicompost improves the soil porosity, aeration, drainage, water holding capacity and enhances soil microbial activity (Sinha et al. 2010). Gopal et al. (2009) noted 80% increase in hydraulic conductivity and six-fold increase in water infiltration following vermicompost application in various soils. This indicated that the vermicompost can be used to mitigate the rampant soil degradation in the communal areas of Zimbabwe as it promotes soil aggregation thereby confers resistance to erosion (Parwada and Van Tol 2018).

The application of vermicompost increases crop growth and yields threefold as compared to the conventional manure (Sinha et al. 2010). This is due to the increased availability and high solubility of nutrients and the presence of plant growth hormones in the vermicompost. Generally, the application of vermicompost modifies the physio-chemical and biological properties of the soil and increases soil biodiversity, thereby reducing environmental risks associated with inorganic fertilizers (Lim et al. 2015). This clearly shows that the vermicompost has other indirect benefits on soil properties, e.g., increased water holding capacity therefore can enhance crop productivity in areas that receive low rainfall.

Soil chemical properties such as pH, electrical conductivity, OM, and nutrient status improved significantly and led to better plant growth and yield owing to vermicompost application (Anoop and Garg 2011). Moreover, the vermicompost can supply plant hormones, enzymes, and vitamins to the soil, thereby improving soil health and crop productivity. Many researchers agree that the conventional compost has higher ammonium content while the vermicompost is higher in nitrates which is the more available form of nitrogen to promote better plant growth performance and yield. Chadwick et al. (2018) noted that the vermicompost generally had higher N availability than the conventional compost on a weight basis and the supply of several other nutrient, e.g., P, K, S, and Mg, was significantly increased by addition of vermicompost as compared to the conventional manure to soil.

The vermicompost has the ability to supply nutrients for a long time while the conventional manure cannot release the required amounts of macro-and-micronutrients to plant in a shorter time after application (Hait and Tare 2011). The applied conventional manure has to decompose first in order to provide a significant quantity of several important plant nutrients. Part of the N from organic matter is converted into plant-available mineral forms such as NH_4^+ and NO_3^- through the process of mineralization (Hanc and Pliva 2013). However, the timing and amount of mineralization often do not coincide with crop need, making in-season fertilization necessary. This lack of synchrony between nitrogen mineralized from the conventional organic manure and crop nitrogen uptake is a major challenge for soil fertility management in organic systems. Although not widely researched on Chadwick et al. (2018)) reported that microorganisms in the worm casts are thought to fix atmospheric N in such quantities that are significant for the earthworm metabolism and as a source of nitrogen for plant growth. The application of vermicompost therefore can modify the physical, chemical, and biological properties of the soil. However, the ideal application rates in different soil types are still unclear.

13.5 Effects of Vermicompost on Organic Vegetable Productivity

Due to its positive effects on the soil, the vermicompost can play a major role in improving growth performance and increase both the quantity and quality yield components of different field crops, vegetables, and flowers. There is a growing demand for organic foods globally, as more people are becoming health conscious and looking for products that have no residual chemicals (Gopal et al. 2009). The consumers are becoming more aware of foods containing synthetic pesticides and fertilizers and are switching to organic foods, which is driving the market growth of organic farming. Further to this, it is generally agreed that organic farming offers more opportunities to developing countries and improves livelihoods, particularly of rural communities. This is because organic farming encourages the use of locally adapted management practices and discourages the reliance on off-farm inputs. The local soil fertility management practices are cheaper and affordable to most farmers therefore increased food productivity. In focusing on organic produce, there is potential to integrate smallholder farmers and rural communities into national exports.

Organic farming is generally understood to be an agricultural method that aims to produce food using natural substances and processes. It is a holistic production management system which promotes and enhances agro-ecosystem health, including biodiversity, biological cycles, and soil biological activity (Abdulraheem et al. 2021). However, in organic farming the use of these methods only is not enough for export markets but the process should be certified to be considered organic. Organic certification is a procedure for verifying that the production process conforms to certain standards. In other words, certification is primarily an acknowledgement that these products have been produced according to organic production standards, e.g., the soil fertility management practices.

13 Role of Vermicompost in Organic Vegetable Production...

Table 13.2 The growth performance of baby spinach (*Spinacia oleracea* L.) under vermicomposted and raw goat manure

Growth parameter	Vermicompost	Goat manure
Days to emergency	3 ± 1.0	5 ± 1.0
Plant height at 2 weeks after planting (cm)	5 ± 0.8	4 ± 0.8
Plant height at 4 weeks after planting (cm)	10.4 ± 2.3	7.9 ± 2.3
Leaf area index at 4WAP ($cm^{-2} g^{-1}$)	8.78 ± 1.5	5.75 ± 1.5
Net assimilation rate at 4WAP ($gg^{-1} day^{-1}$)	0.09 ± 0.03	0.06 ± 0.03
Total dry matter production at 4WAP ($t\ ha^{-1}$)	49.1 ± 7.4	34.6 ± 7.4

Data are means $\pm$ standard error of the means for three replicates. WAP—weeks after planting
Source: Parwada et al. (2020)

Considering that it is a high-grade nutrient-rich organic fertilizer, vermicompost is one of the advocated fertilizers in organic farming. In some studies, farmers were able to increase their crop yields drastically with highly nutrient soil from vermicompost (Hait and Tare 2011). Vermicompost fertilizer contains readily available macronutrients and micronutrients needed for plant growth. Usually, the available macronutrient should be in the range of N (0.1–0.5%), P (0.08–0.5%), K (1.5–3.0%) for normal plant growth and can be wholly supplied by vermicompost (Gajalakshmi et al. 2002). Researchers have reported that vermicompost can increase the solubility nutrients in soil, thereby promoting high plant growth performance and yield (Sinha et al. 2010; Chadwick et al. 2018). Available phosphorus improves vegetable quality, increased root growth, and reduction in crop maturity time (Anoop and Garg 2011). Some macronutrients such as Ca and Mg are used in smaller amounts by the plants though equally important to profitable vegetable production as the N, P, and K. The exchangeable Ca in soils required for ideal growth and performance in most vegetables ranges from 12% to 75% of CEC and range from 4–20% of CEC for the exchangeable Mg (Abdulraheem et al. 2021). The Mg is involved in numerous physiological and biochemical processes activating more enzymes than any other mineral nutrient, thus making a significant contribution to vegetable growth and development. Calcium has also a wide range of roles in physiological and biochemical processes within plants that includes the adjustments of ethylene responses in plants, fruit ripening, flower senescence, and flower abscission (Anoop and Garg 2011). Parwada et al. (2020) observed and enhanced growth performance of baby spinach with the addition of vermicompost goat manure compared to the conventional goat manure (Table 13.2).

There was a 40% reduction in days to emergency and 41.9% increase in total dry matter production at 4WAP of the baby spinach under vermicompost compared to the conventional goat manure (Table 13.2). These observed differences in growth performance could be due to the increased concentration of the available plant nutrients. The micronutrients such as Fe, Cu, Zn, Mn, Mo, and B are used in the formation of chlorophyll, cell division and growth, carbohydrate formation, as well as the maintenance of the plant's enzyme system (Fudzagbo and Abdulraheem 2020). The vermicompost can supply these nutrients in adequacy, therefore an ideal fertilizer for use in organic crop production. Vermicompost has a large surface area which offers several micro sites for nutrient retention and exchange and microbial activity (Anoop and Garg 2011). Hence rich in microbial populations and diversity particularly fungi, bacteria, and actinomycetes (Mupambwa et al. 2017). Application of vermicompost in soil can build up OC, raise the nutrient status, enhance cation exchange capacity, microbial activities, microbial biomass carbon and enzymatic activities all contributing to high plant growth performance and yield.

Studies have revealed that vegetables respond to vermicompost application like hormonal induced activity because it contains high levels of nutrients, humic acids, and humates (Gajalakshmi et al. 2002). Other researchers noted an increase in vegetable growth following vermicompost application regardless of already receiving optimal nutrition (Hanc and Pliva 2013; Lim et al. 2015). The vermicompost also positively influenced the vegetative growth, stimulation of shoot growth and root development in vegetables (Azarmi et al. 2008). Application of vermicompost was used to alter the morphology of vegetables such as increased leaf area and root branching, stimulation of flowering, raised the biomass of flowers ultimately increased fruit yield (Federico et al. 2007; Eswaran and Mariselvi 2016; Abdulraheem et al. 2021).

The vermicompost enhanced seed emergency by 10% in *Vigna radiate* L. (Bhattacharya et al. 2019) and by 40% in *Spinacia oleracea* L. (Parwada et al. 2020) enhanced seedling growth and increased vegetable productivity in general (Eswaran and Mariselvi 2016). Bhattacharya et al. (2019) noted, earlier and improved emergency in vermicompost treated soil than soil without vermicompost. Numerous studies pointed to slow and gradual release of N through vermicompost which increased the concentration of carotene in a wide range of fruits and vegetables, e.g., carrots (Azarmi et al. 2008). Federico et al. (2007)) concluded that the increased N, P, and K content in tomato leaves was caused by foliar application of a liquid extract from vermicompost. Yield and fruit quality increase in tomatoes and were positively correlated to uptake of macronutrients N, P, and K from vermicompost that increased chlorophyll production in the leaves (Abdulraheem et al. 2021).

Gajalakshmi et al. (2002) observed higher content of Ca and vitamin C but less Fe in tomato fruits grown on vermicompost compared with those under inorganic fertilization. Application of vermicompost to soil also led to increased concentrations of P, Ca, Mg, Cu, Zn, and Mn in shoot tissues of red clover and cucumber. Vermicompost increased stem length by 11 mm and diameter by 40 mm in tomato plants even at rates as low as 25% (Federico et al. 2007). An increase in leaf numbers, shoot lengths, shoot and root dry mass of tomato seedlings was noted with 100% application rate of pig manure vermicompost (Azarmi et al. 2008). Other studies have reported that vermicompost contains certain growth promoting hormones such as auxins, cytokinins, and gibberellins, which are secreted by earthworms (Bhattacharya et al. 2019). Many researchers evidenced plant growth promotion following vermicompost inclusion in vegetable production (Mupambwa et al. 2017; Fudzagbo and Abdulraheem 2020), suggesting that application of hormones rich vermicompost into the soil results in better plant growth performance and overall development of vegetables.

13.6 Possible Challenges of Vermicompost Use in Organic Vegetable Production

Vermicomposting is regarded as a clean, sustainable, and zero-waste approach to manage organic wastes but there are still some constraints in popularization of the vermicompost. For instance, this technology has not yet been diffused across the smallholder farmers in Zimbabwe hence limited knowledge about the technology. In organic farming, there is often an assumption that organic equates to safer but current results from one of a registered organic insecticide Spinosad concludes this is not the case (Viani 2022). The Spinosad is a registered organic insecticide for use in over 80 countries but was recently found to possess a far greater risk to beneficial insects than previously thought (Martelli et al. 2022). Very low concentrations of this organic insecticide caused profound effects on beneficial insect species, including vision loss and neurodegeneration in some *Drosophila* species. These new findings suggest that there could be a possibility of generalizing the safety of vermicompost as an organic fertilizer. There could be some unknown negative effects of vermicompost both to the soil and plants which may be more harmful compared to the inorganic fertilizers. Therefore, to clear such doubts, more rigorous research related to its effects on soil and plant quality are required. In addition, there is limited practical application of vermicompost in vegetable production and this requires more attention too.

Vermicompost technology will reduce production costs in horticulture and its use has increased in recent years due to high costs of inorganic fertilizers. The vermicompost could be an effective substitute of inorganic fertilizers in vegetable production. Nevertheless, there is lack of awareness and proper knowledge regarding vermicomposting and the use of vermicompost is limiting its practical use among the smallholder farmers. Therefore, it is necessary to guide farmers about vermicomposting and its appropriate use through various training and extension activities. In Zimbabwe, organizations like the Fambidzanai Permaculture Center (FPC) and the Zimbabwe Organic Producer Promotion Association (ZOPPA) are actively training the farmers on organic farming but more research on the production and use of the vermicompost is still needed. Innovative and effective use of vermicompost technologies must be developed and translated to the organic farmers. The vermicomposting procedure is labor intensive and time consuming and hence requires fastidious management so many farmers may avoid it. It also requires close monitoring since the worms are sensitive to pH, temperature, and moisture content which must be controlled during the entire process which can be challenging to most smallholder farmers.

13.7 Conclusion

Application of vermicompost to the soil can "feed the soil" resulting in a healthy soil unlike when inorganic fertilizers are used. These organic soil amendments are cheap and readily available and hence can increase vegetable quantity and quality among the resource-constrained smallholder farmers. Moreover vermicompost, a chemically and biologically rich fertilizer can increase the cation exchange capacity, NPK content, add plant growth hormones, humic acids, promote soil microbial activity, and reduce root pathogens. This overall improves plant growth performance and yield. Organic soil fertility management is a prerequisite for organic vegetable production, therefore vermicompost can be advocated for use in organic vegetable production. However, there are numerous challenges that can limit the use of vermicompost among the smallholder vegetable producers. The challenges range from the vermicompost production to use, hence the need to intensify research, e.g., to establish the optimum application rates in various soils. There is also a need to intensify confirmatory research to ascertain the safety of the vermicompost on both the soil and plants. Generally, the practical use of vermicompost as a fertilizer is still low and farmer training on the production and use of the vermicompost are also paramount among the smallholder vegetable producers in Zimbabwe.

References

Abdulraheem MI, Charles EF, Fudzagbo J (2021) Amaranthus performance as influence by integrated application of goat manure and urea fertilizer. J Soil Sci Plant Physiol 3(1):1–4. https://doi.org/10.36266/JSSPP/134

Anoop Y, Garg VK (2011) Vermicomposting – an effective tool for the management of invasive weed Parthenium hysterophorus. Bioresour Technol 102(10):5891–5895. https://doi.org/10.1016/j.biortech.2011.02.062

Azarmi R, Ziveh PS, Satari MR (2008) Effect of vermicompost on growth, yield and nutrition status of tomato (*Lycopersicum esculentum*). Pak J Biiol Sci 11(14):1797–1802. https://doi.org/10.3923/pjbs.2008.1797.1802

Bhattacharya S, Debnath S, Debnath S, Saha AK (2019) Effects of vermicompost and urea on the seed germination and growth parameters of *Vigna mungo* L. and *Vigna radiata* L. Wilzek. J Appl Natl Sci 11(2):321–326. https://doi.org/10.31018/jans.v11i2.205

Chadwick DR, Cardenas LM, Dhanoa MS, Donovan N, Misselbrook T, Williams JR, Thorman RE, McGeough KL, Watson CJ, Bell M, Anthony SG, Rees RM (2018) The contribution of cattle urine and dung to nitrous oxide emissions: quantification of country specific emission factors and implications for national inventories. Sci Total Environ 635:607–617. https://doi.org/10.1016/j.scitotenv.2018.04.152

Eswaran N, Mariselvi S (2016) Efficacy of Vermicompost on growth and yield parameters of *Lycopersicum esculentum* (Tomato). Int J Sci Res Publ 6(1):95–108

Federico A, Borraz JS, Molina JAM, Nafate CC, Archila MA, Oliva LM (2007) Vermicompost as a soil supplement to improve growth, yield and fruit quality of tomato (*Lycopersicum esculentum*). Bioresour Technol 98(15):2781–2786

Fudzagbo T, Abdulraheem MI (2020) Earthworms strongly modify microbial biomass and activity triggering enzymatic activities during vermicomposting independently of the application rates of pig slurry. Sci Total Environ 385:252–261

Gajalakshmi S, Ramasamy EV, Abbasi SA (2002) High-rate composting –vermicomposting of water hyacinth (*Eichhornia crassipes*, Mart Solms). Bioresour Technol 83:235–239

Gopal M, Gupta A, Sunil E, Thomas VG (2009) Amplification of plant beneficial microbial communities during conversion of coconut leaf substrate to vermicompost by *Eudrilus* sp. Curr Microbiol 59:15–20

Habetamu G, Gebreslassie H, Hirut B (2021) Evaluation of earthworm multiplication bedding materials for effective Vermicompost production at Jimma, Southwestern Ethiopia. Sci Develop 2:57–61. https://doi.org/10.11648/j.scidev.20210203.15

Hait S, Tare V (2011) Optimizing vermistabilization of waste activated sludge using vermicompost as bulking material. Waste Manag 31:502–511

Hanc A, Pliva P (2013) Vermicomposting technology as a tool for nutrient recovery from kitchen bio-waste. J Mater Cycles Waste Manage 15:431–439

Lim SL, Wu TY, Lim PN, Shak KP (2015) The use of vermicompost in organic farming: overview, effects on soil and economics. J Sci Food Agric. 95(6):1143–1156. https://doi.org/10.1002/jsfa.6849. Epub 2014 Aug 26. PMID: 25130895

Martelli F, Hernandes NH, Zuo Z, Wang J, Wong C, Karagas NE, Roessner U, Rupasinghe T, Robin C, Venkatachalam K, Perry T, Batterham P, Bellen HJ (2022) Low doses of the organic insecticide spinosad trigger lysosomal defects, elevated ROS, lipid dysregulation, and neurodegeneration in flies. eLife 11:e73812. https://doi.org/10.7554/eLife.73812

Mupambwa HA, Ncoyi K, Mnkeni PNS (2017) Potential of chicken manure vermicompost as a substitute for pine bark based growing media for vegetables. Int J Agric Biol 19:1007–1011

Parwada C, Chigiya V, Ngezimana W, Chipomho J (2020) Growth and performance of baby spinach (*Spinacia oleraceae* L.) grown under different Organic Fertilizers. *Inter. J. Agron.* 2020, Article ID 8843906, 6 pages https://doi.org/10.1155/2020/8843906

Parwada C, Chipomho J, Mandumbu R (2022) In-field soil conservation practices and crop productivity in marginalized farming areas of Zimbabwe. In: Mupambwa HA, Nciizah AD, Nyambo P, Muchara B, Gabriel NN (eds) Food security for African smallholder farmers. Sustainability sciences in Asia and Africa. Springer, Singapore. https://doi.org/10.1007/978-981-16-6771-8_5

Parwada C, Van Tol J (2018) Effects of litter quality on macro aggregates reformation and soil stability in different soil horizons. Environ Dev and Sustain 21:1321–1339

Rostami R, Nabaei A, Eslami A (2009) Survey of optimal temperature and moisture for worms' growth and operating vermicompost production of food wastes. Health Environ 1(2):105–112

Rostami R, Nabaei A, Eslami A, Najafi Saleh H (2010) Survey of E. foetida population on pH, C/N ratio and process's rate in vermicompost production process from food wastes. Ofogh-e-Danesh 35(52):93–98

Savci S (2012) Investigation of the effects of chemical fertilizers on environment. APCBEE Procedia 1(2012):289–292

Sinha R, Agarwal S, Chauhan K, Valani D (2010) The wonders of earthworms & its vermicompost in farm production: Charles Darwin's 'friends of farmers', with potential to replace destructive chemical fertilizers. Agric Sci 1:76–94. https://doi.org/10.4236/as.2010.12011

Sørensen P, Weisbjerg M, Lund P (2003) Dietary effects on the composition and plant utilization of nitrogen in dairy cattle manure. J Agric Sci 141(1):79–91. https://doi.org/10.1017/S0021859603003368

Svotwa E, Baipai R, Gwatibaya S, Tsvere M, Jiyane S (2008) Socioeconomic trends and constraints in organic farming in the smallholder farming sector of Zimbabwe. J Sustain Dev in Africa 10:214–228

Viani A (2022) Organic insecticides more damaging to non-target insects than synthetic counterparts (2022, February 24) Retrieved 28 February 2022 from https://phys.org/news/2022-02-insecticides-non target-insects-synthetic-counterparts.html

Chapter 14
Co-application of Vermicompost with Other Amendments for the Improvement of Infertile/Degraded Soils

Akinson Tumbure, Charity Pisa, and Pardon Muchaonyerwa

Abstract Soil degradation and fertility decline are serious challenges to sustainable agriculture in Sub-Saharan Africa. While organic residue application is a good practice to replenish soil nutrients and improve soil quality, the margins by which soil attributes change per season are usually low unless huge quantities are applied. On the other hand, different types of organic amendments, including vermicompost, can only optimally supply a limited range of soil nutrients. Co-application of vermicompost with other soil amendments provides a good opportunity for African farmers to address these challenges and improve the productivity of their soils. This chapter explores the changes in soil properties after co-application of vermicompost with other amendments. The benefits and limitations of co-applying vermicompost with mineral fertilisers, biochar, inorganic minerals and microbial inoculants are discussed. The paper reveals that there are more benefits than limitations in co-application as the practice provides greater soil fertility compared to sole applications of vermicompost or other amendments.

Keywords Soil amendments · Co-application · Biochar · Microbial inoculants · Soil fertility · Smallholder farmers

14.1 Introduction

Soil degradation, including soil erosion, has serious negative effects on soil, water and air quality, with erosion causing loss of soil particles and associated nutrients and organic matter. Poor soil fertility and inappropriate management strategies result

---

A. Tumbure (✉) · C. Pisa
Faculty of Earth and Environmental Sciences, Marondera University of Agricultural Sciences and Technology, Marondera, Zimbabwe

P. Muchaonyerwa
School of Agricultural, Earth and Environmental Sciences, University of KwaZulu-Natal, Scottsville, Pietermaritzburg, South Africa

© The Author(s), under exclusive license to Springer Nature Singapore Pte Ltd. 2023
H. A. Mupambwa et al. (eds.), *Vermicomposting for Sustainable Food Systems in Africa*, Sustainable Agriculture and Food Security,
https://doi.org/10.1007/978-981-19-8080-0_14

in low agricultural productivity, food insecurity in Sub-Saharan Africa (Zingore et al. 2015). Improved management strategies, including the use of inorganic fertilisers, are required in order to improve crop productivity (ten Berge et al. 2019). While these fertilisers make nutrients available, they are often too expensive for the majority farmers, especially those in the smallholder sector, in addition to their low contribution to soil organic matter. There is a drive to encourage farmers to apply organic amendments for improved and environmentally sustainable agricultural production, globally.

The leading motivation for application of organic material has been to improve or maintain soil fertility and agricultural sustainability (Demiraj et al. 2018; Nair et al. 2017; Pisa et al. 2020). This is because, use of organic materials can have several effects that lead to conditioning the soil, improving soil water-holding capacity, soil structure and improving availability of soil nutrients. Effects of organic materials on plant growth and yield are, however, highly variable (Libutti et al. 2020) and the variability is largely attributed to the quality and quantity of materials applied and the nature of soils used. Vermicompost, the end product of the decomposition of organic waste by earthworms and mesophilic microbes, is receiving increased attention because of its potential to improve soil structure and plant nutrients (Abduli et al. 2013; Blouin et al. 2013), and to suppress pathogens (Jack 2010).

During vermicomposting, the organic wastes pass through the gut of earth worms, where their composition is changed and the availability of both macronutrients, including N and P, and micro-nutrients, is increased while the C/N ratio is reduced (Atiyeh et al. 2001; Messiga et al. 2022). Vermicompost is rich in fungi, bacteria and actinomycetes, enzymes and hormones that enhance plant growth and reduce plant diseases (Bhat et al. 2018; Hussain et al. 2017). Microbes in the vermicompost produce plant growth hormones such as auxins, cytokinins and humic substances (Atiyeh et al. 2001; Muscolo et al. 1999), which play a major role in plant growth and development. A number of researchers have reported stimulation of soil microbial growth and activity on application of vermicompost and consequently improved mineralisation of soil plant nutrients.

Atiyeh et al. (2001) reported greater dehydrogenase enzyme activity in vermicompost when compared to commercial medium. In low input agricultural systems, vermicompost has been used to substitute inorganic fertilisers (Bhat et al. 2018). However, the composition of vermicompost depends on the feedstocks and vermicomposting conditions used, which affect the proportions of available nutrients. As such, co-application of vermicompost with other soil amendments could optimise availability of plant essential nutrients and enhance crop productivity. The objective of this chapter is to elucidate the opportunities of improving chemical, physical and biological quality of infertile soils through co-applying vermicompost with other substances.

14.1.1 The General Vermicomposting Process and Typical Vermicompost Characteristics

Vermicompost can be produced from various bio-wastes which include livestock dung, human sewage sludge, plant residues/trimmings and household wastes to name a few. If need be, the bio-wastes are shredded to appropriate sizes, added to containers (called vermi-reactors) and two or more types of wastes can be blended together. A specific number of earthworms (usually *Eisenia fetida*) are added to the vermi-reactors after the feedstock has undergone an initial short-term decomposition period. Moisture content is usually maintained around 70% w/w and the worms are protected from sunlight. The vermi-degradation process is a result of earthworm ingestion of the added biomass which then undergoes rapid degradation under the action of various microorganisms, enzymes and hormones found in the earthworms gut (Chatterjee et al. 2021). This vermi-degradation process generally takes at least 90 days for vermicompost to mature. Examples of typical N, P and organic C content of the final vermicompost that is made from various feedstocks are given in Table 14.1.

Depending on the feedstock, typical concentration of P and N in vermicompost is usually around 0.5 and 1.5%, respectively (Table 14.1). These values mean that typical vermicompost may need to be applied in quantities upwards of 20 t ha^{-1} to supply sufficient nutrients (e.g. P) to a maize crop. There is therefore merit in looking into combinations of vermicompost and other materials that can be used to supply soil nutrients and improve soil quality. However, some external challenges still exist that could hinder the adoption and upscaling of vermicomposting technologies.

Table 14.1 Selected characteristics of vermicompost made from various feedstocks

Feedstock materials	pH	Total organic C (%)	Total N (%)	Total P (%)	References
Rabbit, cow, pig dung, poultry litter	6.4–9.0	12.1–28.7	1.3–2.5	0.22–1.06	Chatterjee et al. (2021)
Cabbage, cauliflower wastes	7.3–7.5	25–30	1.5–1.8	0.93–1.10	Mago et al. (2022)
Textile mill sludge, cow dung, tea waste	7.1–7.6	15.1–36.5	0.25–1.7	0.27–1.20	Badhwar and Singh (2022)
Dairy sludge, sugarcane press mud	7.3–7.9	26.7–30	0.65–0.69	2.69–2.92	Sharma et al. (2022)
Paper sludge, grass clippings, vegetable waste, cow, sheep manure	6.6–8.4	–	1.1–2.0	0.23–0.4	Raimi et al. (2022)

– not reported

14.1.2 Vermicompost Use by Smallholder Farmers in Africa, General Challenges and Upscaling Opportunities

The value of vermicompost for agricultural productivity has been generally appreciated by smallholder farmers. A study by Mainoo et al. (2008) reported that the attitudes of Ghanaian smallholder farmers were generally positive to adoption of vermicomposting technology, although several challenges, such as limited farm resources and lack of vermicomposting knowledge, existed at the time. For instance, Liégui et al. (2021) explains that the main reason why vermicomposting technology has not been fully adopted is that smallholder farmers have poor understanding of the technology in Cameroon. Similarly, low adoption rates are also noted in Ethiopia (Gebrehana et al. 2022), while in Southern Africa, vermicomposting is reportedly done by only a few small-scale farmers (Mugwendere et al. 2015). There is potential for upscaling vermicomposting technologies in African farming systems, which can be approached in two different ways.

The first approach hinges on small-scale on-farm vermicomposting, which is achieved through construction of vermi-reactors at household or farm level using various materials including concrete, plastic, or wood (Shanka 2020). As expected, this approach presents specific farmer challenges, including lack of resources (worms and feedstock), poor knowledge and lack of space (Gebrehana et al. 2022; Liégui et al. 2021; Singh et al. 2020). While many farmers prefer animal wastes as feedstock, sewage sludge and tree litter can also be used to solve the problem of lack of feedstock. When tree litter is used, the final nutritional characteristics of the vermicompost will be influenced by the tree species (Mokgophi et al. 2020). While sewage sludge provides a valuable feedstock for vermicomposting, its handling poses the risk of spreading pathogenic microorganisms pre and post vermi-degradation. This can be solved by using a short-term thermophilic composting (1 week) before vermicomposting, vermifiltering (Furlong et al. 2017) or ammonia sanitisation (Lalander et al. 2015). However, Atanda et al. (2018) reported that the risk of having pathogenic microorganisms in the final vermicompost product is low, based on assessment of 60 vermicompost products from several South African farms.

The second approach is the establishment of vermicomposting companies that centrally collect and process the waste (vermicomposting) and make vermicompost and vermicompost-leachates products available to farmers. This approach has the advantage of being able to tap into endless supplies of municipal and household organic wastes, and that the safety of the vermicompost products can also be maintained and regulated by independent bodies.

14.1.3 The Need to Co-apply Vermicompost with Other Amendments

The prevalence of multi-nutrient deficiencies in soils of sub-Saharan Africa means that one type of organic amendment is usually not enough to correct all the deficiencies, and co-application with another amendment may be required. In Zimbabwe, deficiencies of exchangeable cations in soil have been noted as a result of applications of unfortified organic resources over multiple years (Mtangadura et al. 2017). Organic resources such as vermicompost may have low levels of some nutrients and might need to be applied at high quantities in order to have significant soil effects that translate to improved crop yields. For example, considering that P content can be as low as 0.22% (Table 14.1), to supply 45 kg of P about 20.5 t ha^{-1} of vermicompost would be needed. Also some nutrients present in vermicompost, such as phosphorus, may be largely unavailable, e.g. in the organophosphate form (Parastesh et al. 2019). As such, when vermicompost with considerable mineral N is applied at high rates to meet available P requirements, it results in leaching of N, mainly as NO_3^-, which is detrimental to groundwater quality. High doses of vermicompost have also been reported to increased ammonia (NH_3) and nitrous oxide (N_2O) emissions, which were reduced when the vermicompost was co-applied with biochar (Wu et al. 2019). This suggests that co-application of vermicompost with other amendments may reduce the negative effects of application of vermicompost at high rates. Coupling vermicompost with organophosphate degrading bacteria will be essential in increasing P availability, and therefore reduce the need to apply extremely high rates of the vermicompost.

14.2 Co-applying Vermicompost with Microbial Inoculants

Combining vermicompost and microbial inoculants can synergistically improve soil quality through various chemical and physical pathways. This is true for both microbial inoculants that are added during vermicomposting and those added together with vermicompost at the soil application stage. In soil-related sciences, *Rhizobium, Azospirillum, Bacillus* and *Pseudomonas* are amongst the most researched bacterial genera while *Glomus* is the most researched fungal genus. The *Rhizobium* and *Azospirillum* are important in soil N nutrient cycling due to their ability to biologically fix N (Singh et al. 2021), while *Bacillus* and *Pseudomonas* are essential in breaking down organic and inorganic nutrient sources due to their metabolic diversity. The importance of fungi of the *Glomus* genus cannot be overstated as these form plant-mycorrhizal associations beneficial to plants and are instrumental in the mineralisation of various soil nutrients and uptake of nutrients that would otherwise not be accessible to the crop roots. In addition to supplying nutrients, foliar applications of vermicompost extracts may suppress oomycete pathogens and plant diseases (Jack 2012).

14.2.1 Vermicompost and Arbuscular Mycorrhizal Fungi

Several studies report positive changes in soil properties after co-application of vermicompost and arbuscular mycorrhizal fungi (AM fungi). In a study by Akhzari et al. (2015), AM fungi inoculation with vermicompost significantly lowered soil bulk density compared to when vermicompost only was applied. The lower bulk density is likely to be a result of improvements of soil structure after adding vermicompost and AM fungi. The AM fungi improve soil structure in two major ways (Fig. 14.1).

Firstly, they produce a glycoprotein called glomalin, which aggregates and stabilises soil particles by forming stable bridges with clay (Agnihotri et al. 2021; Benaffari et al. 2022; Hussain et al. 2021). Secondly, the hyphae of AM fungi can also bind soil particles, thereby increasing soil aeration, porosity and soil water-holding capacity (SWHC) (Akhzari et al. 2015). Studies by Benaffari et al. (2022) revealed that AM fungi and vermicompost application resulted in greater glomalin-related soil proteins. An indirect mechanism by which soil structure may also improve after co-application of vermicompost and AM fungi is through the promotion of aquaporin synthesis in plants by AM fungi which enhances water uptake and may result in better root establishment (Benaffari et al. 2022). Better root establishment especially when considering grasses can stabilise soil aggregates (Yavitt et al. 2021). On the other hand, organic carbon from the vermicompost contributes to

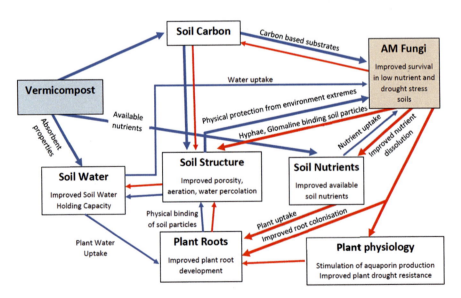

Fig. 14.1 Relationships between vermicompost and arbuscular mycorrhiza (AM fungi) when co-applied in soil and the effects of these relationships on the soil and plant condition. Blue coloured processes are influenced by vermicompost, and red coloured processes are influenced by AM fungi

improved soil structure, soil tilth, soil porosity and enhances root growth (Chaoui et al. 2003).

14.2.2 Vermicompost with Phosphate Solubilising and N Fixing Bacteria Inoculants

Amongst other factors, available nutrient content in vermicompost is greatly influenced by the substrate used in the vermicomposting process (Singh et al. 2021), with some nutrients, though present, being not be available. For example, P could be present in vermicompost largely in the unavailable organophosphate form (Parastesh et al. 2019). In such cases, the dual application of vermicompost with microbial inoculants may promote greater soil nutrient availability compared to the application of vermicompost alone. Studies by Lukashe et al. (2019) reported a complementary influence resulting in improved vermi-degradation and nutrient mineralisation when phosphate solubilising bacteria (PSB) was added during vermicomposting of a mixture of fly ash, cow dung and wastepaper. When added to soil, vermicompost with bacterial inoculants will be able to support a greater variety of microbial processes than when vermicompost alone is added due to improved nutrient availability. The common, drier and poorly fertile soils in Africa limit microbial growth and activity (Tumbure et al. 2022) and co-application of vermicompost and microbial inoculants may significantly improve microbial survival and hence soil microbial diversity.

In a study by Singh et al. (2021), combining *Azospirillum* and vermicompost was shown to improve soil health through favourable membrane activities that bolster microbial and plant tolerance to abiotic stresses. In another study by Khan et al. (2015), the soil organic carbon pool did not change when microbial inoculants were added to vermicompost compared to when only vermicompost was added. However, in the same study, soil available P was significantly higher in treatments that received microbial inoculants consisting of *Pseudomonas monteilii* or *Glomus intraradices* compared to vermicompost addition only. Khosravi et al. (2017) reported that *Pseudomonas fluorescens* produced iron chelating siderophores and solubilised insoluble inorganic and organic phosphates when applied to soil together with vermicompost and phosphate rock. Significant increases in soil available N are also reported after applying vermicompost with PSB + Azotobacter compared to when vermicompost only is applied (Sumit Rai et al. 2014).

Vermicompost and its liquid extract is good at maintaining live bacteria over longer periods of time (Arfarita et al. 2022). This presents one of the major effects by which microorganisms and vermicompost find synergy and beneficial effects of co-application are enhanced. Results from a study by Mal et al. (2021) reported improved available macronutrients and microbial populations when N fixing, and P solubilising bacteria were added to vermicompost. Similar results including an increase in enzymatic activities and plant nutrient uptake were observed by Ghadimi et al. (2021) after the use of an Azotobacter (N fixing) inoculant and vermicompost. Adding vermicompost together with N fixing bacteria has been shown to significantly improve nodulation and nitrogen fixation in legume plants (Mathenge et al. 2019). In a different study, soil organic carbon increases of up to 145, 70 and 128% were observed after addition of sole vermicompost, vermicompost and an N fixing *Bacillus* spp. inoculant and vermicompost and P solubilising *B. megaterium* were used, respectively (Verma et al. 2016). In the same study by Verma et al. (2016), related improvement in microbial populations, soil alkaline phosphatase activity, available nutrients in soil (N, P, K) and an increase in sweet basil yields was also observed. These results are important because they show that using N fixing bacteria with vermicompost has several benefits.

It should also be noted that vermicompost provides its own contribution of beneficial microorganisms too such as N fixing microbes which lead to the fixation of more atmospheric N which in turn enriches the soils' nitrogen status (Lim et al. 2015; Padmavathiamma et al. 2008). This results in improved soil fertility and quality (Gutiérrez-Miceli et al. 2007; Ievinsh 2011; Papathanasiou et al. 2012; Pramanik et al. 2007; Zaller 2007). Co-application of vermicompost with mineral fertilisers has also been shown to have positive effects on crop yields.

14.3   Co-applying Vermicompost and Mineral Fertilisers

The co-application of vermicompost and mineral fertilisers has been seen to improve plant growth and yield. Studies by Moraditochaee et al. (2011) showed that application of $6\,\text{t}\,\text{ha}^{-1}$ vermicompost and $50\text{--}75\,\text{kg}\,\text{N}\,\text{ha}^{-1}$ of inorganic nitrogen fertiliser produced significantly higher eggplant fruit yield ($41.44\,\text{t}\,\text{ha}^{-1}$) compared to the sole application of mineral N and vermicompost. Studies by De Grazia et al. (2006) and Devi et al. (2002) report similar results. Some studies have however reported the co-application of vermicompost, and mineral fertilisers reduced the positive yield effect of vermicompost. For instance, results from a meta-analysis by Blouin et al. (2019) report that combined soil application of mineral fertiliser and vermicompost reduced commercial yield from 38 to 14%. In another study, mineral fertiliser was reported to have significantly reduced the effects of vermicompost when N application exceeded $30\,\text{kg}\,\text{N}\,\text{ha}^{-1}\,\text{year}^{-1}$ (van Groenigen et al. 2014) to the extent that it did not differ from the un-amendment control. This is attributed to the negative effect that the mineral fertiliser has on earthworms (which are introduced to the soil when vermicompost is applied to the soil) which are key organisms in the mineralisation process of SOM. These results indicate that an increase in plant biomass is partially due to the availability of mineral N (and other nutrients) as a result of earthworm activity or vermicompost addition (Aira et al. 2005; Lavelle et al. 1992). This means that the positive effects of earthworms increase with more organic material addition, but high N availability reduces the positive effects of vermicompost.

The co-application of vermicompost with mineral P fertilisers increases P availability. This is because vermicompost reduces P fixation by soil particles as the organic acids in vermicompost compete for the binding sites with orthophosphates and metal complexes. Increased microbial P accumulation also reduces the amount of labile P that eventually get fixed to soil particles. Chiranjeeb et al. (2020) reported increased microbial biomass P after vermicompost plus NPK fertiliser and attributed this to an increase in soil microbial population and reproduction of P accumulating microbes. The accumulation of P by microbes can be beneficial for soil P retention and availability because microbial turnover where P is mineralised and taken up again takes place at a relatively fast rate.

Vermicompost application to low nutrient soils can then be a solution for yield improvement for smallholder farmers who have little or no access to mineral fertilisers. While co-application with mineral fertilisers may improve the effectiveness of vermicompost, some natural minerals may also make a contribution (Table 14.2).

14.4 Co-applying Vermicompost with Natural Minerals

14.4.1 *Vermicompost and Silicates*

Silicate rock minerals are the most abundant mineral class and huge quantities of these in the form of waste rock dust are produced each year from the mining industry (Swoboda et al. 2022). In countries such as India, the material loss as rock dust from silicate rock mining to their final use in various end products is estimated to be 70% (de Souza et al. 2019). In Zimbabwe, silicate minerals such as granite, serpentine and dunite are abundant (Chaumba 2017, 2019; Tumbure et al. 2021). Application of rock powders to agricultural fields has potential to sustainably supply some of the plant nutrient needs. However, nutrient availability from these rocks when unmodified is usually low (Tumbure et al. 2021). Combining rock powder applications with vermicompost may provide a good option to optimise nutrient availability (Table 14.2). The addition of a silica mineral phlogopite to vermicompost was shown to result in the partial weathering of phlogopite to vermiculite, thereby improving the K nutritional value of the vermicompost (Jafari et al. 2021a). Addition of phlogopite enriched vermicompost to soil resulted in further increased rates of K mineralisation as phlogopite was converted to smectite due to root and microbial exudates in the rhizosphere (Jafari et al. 2021b). Other minerals such as the aluminosilicate group of zeolites due to their high porosity and high-water absorption properties have been shown to be effective in alleviating soil water deficit stress when applied with vermicompost (Baghbani-Arani et al. 2021).

Table 14.2 Reported soil physico-chemical changes after co-application of natural minerals and vermicompost

Rock type	Minerals	Application rates	Soil description	Enhanced soil physico-chemical changes	References
Carbonate	Calcite $CaCO_3$	Mixture: multiple mixes of 0, 5.93, 9.57 and 11.96 t lime ha^{-1}, and 0, 5 and 10 t ha^{-1} vermicompost	Nitisol, pH 4.62	Increased labile P and decreased soil P-sorption capacity and was more effective in reducing exchangeable Al and acidity compared to sole applications	Alemu and Melese (2021)
		Mixture: 1:1:2 (w/w) ratio of lime, sepiolite and vermicompost. Mixture applied at 1, 2 and 5% mass of soil	Contaminated calcareous soil	Reduced heavy metal availability (Cd, Pb and Zn), increased availability of soil N, P and K and decreased diversity of soil bacterial and fungal communities	Tan et al. (2022)
Silicate	Phlogopite	Mixture: 4:1 (w/w) ratio of dried cow manure to phlogopite. Mixture applied at 1:6 parts sand (w/w)	Washed quartz sand	Increased soil K availability	Jafari et al. (2021b)
	Steatite	50 t ha^{-1} of vermicompost enriched with steatite powder at 20% (w/w)	Oxisol, pH 5.1	Unfavourably increased soil Cr and Pb availability	de Souza et al. (2019)
	Gneiss powder	Gneiss powder added during vermicomposting at 20% (w/w) Final vermicompost	Oxisol, pH 6.2	Increased available P (Mehlich-1 extractable P), improved K and Ca availability	(Paula De Souza et al. 2018)

(continued)

Table 14.2 (continued)

Rock type	Minerals	Application rates	Soil description	Enhanced soil physico-chemical changes	References
		applied at 20 t ha^{-1}			
	Zeolite, clinoptilolite type	Vermicompost at 2.7 t ha^{-1} Zeolite at 9 t ha^{-1}	Sandy loam, pH 7.35	Improved soil water-holding capacity alleviated plant soil water deficit stress	Baghbani-Arani et al. (2021)
Phosphate rocks (apatite)	Sedimentary carbonate apatite	Rock phosphate at 200 mg kg^{-1} soil Vermicompost at 15 t ha^{-1}	Acid lateritic soils (Aqualfs) pH 5.4	Increased available soil P content more than sole applications Had no effect on available soil N and K	Pramanik et al. (2009)

14.4.2 Vermicompost and Natural Carbonates

Carbonate rich rocks are abundant in many African countries, and they are generally used for correcting soil acidity. Amongst these rocks, calcium carbonate or limestone has the advantage of being a low-cost commodity (Barber 1991). Use of limestone in the majorly acidic soils of the tropics can improve availability of soil P and Mo and reduce Al/Fe toxicity. Limestone can be added during the vermicomposting process or together with vermicompost as a soil co-application. Some studies have reported that adding lime at a rate of 5 g/kg during vermicomposting could be ineffective in improving nutrient content and enzymatic activities of vermicompost (Pramanik et al. 2007). Furthermore, when lime is applied to soil together with vermicompost, several soil quality benefits have also been reported. A study by Alemu and Melese (2021) showed that, while lime alone did not affect available soil P (NH_4Cl extractable P), there was a significant positive effect on available P when combined with vermicompost. The same authors reported that combined soil application of vermicompost and lime was more effective at reducing exchangeable acidity and exchangeable Al and Fe while improving available P. Increase in availability of soil N, P and K and reduction in the bioavailability of soil Cd, Pb and Zn were also reported by Tan et al. (2022) after combined soil application of lime, sepiolite and vermicompost (Table 14.2). While soil P availability is a major limitation to crop production, co-application of vermicompost with rock phosphate could make a substantial contribution.

14.4.3 Vermicompost and Rock Phosphates

Numerous studies have focused on the enrichment of vermicompost with rock phosphates during the vermicomposting process in order to improve its fertiliser value (Busato et al. 2021; Rahbar Shiraz et al. 2019). However, Mupambwa and Mnkeni (2018) highlight that even though positive, increases in available P are noted, the levels are usually much less compared to inorganic fertilisers due to the limited quantity of RP applied. Another limitation results from feedback mechanisms due to accumulation of some available nutrients such as N which limit growth of worms as explained previously and accumulation of RP reaction products such as Ca that is known to greatly limit P dissolution from RP (Tumbure et al. 2022). It is possible that as opposed to addition of RP during vermicomposting, soil co-application of RP with vermicompost during cropping could result in more mineralised nutrients due to plant uptake effects. For example, plant Ca and P uptake would in effect drive further the PR dissolution reaction forward in the presence of vermicompost, thereby enhancing dissolution by low molecular weight organic acids excreted by bacteria hosted in vermicompost. Some studies have reported that the combined soil application of RP and vermicompost results in significantly higher available soil P than sole application of either RP or vermicompost (Table 14.2) in soil.

14.5 Co-applying Vermicompost with Biochar

Soil application of biochar enhances nutrient retention, particularly nitrogen and carbon (Borchard et al. 2012; Cheng et al. 2012; Clough et al. 2013; Doan et al. 2015; Farrell et al. 2014). Biochar application-induced soil nutrient retention may lead to better plant growth (Steiner et al. 2008; Vaccari et al. 2011). Co-application of biochar and vermicompost has been reported to increase biochar stability and reduce OM solubility (Doan et al. 2014; Ngo et al. 2013).

Dissolved organic carbon (DOC) that is added to soil through biochar and vermicompost is key to the biogeochemical processes in the environment due to its reactivity and mobility. Although dissolved organic matter represents only a small fraction of the SOM, the availability of DOC ensures the microbial turnover of nutrients in soils, which is essential for the fertility of soils. Soils with biochar and vermicompost could therefore provide several advantages for crop growth.

Biochar and vermicompost application to soils may suppress soilborne pathogens and consequently increase crop productivity. For instance, Shoaf (2014) reported that co-applying biochar and vermicompost reduced P. *capsici* root infection in two soils. In a different study by Ebrahimi et al. (2021) increased eggplant yield and water use efficiency was reported after co-application of biochar and vermicompost compared to sole applications. The vermicompost and biochar treatment also had the highest water use efficiency when water was applied at 50% plant water requirement.

These examples suggest that the response to applications of biochar and vermicompost varies depending on the environmental conditions,

Biochar influences the leaching of nutrients because it alters the soil's bulk density, porosity, permeability and water retention. It also modifies the chemical properties of the soil, such as the pH, cation and anion exchange capacities, as well as the soil's capacity to absorb soluble organic and inorganic compounds (Becagli et al. 2021; Tumbure et al. 2020). Biochar however increases the carbon to nitrogen ratio, which leads to the immobilisation of N, and reduces its availability with its adsorption on the biochar surface (Gul and Whalen 2016). However, the application of biochar combined with a source of nitrogen, like vermicompost, can compensate for the deficiencies in N caused by immobilisation (Nguyen et al. 2017; Prommer et al. 2014).

Both biochar and vermicompost create optimum conditions for plant growth by improving soil porosity, structure and density. Optimum conditions for survival of beneficial soil microbes are created due to biochar's porous and large surface area qualities. Biochar also increases cation exchange capacity (CEC) and therefore improves nutrient retention (Ebrahimi et al. 2021) because of increased nutrient adsorption on soil particles which is antagonistic to leaching forces (Souri et al. 2019). Both biochar and vermicompost have been reported to increase soil pH (Tumbure et al. 2020; Wang et al. 2021), thus enhancing availability of nutrients, especially phosphorous (Álvarez et al. 2017). Vermicompost on the other hand supplies soil nitrogen, resulting in increased plant growth. For example, Singh and Varshney (2013) noted increased soil mineral N (NH_4^+ and NO_3^-) after vermicompost addition. Vermicompost addition may also result in increased populations of N fixing microorganisms in the rhizosphere and increased N fixation (Wang et al. 2017).

Biochar and vermicompost have been seen to also enhance crop production in contaminated soils. Khosropour et al. (2022) investigated the physio-biochemical attributes of the shrub *Berberis integerrima bunge* when vermicompost and biochar were applied in Cd-contaminated soil. The combined application of biochar and vermicompost were observed to be more effective in improving plant growth, with an increase of 34 and 33%, in fresh and dry weights, respectively, when compared with non-treated plants (control). The results suggest the use of combined biochar and vermicompost has the potential to lessen the toxic effects of Cd in *B. integerrima*. The reduction in the effect of cadmium on the plants is attributed to the reduced availability of cadmium effected by both biochar and vermicompost (Wang et al. 2018). In addition to making nutrients available while reducing mobility of heavy metals, co-application of biochar and vermicompost also increases microbial activity.

14.5.1 Effect of Biochar and Vermicompost on Soil Microbial Abundance

Soil microbes are important in balancing nutrients in degraded agro-ecosystems. Soil acidification and salinity caused by continuous cropping alter the soil microbial community structure and inhibit microbial activities (Rathore et al. 2017). The application of biochar and vermicompost has been shown to increase microbial abundance and diversity, which is largely attributed to the increase of soil dissolved organic C from the biochar and vermicompost. Both organic amendments provide labile C to soil microorganisms (Quilliam et al. 2013; Zhao et al. 2017). Soil conditioning effects such as high porosity, aeration, drainage and water-holding capacity after biochar and vermicompost co-application can also indirectly improve soil microbial abundance (Guo et al. 2020; Wang et al. 2018). Biochar and vermicompost-induced soil physio-chemical changes create an improved soil environment for microbial proliferation (Lv et al. 2020), eventually resulting in increase in soil microbial abundance and diversity.

14.6 Conclusions and Recommendations

This chapter indicates that while the value of vermicompost is very clear, its effects on soil condition, microbial activity, nutrient availability and crop productivity are enhanced by co-application of vermicompost with other soil amendments. Co-application with mycorrhizal fungi improves soil physical properties, while inclusion of other microbial inoculants increases nodulation and N fixation, microbial activity, macronutrient availability and uptake. Availability and uptake of macronutrients and crop yield are improved by co-application of vermicompost with rock phosphate, calcium carbonates biochar, with inconsistent results when using mineral fertilisers. It can be recommended that, in order to get the best results from vermicompost, farmers need to co-apply it with microbial inoculants, natural minerals and biochar, while care needs to be taken when mineral fertiliser is used. An opportunity therefore arises for smallholder African farmers to improve the fertility of degraded soils in much less seasonal time.

References

Abduli M, Amiri L, Madadian E, Gitipour S, Sedighian S (2013) Efficiency of vermicompost on quantitative and qualitative growth of tomato plants. Int J Environ Res 7(2):467–472

Agnihotri R, Pandey A, Bharti A, Chourasiya D, Maheshwari HS, Ramesh A, Billore SD, Sharma MP (2021) Soybean processing mill waste plus vermicompost enhances arbuscular mycorrhizal fungus inoculum production. Curr Microbiol 78(7):2595–2607. https://doi.org/10.1007/s00284-021-02532-7

Aira M, Monroy F, Domínguez J (2005) Ageing effects on nitrogen dynamics and enzyme activities in casts of *Aporrectodea caliginosa* (Lumbricidae). Pedobiologia 49(5):467–473

Akhzari D, Attaeian B, Arami A, Mahmoodi F, Aslani F (2015) Effects of vermicompost and arbuscular mycorrhizal fungi on soil properties and growth ofmedicago polymorphal. Compost Sci Util 23(3):142–153. https://doi.org/10.1080/1065657x.2015.1013585

Alemu BA, Melese A (2021) Inorganic phosphorous, lime and vermicompost induced changes on phosphorus fractions and other properties of acidic soil of Cheha district, Ethiopia. Research Square. https://doi.org/10.21203/rs.3.rs-250222/v1

Álvarez JM, Pasian C, Lal R, López Núñez R, Fernández Martínez M (2017) Vermicompost and biochar as substitutes of growing media in ornamental-plant production. J Appl Hortic 19(3): 205–214

Arfarita N, Imai T, Prayogo C (2022) Utilization of various organic wastes as liquid biofertilizer carrier agents towards viability of bacteria and green bean growth. J Trop Life Sci 12(1):1–10. https://doi.org/10.11594/jtls.12.01.01

Atanda AC, Adeleke RA, Jooste PJ, Madoroba E (2018) Insights into the microbiological safety of vermicompost and vermicompost tea produced by south african smallholder farmers. Indian J Microbiol 58(4):479–488. https://doi.org/10.1007/s12088-018-0748-7

Atiyeh RM, Edwards CA, Subler S, Metzger JD (2001) Pig manure vermicompost as a component of a horticultural bedding plant medium: effects on physicochemical properties and plant growth. Bioresour Technol 78(1):11–20. https://doi.org/10.1016/S0960-8524(00)00172-3

Badhwar VK, Singh C (2022) Vermicomposting of textile mill sludge employing Eisenia fetida: role of cow dung and tea waste amendments. Environ Sci Pollut Res Int 29(13):19823–19834. https://doi.org/10.1007/s11356-021-17185-z

Baghbani-Arani A, Modarres-Sanavy SAM, Poureisa M (2021) Improvement the soil physico-chemical properties and fenugreek growth using zeolite and vermicompost under water deficit conditions. J Soil Sci Plant Nutr 21(2):1213–1228. https://doi.org/10.1007/s42729-021-00434-y

Barber B (1991) Phosphate resources of carbonatites in Zimbabwe. Fertil Res 30:247–278

Becagli M, Guglielminetti L, Cardelli R (2021) Effects of combined biochar and vermicompost solution on leachate characterization and nitrogen balance from a greenhouse tomato (*Solanum lycopersicum*) cultivation soil. Commun Soil Sci Plant Anal 52(16):1879–1893. https://doi.org/10.1080/00103624.2021.1900225

Benaffari W, Boutasknit A, Anli M, Ait-El-Mokhtar M, Ait-Rahou Y, Ben-Laouane R, Ben Ahmed H, Mitsui T, Baslam M, Meddich A (2022) The native arbuscular mycorrhizal fungi and vermicompost-based organic amendments enhance soil fertility, growth performance, and the drought stress tolerance of quinoa. Plants (Basel) 11(3):393. https://doi.org/10.3390/plants11030393

Bhat SA, Singh S, Singh J, Kumar S, Bhawana, Vig AP (2018) Bioremediation and detoxification of industrial wastes by earthworms: vermicompost as powerful crop nutrient in sustainable agriculture. Bioresour Technol 252:172–179. https://doi.org/10.1016/j.biortech.2018.01.003

Blouin M, Hodson ME, Delgado EA, Baker G, Brussaard L, Butt KR, Dai J, Dendooven L, Peres G, Tondoh JE, Cluzeau D, Brun J-J (2013) A review of earthworm impact on soil function and ecosystem services. Eur J Soil Sci 64(2):161–182. https://doi.org/10.1111/ejss.12025

Blouin M, Barrere J, Meyer N, Lartigue S, Barot S, Mathieu J (2019) Vermicompost significantly affects plant growth. A meta-analysis. Agron Sustain Dev 39(4):34. https://doi.org/10.1007/s13593-019-0579-x

Borchard N, Wolf A, Laabs V, Aeckersberg R, Scherer H, Moeller A, Amelung W (2012) Physical activation of biochar and its meaning for soil fertility and nutrient leaching—a greenhouse experiment. Soil Use Manag 28(2):177–184

Busato JG, Ferrari LH, Chagas Junior AF, da Silva DB, Dos Santos Pereira T, de Paula AM (2021) Trichoderma strains accelerate maturation and increase available phosphorus during vermicomposting enriched with rock phosphate. J Appl Microbiol 130(4):1208–1216. https://doi.org/10.1111/jam.14847

Chaoui HI, Zibilske LM, Ohno T (2003) Effects of earthworm casts and compost on soil microbial activity and plant nutrient availability. Soil Biol Biochem 35(2):295–302

Chatterjee D, Dutta SK, Kikon ZJ, Kuotsu R, Sarkar D, Satapathy BS, Deka BC (2021) Recycling of agricultural wastes to vermicomposts: characterization and application for clean and quality production of green bell pepper (*Capsicum annuum* L.). J Clean Prod 315:128115. https://doi.org/10.1016/j.jclepro.2021.128115

Chaumba J (2017) Hydrothermal alteration in the main sulfide zone at Unki Mine, Shurugwi Subchamber of the Great Dyke, Zimbabwe: evidence from petrography and silicates mineral chemistry. Minerals 7(7):127. https://doi.org/10.3390/min7070127

Chaumba J (2019) Evolution of the North West Arm and the Central Sector of Mashava Igneous Complex in south central Zimbabwe from an investigation of its silicate minerals compositions. Precambrian Res 324:109–125. https://doi.org/10.1016/j.precamres.2019.01.019

Cheng Y, Cai Z-C, Chang SX, Wang J, Zhang J-B (2012) Wheat straw and its biochar have contrasting effects on inorganic N retention and N2O production in a cultivated Black Chernozem. Biol Fertil Soils 48(8):941–946

Chiranjeeb K, Prasad SS, Singh SP, Bharati V, Jha S (2020) Effect of household vermicompost and fertilizer on soil microbial biomass carbon, biomass phosphorus and biomass nitrogen in incubation experiment. Int J Curr Microbiol App Sci 9(2):1508–1516. https://doi.org/10.20546/ijcmas.2020.902.174

Clough TJ, Condron LM, Kammann C, Müller C (2013) A review of biochar and soil nitrogen dynamics. Agronomy 3(2):275–293

De Grazia J, Tittonell P, Chiesa A (2006) Nitrogen fertilization of Eggplant (*Solanum melongena* L. var. esculentum) transplants and their impact on crop growth after transplanting. IV International Symposium on Seed, Transplant and Stand Establishment of Horticultural Crops; Translating Seed and Seedling 782

de Souza MEP, Cardoso IM, de Carvalho AMX, Lopes AP, Jucksch I (2019) Gneiss and steatite vermicomposted with organic residues: release of nutrients and heavy metals. Int J Recycl Org Waste Agric 8(3):233–240. https://doi.org/10.1007/s40093-019-0244-z

Demiraj E, Libutti A, Malltezi J, Rroço E, Brahushi F, Monteleone M, Sulçe S (2018) Effect of organic amendments on nitrate leaching mitigation in a sandy loam soil of Shkodra district, Albania. Ital J Agron 13(1):93–102

Devi H, Maity T, Thapa U, Paria N (2002) Effect of integrated nitrogen management on yield and Economics of Brinjal. J Interacademicia 6(4):450–453

Doan TT, Bouvier C, Bettarel Y, Bouvier T, Henry-des-Tureaux T, Janeau JL, Lamballe P, Van Nguyen B, Jouquet P (2014) Influence of buffalo manure, compost, vermicompost and biochar amendments on bacterial and viral communities in soil and adjacent aquatic systems. Appl Soil Ecol 73:78–86

Doan TT, Henry-des-Tureaux T, Rumpel C, Janeau JL, Jouquet P (2015) Impact of compost, vermicompost and biochar on soil fertility, maize yield and soil erosion in Northern Vietnam: a three year mesocosm experiment. Sci Total Environ 514:147–154. https://doi.org/10.1016/j.scitotenv.2015.02.005

Ebrahimi M, Souri MK, Mousavi A, Sahebani N (2021) Biochar and vermicompost improve growth and physiological traits of eggplant (*Solanum melongena* L.) under deficit irrigation. Chem Biol Technol Agric 8(1):1–14. https://doi.org/10.1186/s40538-021-00216-9

Farrell M, Macdonald LM, Butler G, Chirino-Valle I, Condron LM (2014) Biochar and fertiliser applications influence phosphorus fractionation and wheat yield. Biol Fertil Soils 50(1):169–178

Furlong C, Rajapaksha NS, Butt KR, Gibson WT (2017) Is composting worm availability the main barrier to large-scale adoption of worm-based organic waste processing technologies? J Clean Prod 164:1026–1033. https://doi.org/10.1016/j.jclepro.2017.06.226

Gebrehana ZG, Gebremikael MT, Beyene S, Wesemael WML, De Neve S (2022) Assessment of trade-offs, quantity, and biochemical composition of organic materials and farmer's perception towards vermicompost production in smallholder farms of Ethiopia. J Mater Cycles Waste Manag 24(2):540–552. https://doi.org/10.1007/s10163-021-01339-9

Ghadimi M, Sirousmehr A, Ansari MH, Ghanbari A (2021) Organic soil amendments using vermicomposts under inoculation of N2-fixing bacteria for sustainable rice production. PeerJ 9:e10833. https://doi.org/10.7717/peerj.10833

Gul S, Whalen JK (2016) Biochemical cycling of nitrogen and phosphorus in biochar-amended soils. Soil Biol Biochem 103:1–15

Guo X-X, Liu H-T, Zhang J (2020) The role of biochar in organic waste composting and soil improvement: a review. Waste Manag 102:884–899

Gutiérrez-Miceli FA, Santiago-Borraz J, Molina JAM, Nafate CC, Abud-Archila M, Llaven MAO, Rincón-Rosales R, Dendooven L (2007) Vermicompost as a soil supplement to improve growth, yield and fruit quality of tomato (*Lycopersicum esculentum*). Bioresour Technol 98(15): 2781–2786

Hussain N, Abbasi T, Abbasi S (2017) Enhancement in the productivity of ladies finger (*Abelmoschus esculentus*) with concomitant pest control by the vermicompost of the weed salvinia (*Salvinia molesta*, Mitchell). Int J Recycl Org Waste in Agric 6(4):335–343

Hussain S, Sharif M, Ahmad W (2021) Selection of efficient phosphorus solubilizing bacteria strains and mycorrhizea for enhanced cereal growth, root microbe status and N and P uptake in alkaline calcareous soil. Soil Sci Plant Nutr 67(3):259–268. https://doi.org/10.1080/00380768. 2021.1904793

Ievinsh G (2011) Vermicompost treatment differentially affects seed germination, seedling growth and physiological status of vegetable crop species. Plant Growth Regul 65(1):169–181

Jack A (2010) Suppression of Plant Pathogens with Vermicomposts. In: Edwards CA, Arancon NQ, Sherman RL (eds) Vermiculture technology: earthworms, organic wastes and environmental management. CRC Press, Boca Raton, FL

Jack ALH (2012) Vermicompost suppression of Pythium aphanidermatum seedling disease: practical applications and an exploration of the mechanisms of disease suppression. Cornell University

Jafari F, Khademi H, Shahrokh V, Cano AFAZ, Acosta JA, Khormali F (2021a) Biological weathering of phlogopite during enriched vermicomposting. Pedosphere 31(3):440–451. https://doi.org/10.1016/s1002-0160(20)60083-2

Jafari F, Khademi H, Shahrokh V, Faz A, Acosta JA (2021b) Earthworm- and rhizosphere-induced biological weathering of phlogopite. J Soil Sci Plant Nutr 22(1):416–427. https://doi.org/10. 1007/s42729-021-00658-y

Khan K, Pankaj U, Verma SK, Gupta AK, Singh RP, Verma RK (2015) Bio-inoculants and vermicompost influence on yield, quality of Andrographis paniculata, and soil properties. Ind Crop Prod 70:404–409. https://doi.org/10.1016/j.indcrop.2015.03.066

Khosravi A, Zarei M, Ronaghi A (2017) Effect of PGPR, Phosphate sources and vermicompost on growth and nutrients uptake by lettuce in a calcareous soil. J Plant Nutr 41(1):80–89. https://doi. org/10.1080/01904167.2017.1381727

Khosropour E, Weisany W, Tahir NA, Hakimi L (2022) Vermicompost and biochar can alleviate cadmium stress through minimizing its uptake and optimizing biochemical properties in Berberis integerrima bunge. Environ Sci Pollut Res Int 29(12):17476–17486. https://doi.org/10. 1007/s11356-021-17073-6

Lalander CH, Komakech AJ, Vinneras B (2015) Vermicomposting as manure management strategy for urban small-holder animal farms—Kampala case study. Waste Manag 39:96–103. https:// doi.org/10.1016/j.wasman.2015.02.009

Lavelle P, Melendez G, Pashanasi B, Schaefer R (1992) Nitrogen mineralization and reorganization in casts of the geophagous tropical earthworm *Pontoscolex corethrurus* (Glossoscolecidae). Biol Fertil Soils 14(1):49–53

Libutti A, Trotta V, Rivelli A (2020) Biochar, Vermicompost, and compost as soil organic amendments: influence on growth parameters, nitrate and chlorophyll content of swiss chard (*Beta vulgaris* L. var cycla). Agronomy 10(3):346. https://doi.org/10.3390/agronomy10030346

Liégui GS, Cognet S, Djumyom GVW, Atabong PA, Noutadié JPF, Chamedjeu RR, Temegne CN, Kengne IMN (2021) An effective organic waste recycling through vermicomposting technology for sustainable agriculture in tropics. Int J Recycl Org Waste Agric 10(3):203–214. https://doi.org/10.30486/IJROWA.2021.1894997.1080

Lim SL, Wu TY, Lim PN, Shak KP (2015) The use of vermicompost in organic farming: overview, effects on soil and economics. J Sci Food Agric 95(6):1143–1156. https://doi.org/10.1002/jsfa.6849

Lukashe NS, Mupambwa HA, Green E, Mnkeni PNS (2019) Inoculation of fly ash amended vermicompost with phosphate solubilizing bacteria (*Pseudomonas fluorescens*) and its influence on vermi-degradation, nutrient release and biological activity. Waste Manag 83:14–22. https://doi.org/10.1016/j.wasman.2018.10.038

Lv H, Zhao Y, Wang Y, Wan L, Wang J, Butterbach-Bahl K, Lin S (2020) Conventional flooding irrigation and over fertilization drives soil pH decrease not only in the top-but also in subsoil layers in solar greenhouse vegetable production systems. Geoderma 363:114156

Mago M, Gupta R, Yadav A, Kumar Garg V (2022) Sustainable treatment and nutrient recovery from leafy waste through vermicomposting. Bioresour Technol 347:126390. https://doi.org/10.1016/j.biortech.2021.126390

Mainoo N-OK, Barrington S, Whalen JK (2008) Vermicompost as a fertilizer for urban and peri-urban farms: perceptions of farmers in Accra, Ghana. Ghana J Agric Sci 41:219–226

Mal S, Chattopadhyay GN, Chakrabarti K (2021) Microbiological integration for qualitative improvement of vermicompost. Int J Recycl Org Waste Agric 10:157–166. https://doi.org/10.30486/IJROWA.2021.1902019.1087

Mathenge C, Thuita M, Masso C, Gweyi-Onyango J, Vanlauwe B (2019) Variability of soybean response to rhizobia inoculant, vermicompost, and a legume-specific fertilizer blend in Siaya County of Kenya. Soil Tillage Res 194:104290. https://doi.org/10.1016/j.still.2019.06.007

Messiga AJ, Hao X, Dorais M, Bineng CS, Ziadi N, Naeth MA (2022) Supplement of biochar and vermicompost amendments in coir and peat growing media improves N management and yields of leafy vegetables. Can J Soil Sci 102(1):39–52. https://doi.org/10.1139/cjss-2020-0059

Mokgophi MM, Manyevere A, Ayisi KK, Munjonji L (2020) Characterisation of *Chamaecytisus tagasaste*, *Moringa oleifera* and *Vachellia karroo* vermicomposts and their potential to improve soil fertility. Sustainability 12(22):9305. https://doi.org/10.3390/su12229305

Moraditochaee M, Bozorgi HR, Halajisani N (2011) Effects of vermicompost application and nitrogen fertilizer rates on fruit yield and several attributes of eggplant (*Solanum melongena* L.) in Iran. World Appl Sci J 15(2):174–178

Mtangadura TJ, Mtambanengwe F, Nezomba H, Rurinda J, Mapfumo P (2017) Why organic resources and current fertilizer formulations in Southern Africa cannot sustain maize productivity: evidence from a long-term experiment in Zimbabwe. PLoS One 12(8):1–23. https://doi.org/10.1371/journal.pone.0182840

Mugwendere T, Mtaita T, Mutetwa M, Tabarira J (2015) Use of vermicompost as a soil supplement on growth and yield of rape (*Brassica napus*). J Glob Innov Agric Soc Sci 3(1):25–31. https://doi.org/10.17957/jgiass/3.1.701

Mupambwa HA, Mnkeni PNS (2018) Optimizing the vermicomposting of organic wastes amended with inorganic materials for production of nutrient-rich organic fertilizers: a review. Environ Sci Pollut Res Int 25(11):10577–10595. https://doi.org/10.1007/s11356-018-1328-4

Muscolo A, Bovalo F, Gionfriddo F, Nardi S (1999) Earthworm humic matter produces auxin-like effects on *Daucus carota* cell growth and nitrate metabolism. Soil Biol Biochem 31(9):1303–1311

Nair VD, Nair P, Dari B, Freitas AM, Chatterjee N, Pinheiro FM (2017) Biochar in the agroecosystem–climate-change–sustainability nexus. Front Plant Sci 8:2051

Ngo P-T, Rumpel C, Ngo Q-A, Alexis M, Vargas GV, de la Luz Mora Gil M, Dang D-K, Jouquet P (2013) Biological and chemical reactivity and phosphorus forms of buffalo manure compost, vermicompost and their mixture with biochar. Bioresour Technol 148:401–407

Nguyen TTN, Xu C-Y, Tahmasbian I, Che R, Xu Z, Zhou X, Wallace HM, Bai SH (2017) Effects of biochar on soil available inorganic nitrogen: a review and meta-analysis. Geoderma 288:79–96

Padmavathiamma PK, Li LY, Kumari UR (2008) An experimental study of vermi-biowaste composting for agricultural soil improvement. Bioresour Technol 99(6):1672–1681. https://doi.org/10.1016/j.biortech.2007.04.028

Papathanasiou F, Papadopoulos I, Tsakiris I, Tamoutsidis E (2012) Vermicompost as a soil supplement to improve growth, yield and quality of lettuce (*Lactuca sativa* L.). J Food Agric Environ 10(2):677–682

Parastesh F, Alikhani HA, Etesami H (2019) Vermicompost enriched with phosphate–solubilizing bacteria provides plant with enough phosphorus in a sequential cropping under calcareous soil conditions. J Clean Prod 221:27–37. https://doi.org/10.1016/j.jclepro.2019.02.234

Paula De Souza ME, Cardoso IM, De Carvalho AMX, Lopes AP, Jucksch I, Janssen A (2018) Rock powder can improve vermicompost chemical properties and plant nutrition: an on-farm experiment. Commun Soil Sci Plant Anal 49(1):1–12. https://doi.org/10.1080/00103624.2017.1418372

Pisa C, Wuta M, Muchaonyerwa P (2020) Effects of incorporation of vermiculite on carbon and nitrogen retention and concentration of other nutrients during composting of cattle manure. Bioresour Technol Rep 9:100383

Pramanik P, Ghosh GK, Ghosal PK, Banik P (2007) Changes in organic—C, N, P and K and enzyme activities in vermicompost of biodegradable organic wastes under liming and microbial inoculants. Bioresour Technol 98(13):2485–2494. https://doi.org/10.1016/j.biortech.2006.09.017

Pramanik P, Bhattacharya S, Bhattacharyya P, Banik P (2009) Phosphorous solubilization from rock phosphate in presence of vermicomposts in Aqualfs. Geoderma 152(1–2):16–22. https://doi.org/10.1016/j.geoderma.2009.05.013

Prommer J, Wanek W, Hofhansl F, Trojan D, Offre P, Urich T, Schleper C, Sassmann S, Kitzler B, Soja G (2014) Biochar decelerates soil organic nitrogen cycling but stimulates soil nitrification in a temperate arable field trial. PLoS One 9(1):e86388

Quilliam RS, Glanville HC, Wade SC, Jones DL (2013) Life in the 'charosphere'—does biochar in agricultural soil provide a significant habitat for microorganisms? Soil Biol Biochem 65:287–293

Rahbar Shiraz S, Jalili B, Bahmanyar MA (2019) Amendment of vermicompost by phosphate rock, steel dust, and *Halothiobacillus neapolitanus*. Waste Biomass Valorization 11(8):4207–4213. https://doi.org/10.1007/s12649-019-00740-8

Rai S, Rani P, Kumar M, Rai AK, Shahi SK (2014) Effect of integrated use of vermicompost, FYM, PSB and azotobacter on physico-chemical properties of soil under onion crop. Environ Ecol 32 (October—December 2014):1797–1803

Raimi AR, Atanda AC, Ezeokoli OT, Jooste PJ, Madoroba E, Adeleke RA (2022) Diversity and predicted functional roles of cultivable bacteria in vermicompost: bioprospecting for potential inoculum. Arch Microbiol 204(5):261. https://doi.org/10.1007/s00203-022-02864-3

Rathore AP, Chaudhary DR, Jha B (2017) Seasonal patterns of microbial community structure and enzyme activities in coastal saline soils of perennial halophytes. Land Degrad Dev 28(5):1779–1790

Shanka D (2020) Roles of eco-friendly low input technologies in crop production in sub-Saharan Africa. Cogent Food Agric 6(1):1843882. https://doi.org/10.1080/23311932.2020.1843882

Sharma D, Prasad R, Patel B, Parashar CK (2022) Biotransformation of sludges from dairy and sugarcane industries through vermicomposting using the epigeic earthworm Eisenia fetida. Int J Recycl Org Waste Agric 11:165–175. https://doi.org/10.30486/IJROWA.2021.1922034.1196

Shoaf NL (2014) Biochar and vermicompost amendments in vegetable cropping systems: impacts on soil quality, soil-borne pathogens and crop productivity. Purdue University

Singh R, Varshney G (2013) Effects of carbofuran on availability of macronutrients and growth of tomato plants in natural soils and soils amended with inorganic fertilizers and vermicompost. Commun Soil Sci Plant Anal 44(17):2571–2586

Singh A, Omran E-SE, Singh GS (2020) Vermicomposting Impacts on Agriculture in Egypt. In: Omran E-SE, Negm AM (eds) Technological and modern irrigation environment in Egypt: best management practices & evaluation. Springer International, Cham, pp 181–203. https://doi.org/10.1007/978-3-030-30375-4_9

Singh Y, Bhatnagar P, Singh J, Sharma MK, Jain SK, Maurya IB, Sharma YK (2021) Augmentation of plant growth attributes, soil physico-chemical properties and microbial population in custard apple cv. balanagar in response to *Azospirillum brasilense* and vermicompost application in Humid Zone of South Eastern Rajasthan, India. Commun Soil Sci Plant Anal 52(21):2701–2714. https://doi.org/10.1080/00103624.2021.1956517

Souri MK, Naiji M, Kianmehr MH (2019) Nitrogen release dynamics of a slow release urea pellet and its effect on growth, yield, and nutrient uptake of sweet basil (*Ocimum basilicum* L.). J Plant Nutr 42(6):604–614

Steiner C, Glaser B, Geraldes Teixeira W, Lehmann J, Blum WE, Zech W (2008) Nitrogen retention and plant uptake on a highly weathered central Amazonian Ferralsol amended with compost and charcoal. J Plant Nutr Soil Sci 171(6):893–899

Swoboda P, Doring TF, Hamer M (2022) Remineralizing soils? The agricultural usage of silicate rock powders: a review. Sci Total Environ 807(Pt 3):150976. https://doi.org/10.1016/j.scitotenv.2021.150976

Tan C, Luo Y, Fu T (2022) Soil microbial community responses to the application of a combined amendment in a historical zinc smelting area. Environ Sci Pollut Res Int 29(9):13056–13070. https://doi.org/10.1007/s11356-021-16631-2

ten Berge HFM, Hijbeek R, van Loon MP, Rurinda J, Tesfaye K, Zingore S, Craufurd P, van Heerwaarden J, Brentrup F, Schröder JJ, Boogaard HL, de Groot HLE, van Ittersum MK (2019) Maize crop nutrient input requirements for food security in sub-Saharan Africa. Glob Food Sec 23:9–21. https://doi.org/10.1016/j.gfs.2019.02.001

Tumbure A, Bishop P, Bretherton M, Hedley M (2020) Co-pyrolysis of maize stover and igneous phosphate rock to produce potential biochar-based phosphate fertilizer with improved carbon retention and liming value. ACS Sustain Chem Eng 8(10):4178–4184. https://doi.org/10.1021/acssuschemeng.9b06958

Tumbure A, Bishop P, Hedley MJ, Bretherton MR (2021) Increasing phosphorus solubility by sintering igneous Dorowa phosphate rock with recycled glass. J Therm Anal Calorim 145(6):3019–3030. https://doi.org/10.1007/s10973-020-10078-2

Tumbure A, Bretherton MB, Bishop P, Hedley MJ (2022) Phosphorus recovery from an igneous phosphate rock using organic acids and pyrolysis condensate. Sci Afr 15:e01098. https://doi.org/10.1016/j.sciaf.2022.e01098

Vaccari F, Baronti S, Lugato E, Genesio L, Castaldi S, Fornasier F, Miglietta F (2011) Biochar as a strategy to sequester carbon and increase yield in durum wheat. Eur J Agron 34(4):231–238

van Groenigen JW, Lubbers IM, Vos HM, Brown GG, De Deyn GB, van Groenigen KJ (2014) Earthworms increase plant production: a meta-analysis. Sci Rep 4:6365. https://doi.org/10.1038/srep06365

Verma SK, Pankaj U, Khan K, Singh R, Verma RK (2016) Bioinoculants and vermicompost improveocimum basilicumyield and soil health in a sustainable production system. Clean Soil Air Water 44(6):686–693. https://doi.org/10.1002/clen.201400639

Wang X-X, Zhao F, Zhang G, Zhang Y, Yang L (2017) Vermicompost improves tomato yield and quality and the biochemical properties of soils with different tomato planting history in a greenhouse study. Front Plant Sci 8:1978

Wang Y, Xu Y, Li D, Tang B, Man S, Jia Y, Xu H (2018) Vermicompost and biochar as bio-conditioners to immobilize heavy metal and improve soil fertility on cadmium contaminated soil under acid rain stress. Sci Total Environ 621:1057–1065

Wang F, Wang X, Song N (2021) Biochar and vermicompost improve the soil properties and the yield and quality of cucumber (*Cucumis sativus* L.) grown in plastic shed soil continuously cropped for different years. Agric Ecosyst Environ 315:107425. https://doi.org/10.1016/j.agee.2021.107425

Wu D, Feng Y, Xue L, Liu M, Yang B, Hu F, Yang L (2019) Biochar combined with vermicompost increases crop production while reducing ammonia and nitrous oxide emissions from a paddy soil. Pedosphere 29(1):82–94. https://doi.org/10.1016/s1002-0160(18)60050-5

Yavitt JB, Pipes GT, Olmos EC, Zhang J, Shapleigh JP (2021) Soil organic matter, soil structure, and bacterial community structure in a post-agricultural landscape. Front Earth Sci 9:590103. https://doi.org/10.3389/feart.2021.590103

Zaller JG (2007) Vermicompost as a substitute for peat in potting media: effects on germination, biomass allocation, yields and fruit quality of three tomato varieties. Sci Hortic 112(2):191–199

Zhao H-T, Li T-P, Zhang Y, Hu J, Bai Y-C, Shan Y-H, Ke F (2017) Effects of vermicompost amendment as a basal fertilizer on soil properties and cucumber yield and quality under continuous cropping conditions in a greenhouse. J Soils Sediments 17(12):2718–2730

Zingore S, Mutegi J, Agesa B, Tamene L, Kihara J (2015) Soil degradation in sub-Saharan Africa and crop production options for soil rehabilitation. Better Crops 99(1):24–26

Chapter 15
Sustainable Enhancement of Soil Fertility Using Bioinoculants

Mukelabai Florence and Chimwamurombe Percy

Abstract Smallholder farmers are depending upon subsistence agriculture on very poor sandy soils for their living, that lack adequate amounts of nutrients to support the ideal crop growth such as nitrogen which is the most limiting nutrient for growth of leguminous plants such as common beans, soya beans, cowpeas, and garden peas, this is because the elemental N present in the soil cannot support crop growth. Therefore, this chapter will be looking at the sustainable enhancement of soil fertility using bioinoculants. Opportunely, soil microorganisms precisely bacteria called rhizobia are able to colonize the rhizosphere, infect legume roots, and biologically fix N in soil by forming a symbiotic bond with the plant and converting free N into ammonia for the plants to use, hence making soils fertile for plant growth. From the many studies done, it has been observed and can be concluded that leguminous cultivars respond very well to inoculant treatments compared to chemical N fertilizers. Inoculants are more cost effect for small-scale farmers compared to the expensive chemical N fertilizers, thereby proving that inoculants are able to sustain and enhance soil fertility.

Keywords Nitrogen fixation · Legume · Bioinoculants · Rhizosphere · Endophytic microbes

15.1 Introduction

Inadequate economic resources, peril of crop failure, and inconsistency in yields are among the reasons that contribute to farmers' hesitancy to accept new technologies in Africa's low rain-fed agricultural systems (Ogada et al. 2010). Risks will increase in many parts of Sub-Saharan Africa (SSA) because of climate change and SSA is

---

M. Florence · C. Percy (✉)
Department of Biology, Chemistry and Physics, School of Natural and Applied Sciences, Namibia University of Science and Technology, Windhoek, Namibia

© The Author(s), under exclusive license to Springer Nature Singapore Pte Ltd. 2023
H. A. Mupambwa et al. (eds.), *Vermicomposting for Sustainable Food Systems in Africa*, Sustainable Agriculture and Food Security,
https://doi.org/10.1007/978-981-19-8080-0_15

expected to suffer a decline in suitable land area for cultivation of crops (Lane and Jarvis 2007).

In addition, the rise in frequency and strength of cyclical droughts has caused a reduction in some of the most important food crops, which was studied according to Hase (2013). Low agricultural productivity leading to increased food insecurity has also been caused by a lack of incentives to farmers for engaging in optimum land management policies that would speed up the need for technological change, leading to better productivity (Ali et al. 2017). Farmers are trapped in a vicious circle of poverty and hunger as the soils continue to degrade unabated due to the fact that very few farmers use farming practices that physically conserve the soil (Taapopi et al. 2018).

Smallholder farmers depend upon subsistence agriculture on very poor sandy soils for their living that lack adequate amounts of nutrients to support the ideal crop growth such as Nitrogen (N) which is the most limiting nutrient for growth of leguminous plants such as common beans, soya beans, cowpeas and garden peas, this is because the elemental N present in the soil cannot support crop growth (Reinhold-Hurek et al. 2015), and therefore must be converted from N to ammonia for the plant to utilize it.

N is required and utilized in high quantities by plants, due to the fact that it is the basic component of many chemical compounds, including proteins and nucleic acids. Pollution to the underground water as well as the soil is caused by synthetic N chemical fertilizers according to Farahvash et al. (2010). Nitrogen fertilizers are also expensive and cause a rise in the financial production of crops. Hence, organic and cheaper means need to be further explored and reduce these adverse effects and costs of obtaining N (Farahvash et al. 2010).

Opportunely, soil microorganisms precisely bacteria called rhizobia are able to colonize the rhizosphere, infect legume roots and biologically fix N in soil by forming a symbiotic bond with the plant and converting free N into ammonia for the plants to use, hence making soils fertile for plant growth (Mohammadi and Sohrabi 2012; Eskin 2012). The key factor to sustain agricultural production is Biological Nitrogen Fixation (BNF). Therefore, unless we carefully make use of BNF, increasing crop yield cannot be fulfilled.

Legumes are one of the most diverse plants on earth spread wide in tropics and temperate zones (Sprent and James 2007). Legumes belong to the superfamily of angiosperms (*Leguminosae/Fabaceae*) from the order *Fabales* and the clade *eurosid* (Doyle and Luckow 2003; Tran and Nguyen 2009). Legumes are able to grow in degraded soils because they are able to fix N by forming a symbiotic relationship with rhizobia (Wong 2003; Freitas et al. 2004).

Legumes not only provide traditional diets throughout the world, but also have multiple benefits to both the soil and other crops through intercropping (Stajković et al. 2011). Among the vast and potential uses of grain legumes like soybean, cowpea, common bean, and peas as human food, animal feed, and soil fertility enhancer, they are also able to grow different agro-ecological environments (Hendawey and Younes 2013).

The collaboration between plants and soil microorganisms mostly happens in the rhizosphere (Marschner et al. 2011). The legume-rhizobial symbiosis has a large impact on success of legumes, the atmospheric N the organisms fix can be more than the fertilizer N an average farmer can afford to buy and apply according to a study done by Rumjanek. Therefore, legume-rhizobia symbiosis can provide an easy and inexpensive way to enhance soil fertility and improve crop production.

About 20 million tons of atmospheric nitrogen is transformed into ammonia which is 50–70% of the world BNF by root nodule rhizobia. The success of a symbiotic relationship between legumes and rhizobia is determined by the high fixed nitrogen in hosts. Hosts are able to potentially obtain different rhizobia when invading new habitats. However, host range expansion may be limited by the symbiont distribution. Therefore, this chapter will be looking at the sustainable enhancement of soil fertility using bioinoculants.

15.2 Effective Bioinoculants and Their Uses in Soil Fertility Management

Microbial inoculants also known as soil inoculants or bioinoculants or biofertilizers are agricultural amendments that use beneficial rhizospheric or endophytic microbes to promote plant health. Bioinoculants are composed principally of fungal and/or bacterial isolates, and infrequently comprise of other abiotic additives such as nutrients, or inorganic/organic carriers. Numerous microbes involved form symbiotic relationships with the target crops where both parties benefit (mutualism). Microbial inoculants not only promote plant growth by providing essential nutrients, but also promote plant growth by stimulating plant hormone production.

Microbial biofertilizers are mostly made of bacteria, fungi, and cyanobacteria, of which most of them are able to form symbiotic relationships with plants. The most imperative forms of microbial fertilizers, based on their nature and occupation, are those that provide nitrogen and phosphorus. The following are examples:

15.2.1 Nitrogen-Fixing Biofertilizers

They convert nitrogen into ammonia an organic compound in plants. They are subdivided into three groups called free-living, symbiotic, and associative biofertilizers.

(a) *Free-living biofertilizers.* Free-living nitrogen-fixing bacteria do not need to generate a symbiotic relationship with plants to survive and replicate. These are essential due to the fact that numerous plants such as corn are not able to make a symbiotic relationship with nitrogen-fixing bacteria. These natural microbes exist in small concentrations relative to the overall microbial population within the soil. Examples of such bacteria include *Anabaena, Azotobacter, Beijerinckia, Derxia, Aulosira, Tolypothrix, Cylindrospermum, Stigonema, Clostridium, Klebsiella, Nostoc, Rhodopseudomonas, Rhodospirillum, Desulfovibrio, Chromatium,* and *Bacillus polymyxa.*

(b) *Symbiotic biofertilizers.* Soil microorganisms specifically bacteria called *rhizobia* are able to form symbiotic relationships with plants by colonizing the rhizosphere, infecting legume roots, and biologically fixing nitrogen in soil. Therefore, making soils fertile for plant growth great examples of *Rhizobia* include *Rhizobium, Bradyrhizobium, Sinorhizobium, Azorhizobium, Mesorhizobium, Allorhizobium, Frankia, Anabaena azollae,* and *Trichodesmium.*

(c) *Associative bacteria* (blue cells) are in close interaction with the plant surface and sometimes can be found within plant tissues. Associative N fixation (ANF) is the process by which dinitrogen gas is converted into ammonia by bacteria in casual association with plants, meaning they have a less intimate association with roots. Examples include *Azospirillum* spp. (*A. brasilense, A. lipoferum, A. amazonense, A. halopraeferens,* and *A. irakense*), *Acetobacter diazotrophicus, Herbaspirillum* spp., *Azoarcus* spp., *Alcaligenes, Bacillus, Enterobacter, Klebsiella,* and *Pseudomonas.*

15.3 Phosphorus Acting Microbes

These are equally subdivided into two groups, namely the phosphate solubilizing and phosphate mobilizing.

(a) *Phosphate solubilizing* microorganisms (PSMs) are a group of beneficial microorganisms capable of hydrolyzing organic and inorganic phosphorus compounds from insoluble compounds. Examples include *Bacillus megaterium var. phosphaticum, B. subtilis, B. circulans, B. polymyxa, Pseudomonas striata, Penicillium* spp., *Aspergillus awamori, Trichoderma, Rhizoctonia solani, Rhizobium, Burkholderia, Achromobacter, Agrobacterium, Micrococcus, Aerobacter, Flavobacterium,* and *Erwinia.*

(b) *Phosphorus-mobilizing bacteria* (PMB) are beneficial bacteria that effectively mobilize P through solubilization of sorbed P pools and mineralization of organic P compounds which are otherwise not readily available to the plant. Examples include *Arbuscular mycorrhiza* (*Glomus* sp., *Gigaspora* sp., *Acaulospora* sp., *Scutellospora* sp., and *Sclerocystis* sp.), *ectomycorrhiza* (*Laccaria* spp., *Pisolithus* spp., *Boletus* spp., *Amanita* spp.), *ericoid mycorrhizae* (*Pezizella ericae*), and *orchid mycorrhiza* (*Rhizoctonia solani*).

15.4 Micronutrients Biofertilizers

These are grouped in the potassium solubilizing, silicate and zinc solubilizing biofertilizers. Soil-dwelling microorganisms can also be used as biofertilizers to make available many other types of nutrients other than nitrogen and phosphorus such as potassium, zinc, iron, and copper. Certain rhizobacteria can solubilize insoluble potassium forms, which is another important nutrient for plant growth. Large biomass yields due to improved potassium uptake have been detected with *Bacillus edaphicus* (for wheat), *Paenibacillus glucanolyticus* (for black pepper), and *Bacillus mucilaginosus* in co-inoculation with the phosphate-solubilizing *Bacillus megaterium* (for eggplant, pepper, and cucumber) (Etesami et al. 2017). Another important mineral is zinc, and is present in very low concentrations in the Earth's crust, because of this it is externally applied as the expensive soluble zinc sulfate to overcome its insufficiencies in the plant. Luckily some microbes such as *Bacillus subtilis, Thiobacillus thiooxidans,* and *Saccharomyces* spp. are able to solubilize the insoluble inexpensive zinc compounds like zinc oxide, zinc carbonate, and zinc sulfide in soil. Equally, other microorganisms can hydrolyze silicates and aluminum silicates by providing protons (which causes hydrolysis) and organic acids (that make complexes with cations and bring them back into a dissolved state) to the medium while metabolizing, and this is extremely advantageous to the plants. For example, an increase in rice growth and grain yield because of an increased dissolution of silica and nutrients from the soil was observed using a silicate-solubilizing *Bacillus* sp. combined with siliceous residues of rice straw, rice husk, and black ash according to a study done by Cakmakci et al. (2007).

15.5 Growth Promoting Rhizobium

Aside from nitrogen-fixing and phosphorus-solubilizing microbes, there are microbes that promote plant growth by synthesizing growth-promoting chemicals. For instance, rhizospheric *Bacillus pumilus* and *Bacillus licheniformis* were observed to produce substantial quantities of physiologically active plant hormone gibberellin. Examples of such microorganisms include *Agrobacterium, Achromobacter, Alcaligenes, Arthrobacter, Actinoplanes, Azotobacter, Bacillus, Pseudomonas fluorescens, Rhizobium, Bradyrhizobium, Erwinia, Enterobacter, Amorphosporangium, Cellulomonas, Flavobacterium, Streptomyces,* and *Xanthomonas.*

15.6 Compost Biofertilizers

Compost is a rotting, breakable, murky material that makes a symbiotic food web within the soil, which consists of approximately 2% (w/w) of phosphorus, potassium, and nitrogen, along with microorganisms, earthworms, and dung beetles. Compost is produced from an extensive diversity of materials like straw, leaves, cattle-shed bedding, fruit and vegetable wastes, biogas plant slurry, industrial wastes, city garbage, sewage sludge, factory waste, etc. Microbial organic solid residue oxidation causes the creation of humus-containing material, that can be used as an organic fertilizer that sufficiently aerates, aggregates, buffers, and keeps the soil moist, aside from providing advantageous minerals to the crops and increasing soil microbial diversity. Compost is synthesized by many different decomposing microorganisms such as *Trichoderma viride, Aspergillus niger, A. terreus, Bacillus* spp., Gram-negative bacteria *such as Pseudomonas, Serratia, Klebsiella,* and *Enterobacter*, etc. that are able to plant cell wall-degrading cellulolytic or lignolytic and other activities, besides having proteolytic activity and antibiosis (by production of antibiotics) that suppresses other parasitic or pathogenic microorganisms. Another important type (vermicompost) contains earthworm cocoons, excreta, microorganisms (like bacteria, actinomycetes, fungi), and different organic matters, which provide nitrogen, phosphorus, potassium, and several micronutrients, and efficiently recycles animal wastes, agricultural residues, and industrial wastes cost-effectively and uses low energy.

15.7 Factors Affecting Compost

There are mainly four factors that control and affect the composting process, namely nutrition (carbon: nitrogen ratio of the material), moisture content, oxygen (aeration), and temperature.

15.7.1 Carbon to Nitrogen Ratio

In order for decomposition to occur the microorganisms responsible for decomposition of organic matter require a specific amount or ration of carbon and nitrogen as a nutrient to grow and reproduce. Microbes work efficiently if carbon: nitrogen ratio is 30:1. If the carbon ratio surpasses 30, the rate of composting reduces due to reduced growth and reproduction of the microbes. Therefore, C: N ratios that are as low as 10:1 or as high as 50:1 will cause a decrease in decomposition of the organic waste material.

15.7.2 Moisture Content

Ideally the percentage of the moisture content is supposed to be 60% (El-Haggar et al. 1998). The original moisture content must range within 40–60% reliant on the components of the mixture. A simple reduction of moisture by less than 40% will cause microbial activity to slow down and became dormant. An increase of the moisture content above 60% will cause decomposition to slow down and an odor from the anaerobic decomposition will be emitted.

15.7.3 Oxygen (Aeration)

A nonstop supply of oxygen through aeration is extremely vital to guarantee aerobic fermentation (decomposition). Good aeration is required to control the environment for biological reactions and achieve the optimal efficiency. Diverse procedures can be used to perform the required aeration according to the composting techniques. The most frequent forms of composting techniques are: vermin-composting, forced composting natural composting, and passive composting.

15.7.4 Temperature

Heat is produced by the activity of bacteria and other microorganisms while decomposing (oxidize) organic material. The heat produced ranges from 32 to 60 ° C for efficient compost to occur, but if the temperatures rise or decrease outside this range, microbial activity will slow down, or might be destroyed.

15.8 Nitrogen Fixation in Legumes and Its Benefits

All organisms capable of fixing nitrogen into the soil come from the biological group called prokaryotes. Nitrogen-fixing organisms come from the family *rhizobiaceae* (*Rhizobium, Sinorhizobium, Bradyrhizobium, Mesorhizobium and Azorhizobium*, collectively called rhizobia) and all are able to form a symbiosis with leguminous plants (Zahran 2001) and non-leguminous trees (e.g., *Frankia*).

All organisms that reduce N to ammonia do so with the aid of an enzyme complex called nitrogenase (Zahran 2001). The nitrogenase enzymes are irreversibly inactivated by oxygen, and the process of nitrogen fixation uses a large amount of energy (Zahran 2001). Acetylene reduction assay is used to measure nitrogenase activity, which is economical and sensitive (Zahran et al. 2012).

Rhizobial nodulation at the genetic level is done by the nodulation genes called *nod*, *noe*, and *nol* (Steenkamp et al. 2008). The structure and formation of nodules genes is regulated and controlled by Nod factors (NFs), which represent the primary signaling molecules implicated in host infection and nodule organogenesis (Steenkamp et al. 2008).

15.9 Benefits of Nitrogen Fixation in Legumes

The most limiting nutrient for growth of leguminous plants like common beans, soya beans, cowpeas, and garden peas is N (Howieson and Committee 2007). N is vital in plant cells for synthesis of enzymes, proteins, chlorophyll, deoxyribonucleic acid (DNA), and ribonucleic acid (RNA), thus extremely important for plant growth and synthesis of food and feed (Matiru and Dakora 2004).

The major producers of leguminous plants in Africa are small-scale farmers and they hardly use fertilizers due to their low income; that is why this crop is highly dependent on nitrogen fixation. The success of legumes is impacted largely by the legume-rhizobial symbiosis, hence atmospheric nitrogen fixed by the organisms is more than the fertilizer nitrogen an average farmer can afford to buy and apply. Thus, legume-rhizobia symbiosis provides an easy and low-cost way of improving soil fertility and crop production (Roychowdhury et al. 2013).

15.10 Studies Done on the Use of BNF and Its Effects on Crop Yields

According to study done by Mukelabai, cowpeas are largely grown by smallholder farmers in Namibia and other parts of Sub-Saharan Africa under low input agricultural systems with little to no fertilizer application; that is why biological nitrogen fixation in traditional cropping system is of great importance for system sustainability. The cowpea residue is typically incorporated into the soil, and therefore the high N and P content in the shoots resulting from enhanced plant growth and nitrogen fixation could provide additional residual N and P for subsequent crops (Giller 2002).

The effectiveness of the inoculant strains in fixing nitrogen was proved in the production of higher grain yield per hectare as well as an increase in the pod number per hectare in inoculated plants when compared to the non-inoculated plants (Fig. 15.1) according to a study done by Mukelabai. The result obtained were similar with a report by Martins et al. (2003) which demonstrated that inoculation of cowpeas increased the grain yield. Mukelabai showed that there was an increase in the grain yield by 61% due to the use of an inoculant. Likewise, Martins et al. (2003) saw a significant increase in grain yield of up to 30% (533–693 kg/ha).

15 Sustainable Enhancement of Soil Fertility Using Bioinoculants

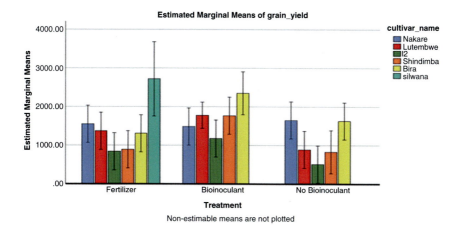

Fig. 15.1 Graph of the two-way Anova analysis of grain yield among the three treatments

Table 15.1 The bioinoculants that were used in cowpea test fields by Mukelabai

Strain/inoculant	Genus	Host
1–7	*Bradyrhizobium*	Cowpea (6)
14–3	*Bradyrhizobium*	Cowpea (6)

Mixing of both phosphate and the bio-inoculum increased the grain weight by 61% while the non-inoculated field or negative control only containing phosphate only increased the grain weight by 35% according to Mukelabai. Onduru et al. (2008) correspondingly described a comparable positive interaction between the inoculant and phosphate fertilizer for cowpea grain yield which led to a 54% increase in grain yield compared with control field.

The reason for the increase in grain yield can be attributed to the important role that phosphate plays in both nodulation, nitrogen fixation and plant growth processes through enhanced root development and root hair formation (Nielsen et al. 2001; Nziguheba et al. 2016), nodule initiation, growth and as energy source for N fixation process that have direct effects on nitrogenase activity in nodules and photosynthesis (Hogh-Jensen et al. 2002). Hence, the combination of phosphate fertilizer to N fixing legumes on phosphate lacking soils further improved nitrogen fixation and crop yields.

In a study that was conducted in Omusati region by Mukelabai it was observed that in spite of high numbers of indigenous rhizobia population sizes across the study location, inoculation with the strains 1–7 and 14–3 (Table 15.1) increased the grain weight, yields per hectare and the number of pods per hectare in the cultivars used. This study's result corresponds to a study done by Gachande and Khansole (2010) who also identified the Rhizobium *Bradyrhizobium japonicum* based on its morphological, cultural, biochemical characteristics and its benefits in host specificity.

Small-scale farmers that are the main legume producers in Africa hardly use fertilizers during legume production due to their low income; Therefore, the crop is largely dependent on fixed nitrogen from symbionts. Unfortunately, most time, the native nitrogen fixers are competitive to inoculants, but they might not be the best or even an efficient strain and could possibly be unsuited to the host plant.

Therefore, we cannot rely on native nitrogen fixers without acquiring prior information on its competence and compatibility with host legume, because it can lead to crop production failure. Hence identification of rhizobia strains in African farmers' fields using specific hosts and testing their efficiency in nitrogen fixation and effectiveness in cross inoculation is crucial for improvement of legume production.

The host specificity leads to a perfect match between legume and rhizobia resulting into effective nodule (deep red inside) formation and nitrogen fixation. If cross inoculation with no perfect match has occurred, ineffective nodules (green or white inside) or no nodules may be formed, and nitrogen fixation does not occur (Gwata et al. 2004).

Hence host specificity may have played a role and that is why this study only acquired a small amount of rhizobial strains while the other study managed to collect and sequenced a larger number of rhizobial strains of up to 75 isolates due to the different types of crops they used to trap different isolates. The indole production test was negative for all the isolates; this result was similar to that observed by Monica et al. (2015).

15.11 Conclusions

Inoculants are more cost effect for small-scale farmers compared to the expensive chemical N fertilizers. Therefore, proving that inoculants are able to sustain and improve soil fertility in degraded, nutrient deprived and desolate areas of the world. Furthermore, these bioinoculants can be used to act as amendments that enhance the biodegradation process. Research toward this is required to also allow for the isolation of natural and locally adapted microbial cocktails that can be used as bio-fertilizer.

References

Ali S, Liu Y, Ishaq M, Shah T, Abdullah, Ilyas A, Din I (2017) Climate change and its impact on the yield of major food crops: evidence from Pakistan. Foods 6:39

Cakmakci R, Dönmez MF, Erdoğan Ü (2007) The effect of plant growth promoting rhizobacteria on barley seedling growth, nutrient uptake, some soil properties, and bacterial counts. Turk J Agric For 31:189–199

Doyle JJ, Luckow MA (2003) The rest of the iceberg. Legume diversity and evolution in a phylogenetic context. Plant Physiol 131:900–910

El-Haggar SM, Hamoda MF, Elbieh MA (1998) Composting of vegetable waste in sub-tropical climates. Int J Environ Pollut 9:411–420

Eskin N (2012) Colonization of *Zea mays* by the nitrogen fixing bacterium. *Gluconacetobacter diazotrophicus*

Etesami H, Emami S, Alikhani HA (2017) Potassium solubilizing bacteria (KSB): mechanisms, 507 promotion of plant growth, and future prospects, a review. J Soil Sci Plant Nutr 17:897–911

Farahvash F, Mirshekari B, Niosha Z (2010) Effects of biofertilizers (*Azotobacter* and nitroxine) and different rates of chemical fertilizers on some attributes of cowpea (*Vigna unguiculata* (L.) Walp.). J Food Agric Environ 8 (2):664–667

Freitas H, Prasad M, Pratas J (2004) Plant community tolerant to trace elements growing on the degraded soils of são domingos mine in the South East of Portugal: environmental implications. Environ Int 30:65–72

Gachande BD, Khansole GS (2010) Morphological, cultural and biochemical characteristics of *Rhizobium japonicum syn* and *Bradyrhizobium japonicum* of soybean. Biosci Discov 2(1):1–3

Giller KE (2002) Targeting management of organic resources and mineral fertilizers: can we match scientists' fantasies with farmers' realities? In: Vanlauwe B, Diels J, Sanginga N, Merckx R (eds) Integrated plant management in Sub Saharan Africa: from concept to practice. CAB International, Wallingford, pp 155–171

Gwata E, Wofford DS, Boote K, Mushoriwa H (2004) Determination of effective nodulation in early juvenile soybean plants for genetic and biotechnology studies. Afr J Biotechnol 2:417–420

Hase F (2013) Facilitating conservation agriculture in namibia through understanding farmers' planned behaviour and decision making. MSc Thesis. Swedish University of Agricultural Sciences

Hendawey M, Younes A (2013) Biochemical evaluation of some faba bean cultivars under rainfed conditions at El-Sheikh Zuwayid. Ann Agric Sci 58:183–193

Hogh-Jensen H, Schjoerring JK, Soussana JF (2002) The influence of phosphorus deficiency on growth and nitrogen fixation of white clover plants. Ann Bot 90:745–753

Howieson J, Committee GRP (2007) Technical issues relating to agricultural microbial genetic resources (AMiGRs), including their characteristics, utilization, preservation and distribution: draft information paper

Lane A, Jarvis A (2007) Changes in climate will modify the geography of crop suitability: agricultural biodiversity can help with adaptation. SAT eJournal ICRJSA T 4:1–12

Marschner P, Crowley D, Rengel Z (2011) Rhizosphere interactions between microorganisms and plants govern iron and phosphorus acquisition along the root axis-model and research methods. Soil Biol Biochem 43:883–894

Martins LMV, Xavier GR, Rangel FW, Ribeiro JRA, Neves MCP, Morgado LB, Rumjanek NG (2003) Contribution of biological fixation to cowpea: a strategy for improving seed yield in the semi-arid region of Brazil. Biol Fertil Soils 38:333–339

Matiru VN, Dakora FD (2004) Potential use of rhizobial bacteria as promoters of plant growth for increased yield in landraces of African cereal crops. Afr J Biotechnol 3:1–7

Mohammadi K, Sohrabi Y (2012) Bacterial biofertilizers for sustainable crop production: a review. J Agric Biol Sci 7:307–316

Monica N, Roxana V, Carmen P, Ioan R, Rodica P (2015) Isolation and biochemical characterization of *Rhizobium leguminosarum* bv. *Trifolii* and *Sinorhizobium meliloti* using API 20 NE and API 20 E. Bull USAMV Ser Agric 72(1):174–178

Nielsen KL, Eshel A, Lynch JP (2001) The effect of phosphorus availability on carbon economy of contrasting common bean (*Phaseolus vulgaris* L.) genotypes. J Exp Bot 52:329–339

Nziguheba G, Zingore S, Kihara J, Merckx R, Njoroge S, Otinga A (2016) Phosphorus in smallholder farming systems of sub-Saharan Africa: implications for agricultural intensification. Nutr Cycl Agroecosyst 104:321–340

Ogada MJ, Nyangena W, Yesuf M (2010) Production risk and farm technology adoption in the rainfed semi-arid lands of Kenya. Afr J Agric Resour Econ 4:159–174

Onduru D, De Jager A, Muchena F, Gachini G, Gachimbi L (2008) Exploring potentials of rhizobium inoculation in enhancing soil fertility and agroeconomic performance of cowpeas in sub-saharan Africa: a case study in semi-arid Mbeere, Eastern Kenya. Am Eurasian J Sustain Agric 2:185–197

Reinhold-Hurek B, Bünger W, Burbano CS, Sabale M, Hurek T (2015) Roots shaping their microbiome: global hot spots for microbial activity. Annu Rev Phytopathol 53:403–424

Roychowdhury R, Banerjee U, Sofkova S, Tah J (2013) Organic fanning for crop improvement and sustainable agriculture in the era of climate change. Online J Biol Sci 13:50–65

Sprent JI, James EK (2007) Legume evolution: where do nodules and mycorrhizas fit in? Plant Physiol 144:575–581

Stajković O, Delić D, Jošić D, Kuzmanović Đ, Rasulić N, Knežević-Vukčević J (2011) Improvement of common bean growth by co-inoculation with *Rhizobium* and plant growth-promoting bacteria. Rom Biotechnol Lett 16:5919–5926

Steenkamp ET, Stepkowski T, Przymusiak A, Botha WJ, Law IJ (2008) Cowpea and peanut in southern Africa are nodulated by diverse *Bradyrhizobium* strains harboring nodulation genes that belong to the large pantropical clade common in Africa. Mol Phylogenet Evol 48:1131–1144

Taapopi M, Kamwi JM, Siyambango N (2018) Perception of farmers on conservation agriculture for climate change adaptation in Namibia. Environ Nat Resour Res 8(3):33–34

Tran LSP, Nguyen HT (2009) Future biotechnology of legumes. In: Emerich WD, Krishnan H (eds) Nitrogen fixation in crop production. The American Society of Agronomy, Crop Science Society of America and Soil Science Society of America, Madison, pp 265–308

Wong M (2003) Ecological restoration of mine degraded soils, with emphasis on metal contaminated soils. Chemosphere 50:775–780

Zahran HH (2001) Rhizobia from wild legumes: diversity, taxonomy, ecology, nitrogen fixation and biotechnology. J Biotechnol 91:143–153

Zahran HH, Abdei-Fattah M, Yasser MM, Mahmoud AM, Bedmar EJ (2012) Diversity and environmental stress responses of rhizobial bacteria from Egyptian grain legumes. Aust J Basic Appl Sci 6:571–583

Chapter 16
The Potential of Vermicomposts in Sustainable Crop Production Systems

M. T. Mubvuma, I. Nyambiya, K. Makaza, P. Chidoko, C. P. Mudzengi, E. Dahwa, X. Poshiwa, R. Nyamusamba, A. M. Manyanga, M. Muteveri, and H. A. Mupambwa

Abstract The application of vermicompost in semi-arid regions of Southern Africa has been reported to improve seed germination, and enhance seedling growth and development in several horticultural crops, cereals and fruit trees. The study had the objectives to explore the potential of vermicomposting technology in sustainable crop production systems for enhanced adoption. Vermicomposting as a soil amendment has been found to reduce crop irrigation water needs, improve crop water use efficiency, increase leaf area index and radiation use efficiency. Several crop yield growth parameters that include plant height, number of flowers per plant, number of pods per plant, number of fruits per plant, and final crop yield have been reported to improve in situations where vermicompost has been applied, particularly in problem soils. Vermicomposting technology also functions as a cultural method of managing crop insect pests and weeds. In addition, application of vermiwash (aqueous vermicompost extract) into the soil or spraying on surface of plants has been reported to significantly suppress pathogens and pests. Moreover, vermicomposting has been reported to reduce soil salinity and sodicity by as much as 37% and 34%. The liquid vermiwash from vermicompost contains hormones (cytokines, auxin) that facilitate crop growth and development, decomposer bacteria that suppress pathogens,

---

M. T. Mubvuma (✉) · K. Makaza · P. Chidoko · R. Nyamusamba · A. M. Manyanga
Department of Soil and Plant Sciences, Gary Magadzire School of Agriculture, Great Zimbabwe University, Masvingo, Zimbabwe
e-mail: mmubvuma@gzu.ac.zw

I. Nyambiya · M. Muteveri
Department of Physics, Geography and Environmental Sciences, School of Natural Sciences, Great Zimbabwe University, Masvingo, Zimbabwe

C. P. Mudzengi · E. Dahwa · X. Poshiwa
Department of Livestock, Wildlife and Fisheries, Gary Magadzire School of Agriculture, Great Zimbabwe University, Masvingo, Zimbabwe

H. A. Mupambwa
Sam Nujoma Marine and Coastal Resources Research Center, University of Namibia, Henties Bay, Namibia

© The Author(s), under exclusive license to Springer Nature Singapore Pte Ltd. 2023
H. A. Mupambwa et al. (eds.), *Vermicomposting for Sustainable Food Systems in Africa*, Sustainable Agriculture and Food Security,
https://doi.org/10.1007/978-981-19-8080-0_16

amylase, and cellulose that stabilize physical, chemical and biological properties of soil. Vermicompost is cheap, affordable, climate smart, and is ecologically friendly. Although vermicomposting has been regarded as a clean, sustainable, and zero-waste approach to manage organic wastes, there are still some constraints in the popularization of this technology. The major concern amongst the farmers is difficulty in integrating vermicomposting technology into the existing conventional farming systems, and lack of technical know-how. However, with improvement in awareness, there is great potential of vermicomposting technology to play a major role in sustainable crop production systems.

Keywords Vermicompost · Crop production · Radiation use efficiency

16.1 Introduction

Vermicomposting is a non-thermophilic process where earthworms and microbes biodegrade and transform organic waste products into nutrient rich, humified organic fertilizer (Thakur et al. 2021). The process is bio-oxidative and involves natural decomposition process where earthworms will feed on plants, animal or sledge waste material under mesophilic conditions, digest it and excrete fertile humus-like substances called vermin-compost (Patwa et al. 2020; Ahmed and Deka 2022). This technology is economic, eco-friendly, and is a sustainable way for recycling waste from agriculture and agro food industry (Sherathiya Pinalben and Patel 2022). There are two main methods of vermicomposting (bed method and pit method). The bed method involves arranging the organic matter in beds, whilst pit method is when organic waste is arranged in cement pits. The pits often lack aeration and oxygen and this reduces quality of the fertilizer.

Vermicompost has several advantages that include being eco-friendly (Ahmed and Deka 2022), ability to enrich soil fertility levels, being organic, being cheap and its capability to solve problem of waste disposal in a sustainable way (Kiyasudeen et al. 2020). However, vermicompost has problems of requiring longer period of time for transforming the organic waste products into fertilizer. Besides, the method needs moisture content supply which may be difficult to sustain in areas where water availability is a challenge. In addition, the technology requires proper supply of oxygen so as to achieve a good quality fertilizer; however, oxygen levels are often low in pit vermicomposting.

Previous studies have shown a wide application of vermicomposting technology in agriculture and food products waste management. Soto et al. (2022) reported on importance of using tomato waste in vermicomposting whilst Kakati et al. (2022) studied the use of potato waste. Rupani et al. (2022) worked on vermicomposting using palm oil mill waste. All the above cited authors seem to concur on the efficiency of earthworms in managing waste and are quick to conclude that vermicompost is an indispensable fertilizer for plant growth. However, despite all the advantages of vermicompost, very few farmers are using the technology at farm level. There are problems of knowledge gap on how farmers may adopt the

16 The Potential of Vermicomposts in Sustainable Crop Production Systems 263

technology (Vuković et al. 2021). Besides, there is scanty research on the potential of vermicomposts in sustainable crop production systems. Very few studies have reported the potential of vermicomposting technology in sustainable crop production systems, especially under smallholder farming systems. In situations where vermicompost technology has been reported in literature, the effect of its use on sustainable crop production has remained an inexhaustible research gap. In addition, there is limited literature on the use of vermicomposting in seedling production and development (Kamatchi et al. 2018). Moreover, there is scanty information on the effect of vermicomposting on crop physiological parameters such as water use efficiency, radiation capture and its utilization. Information on the response of crop phenological parameters after application of vermicompost has also received minimal attention. Furthermore, the bio-chemistry and physico-chemical properties of vermicompost and how they promote favourable plant root nutrient absorption have scantly been published. Besides, the effects of vermicompost on sustainable weed management, crop arthropod pest and disease management has received limited attention in literature. In Southern Africa, the use of this technology is still limited amongst farmers mainly due to lack of information on optimized production of these organic nutrient sources.

One of the solution for low knowledge gap amongst farmers is by organizing various training and extension activities (Sharma and Garg 2019). Thus this chapter consolidates available literature on the potential of vermicomposting technology in sustainable crop production systems for enhanced adoption of this technology by smallholder farmers.

16.2 Vermicomposting Research in Southern Africa Region

Vermicomposting is a technology that is not very new but has started gaining momentum and recognition in agriculture as a (Coulibaly et al. 2018) sustainable agriculture technology though, not much research has been done on this technology, particularly in the Southern Africa region. However as part of climate-smart agriculture, the body of knowledge around vermicomposting as a tool for improving crop production is steadily growing (Murungweni et al. 2016).

Basic research conducted in Southern Africa has revealed the existence and diversity of earthworms used in vermicomposting. The lifestyle of the African nightcrawler *Eudrilus eugeniae* (Oligochaeta) (Vujoen and Reinecke 1989), its moisture preferences, production and reproduction has been studied in Africa, and has improved the knowledge of vermicomposting in local soils (Viljoen and Reinecke 1990). Other published material includes research on the lifestyle of the European earthwork on African soil (*Dendrobaena veneta* (Oligochaeta) (Viljoen et al. 1991), and a review providing detail on the taxonomy of the African nightcrawler as a vermicomposting agent (Blakemore 2015). The vermicomposting work in Southern Africa appears to build on this research, which has taken place over the past 100 years on the nature of composting earthworm. Some studies have been completed in South Africa with the aim of determining the influence of such material as fly ash generated as waste from power stations on the vermicomposting process. This material has been mediated using earthworms for the possible production of organic fertilizers (Mupambwa 2015). Several studies are being undertaken to investigate the optimized parameters for preparing vermicompost such as the earthworm stocking density, C/N ratio, inorganic material incorporation ratios; moisture content; amongst others (Mupambwa and Mnkeni 2016; Mupambwa et al. 2017).

The practice of vermicomposting has thus gained momentum especially in the twenty-first century (Adhikary 2012) with a number of organizations and individuals beginning to undertake worm farming for the purpose of commercial vermicomposting (Furlong and Innovation 2016) for organic crop production. In a study conducted in the hot region of KwaZulu-Natal, different temperature regimes were used to test the resilience of the tomato crop treated with vermicompost leachate. The results have demonstrated the capacity of vermicomposting to improve total chlorophyll, sugars and proline amounts (Coulibaly et al. 2018) and as such to improve water and radiation use efficiencies in plants.

In soils contaminated with metals from mine tailings done in Zambia, manuring and vermicomposting have been used as methods to immobilize metals such as copper, cadmium, nickel, lead and zinc so that they do not affect plants (Kaninga et al. 2020). The sequestration of unwanted metal excess in saline drylands is a tool that can be used for soil remediation as demonstrated in the experimental mixture of various proportions of igneous rock phosphate with manure/paper compositions (Mupondi et al. 2018). The increase in the amount of bioavailable and absorbable phosphorus (P) is important in dryland regions as phosphorus is important and needed not just for making plant cell walls, but also for the electron transport systems which utilizes water and radiation (Kaninga et al. 2020). One review has looked at the potential of using vermicomposting as climate-smart agriculture in the cool Namib desert. This region is amenable to producing crops without excessive loss of water through transpiration in spite of being a desert (Mupambwa et al. 2019). More recent experimental studies on species abundance of earthworms over four cropping seasons have also been undertaken in the dryland Eastern Cape region of South Africa as part of conservation agriculture practices (McInga et al. 2020). Our chapter thus presents information on the potential use of vermicomposts in sustainable farming systems, thus highlighting options that are available to encourage adoption of this technology mostly by smallholder resource poor farmers.

16.3 Effect of Vermicompost on Seed Germination, Seedling Growth and Development

Application of organic amendment in crop production is increasingly being adopted as an environmentally, economically and agronomically sound practice as validated by many studies.

Compared to other organic amendments, the vermicompost amendment is increasingly being utilized across a wide range of intensive agricultural landscapes. Vermicompost is a much finer structure than composts and contains readily available forms of nutrients making this amendment the most suitable media for seed germination and support for initial seedling growth and development (Arancon et al. 2019; Ebrahimi et al. 2020).

The effect of vermicompost on germination is a function of the interaction between the crop being grown and the concentration of the vermicompost. For example, cabbage, radish and Swedish turnip exposed to lower levels of vermicompost (20%) resulted in germination promotion which was beneficial. Moderate levels of vermicompost (30–50%) resulted in a 50% reduction in germination (Kamatchi et al. 2018). In another study, legumes such as beans and peas were highly inhibited at a vermicompost concentration of 20%, with a linear increase in germination inhibition with vermicompost concentration (Ceritoglu et al. 2021). It is therefore important to make sure that seedlings are not grown in high concentrations of vermicompost. It is also important to conduct studies on the effect of vermicompost on individual crops since the effect is crop-specific (Gashaw and Woldetsadik 2020). The effect of vermicompost on seed germination can also be realized in early seedling growth and development (Ievinsh 2011). However, the effect of vermicompost on seedling growth and development is affected by the type of vermicompost. A high dose of vermicompost can affect seedling development which is expected as the young seedlings are tender. Various mechanisms were identified behind the dualistic nature of vermicompost on seedlings. Firstly, vermicompost contains a lot of nitrate and ammonium which are useful in breaking seed dormancy (Hussain and Abbasi 2018), promoting seed germination. On the other hand, the inhibitory effect may be attributed to excess supply of nutrients, salinization from the vermicompost which slows down water uptake resulting in inhibition (Moles et al. 2019) and also the presence of phenolic compounds which suppress seedling growth (Tanase et al. 2019). Considering the fragility of seedlings, and the need to avoid seedling losses, it is critical that all amendments be tested for their nutrition profile, Ph and EC before application to avoid any unfavourable effects on crop seeds.

In general, the germination and growth of plants are significantly increased in vermicompost to a certain concentration level where inhibition starts. These studies were earlier on corroborated by Atiyeh et al. (2002) who reported significant benefits of vermicompost in promoting germination and initial seedling growth.

16.4 Effect of Vermicompost on Crop Physiological and Phenological Parameters

Research shows the ability of vermicomposting to improve the amount of chlorophyll a and b cell in plants and therefore resulting in improved photosynthetic activity within a plant. The improvement of chlorophyll ensures better use of the available radiation within the plant, and may result in better resilience of crops in dry land regions (Vuković et al. 2021). In a study conducted in China, the use of vermicomposting has been demonstrated to increase the available organic matter to the soil by 38.0–47.9%, whilst making nitrogen available by 92.0–119.0%, and immobilizing phosphorous by about 527.8–595.3% (Li et al. 2020a, b). The higher enzyme activity was noted as the reason for this significant effect that vermicomposting has on soil fertility (Li et al. 2020a, b). This is quite remarkable in building resilience to heat stress in dry land areas where water is scarce and as such ensuring crops make efficient use of the available radiation. Experimental studies have shown that vermicomposting has the potential to increase photosynthetic quantum yield and as such improve crop yields (Younas et al. 2021).

Furthermore, powder and granular vermicompost as a soil amendment has been found to reduce crop irrigation water needs in grass (Rahimian et al. 2021) and improve crop water use efficiency in eggplant (Ebrahimi et al. 2021). Moreover, in a study where vermicompost made from residues of six desert plants, leaf area index, nitrogen use efficiency and radiation use efficiency were observed to increase significantly (Younas et al. 2021). Indeed, a positive crop response to sludge vermicompost fertilizer has been reported in crop root growth and development in ginseng plants (Eo and Park 2019). In addition, several crop yield growth parameters that includes plant height, number of flowers per plant, number of pods per plant, number of fruits per plant and final crop yield have been reported to improve in situations where vermicompost has been applied (Ahirwar and Hussain 2015; Bellitürk et al. 2017), particularly in problem soils (Fig. 16.1a, b).

Blouin et al. (2019) reported a 26% increase in commercial crop yield of *Cucurbitaceae and Asteraceae* family plants, 78% increase in root biomass after application of vermicompost made from cattle manure. In support of this, Younas et al. (2021) reported massive increase in maize plant height, stem diameter, cob length, number of grains/cob, and 1000 grain weight and grain yield of maize crop that had been treated with vermicompost. Also, Wang et al. (2021) suggested that a joint application of vermicompost and biochar may improve cucumber yield and quality. Moreover, Lin et al. (2021) reported economic benefits and internal quality of tobacco when the crop is treated with vermicompost.

(a) (b)

Fig. 16.1 *Sorghum bicolor* (L. Maonche) fertilized with vermicompost (**a**), and *Eleusine coracana* (**b**) fertilized with vermicompost

16.5 Effect of Vermicompost on Sustainable Weed Management

Managing weeds is one of the biggest challenges of farming in Africa. Poor soils and lack of nutrients exacerbate weed management for many farmers in Africa, especially the small-scale farmers who still rely on mechanical weed management methods.

Timing and source of nutrients for the crop in relation to weed germination and growth is crucial to the competitive ability of the crop (Little et al. 2021). Planting of crops is often delayed by access to fertilizers, which are normally in short supply when the season starts. Farmers either plant without applying basal fertilizers or wait until they have the basal fertilizer and use it during planting. Weeds are generally more efficient in extracting nutrients than the crop. When soils lack enough nutrients for the crop, weeds become more competitive. Application of vermicompost will reduce the need for inorganic fertilizers. Vermicompost has nutrients that are immediately available to the crop, which gives the crop a competitive advantage over the weeds. The crop can grow and close its canopy faster, which reduces sunlight that reaches the weeds growing beneath or within the crop canopy. *Striga* is a parasitic weed that extracts water and nutrients from the host plant. In Africa, *Striga* caused great economic losses in several crops especially maize, sorghum and cowpeas. *Striga* grows well in unhealthy and infertile soils. Increasing organic matter and nitrogen reduces the negative impacts of *Striga*. Application of vermicompost at 1 t/ha together with 46 kg N/ha reduced *Striga* infestation and gave optimum sorghum grain and stover yield (Biri et al. 2016).

Application of vermicompost increases activity by soil microorganisms, and the microorganisms in the soil or on the soil can predate weeds, which reduce weed seedbank and weed growth potential. Addition of vermicompost improves soil structure, soil water infiltration and soil aeration. Some weeds are more competitive in poorly drained soils and compacted soils where crop growth is challenged. In some studies, weeds were found to germinate better in tractor-tyre tracks compared to weeds emerging in non-tracked areas (Jurik and Zhang 1999). Ecological weed management enhanced by the use of vermicompost has great potential in sustainable management of weeds.

16.6 Effects of Vermicompost on Crop Arthropod Pests and Disease Management

Besides acting as an excellent sustainable organic source of plant nutrients, vermicomposts are also to have allelopathic like effects against biotic stresses like arthropod pests, nematodes, mites and pathogenic diseases caused by fungi, bacteria and viruses (Table 16.1). Application of vermicompost or vermiwash (aqueous vermicompost extract) into the soil or spraying on surface of plants significantly

Table 16.1 Chemistry of vermiwash/vermicompost and its benefits

Contents of vermiwash/vermicompost	Benefits of vermiwash/vermicompost	References
Decomposer bacteria	Suppress pathogen	Mondal et al. (2015)
Hormones (Cyto-kines, auxin)	Facilitate plant root and shoot growth	Das et al. (2014) and Tripathi et al. (2005)
Amino acid and mucosal secretion	Suppress pathogen and pest	Liu et al. (2004) and Valembois et al. (1984)
Vitamins	Facilitates growth and development of plant	Das et al. (2014) and Tripathi et al. (2005)
Phosphatase	Stabilize physical, chemical and biological properties of soil as well suppress pathogen and solubilize phosphorus	Ghosh (2018)
Actinomycetes	Suppress pathogen	Das et al. (2014), Tripathi et al. (2005), and Balachandar et al. (2018a, b)
Amylase, cellulose	Stabilize physical, chemical and biological properties of soil as well applied in carbon turnover by degrading organic matter	Das et al. (2014) and Tripathi et al. (2005)
Varieties of micronutrientsNi, Mg, Fe, Ca, K	Facilitate the growth and productivities of plant	Ndegwa and Thompson (2000)

suppresses pathogens and pests. It therefore promotes plant health and disease resistance for accelerated plant development. The use of vermicompost in crop production in the semi-arid regions has great potential to reduce the use of synthetic pesticides and fertilizers, thereby significantly reducing crop production costs especially for the resource constrained dryland communal farmers in the semi-arid tropics. Synthetic insecticides are not only expensive to the smallholder farmers but are a hazard to human health and the environment. The use of vermicompost is therefore an attractive alternative but a cheap and eco-friendly bio-pesticide option for sustainable and organic integrated pest management for the smallholder sector in the semi-arid dryland agriculture.

In addition, vermicompost can act against insect and mite pests of economic importance in agriculture. Abou El Atta and Habash (2021) reported a 96.7% reduction in tow-spotted spider mite (*Tetranychus urticae*) after a 50 g application of vermicompost tea. Furthermore, vermicompost tea has been reported to suppress pests that include aphids (*Myzus persicae*), mealie bugs (*Pseudococcus* spp.), cabbage white caterpillars (*Pieris brassicae*) and nematode in tomatoes (Edwards et al. 2007). Application of 0.5 kg of vermicompost was reported to reduce problems of citrus leaf miners in citrus seedlings (Ullah et al. 2019). Vermicompost or vermiwash (aqueous extract of vermicompost) can be combined with other plant-based (botanical) pest control methods to give a synergistic repellent effect against

16 The Potential of Vermicomposts in Sustainable Crop Production Systems

thrips and mites. Vermiwash, if applied as a folia spray in crops in the field has a repellent effect against arthropod pests. The mechanism of arthropod pest suppression is due to changes nutrient status and/or the chemical composition of the plant tissues which affect the palatability to the pest. Earthworms secrete coelomic fluid in the form of mucus which possesses insecticidal, anti-microbial bio-active compounds (Nadana et al. 2020).

Moreover, application of vermicompost or vermiwash can suppress plant parasitic nematode populations especially the root knot nematode *Meloidogyne hapla* in various crops. The possible explanation may be due to the build-up of nematode trapping fungi or fungi that attack and destroy nematode cysts during vermicomposting. Earthworms normally derive their nutrition from fungi and therefore promote fungal activity in soils through their casts.

Vermicompost contains essential anti-microbial and anti-insect chemical substances. The chemical substances have been reported to suppress soil-borne plant pathogens such as *Plasmodiophora brassicae*, *Phytophthora nicotianae*, *Fusarium lycopersici*, *fusarium oxysporium* and *Rhizoctonia solani*. For specific crops, vermicompost can suppress Pythium spp. on cucumbers, *Fusarium graminearum* in wheat, Rhizoctonia spp. on radishes, Verticillium spp. on strawberries and Phomopsis spp. and *Sphaerotheca fuliginea* on grapes in the field. There are two possible mechanisms of pathogen suppression, namely "specific suppression" and "general suppression" (Edwards et al. 2002). Specific suppression is mediated by competition from a limited range of species of microorganisms and general suppression is based on a much wider diversity of microorganisms. General and specific suppression mechanisms include microbial antagonism, nutrient release, induced host resistance and abiotic inhibitory factors of disease suppression. Vermicomposts are richer in microbial communities as compared to traditional thermophilic composts hence the greater potential for pathogen suppression based on microbial competition or microbial antagonism.

16.7 Chemistry of Vermi-Leachate/Vermicompost and Its Benefits

The physico-chemical properties of vermicomposts are a function of the substrates which are used to produce them. As such the material used for vermicomposting needs to be carefully optimized. A range of substrates and the resultant soil nutrients have been described by Pathma and Sakthivel (2012). This work reveals that the common and important matrix of vermicompost consists of humic acid which increases from 40 to 60% on vermicomposting compared to normal compost. Because of the reduction in pH when several organic materials are used, a carefully optimized procedure can be used to neutralize soils which have higher than normal amounts of lime content (Ilker et al. 2016). Beside the reduction in pH because of the increase in humic and other organic acid, one study has reported an increase in the content of heavy metals from cow dung and biogas slurry vermicomposting (Yadav et al. 2013). Further investigation is necessary to establish the level of bioaccumulation of these heavy metals in the plants since they are detrimental to health.

Vermicomposting has been demonstrated to increase the amounts of soluble nitrogen in the form of nitrates (Bachman and Metzger 2007). In addition to improve cation exchange in the soil, next generation sequencing has revealed the prevalence of *Bacillus, Pseudomonas and Paenibacillus* microbiome in processed vermicompost which is important for plant health (Hashimoto et al. 2021). Research shows a general improvement in the micronutrient and macronutrients in growth of crops such as tomato (Azarmi et al. 2008), the reduction of harmful trace metals by as much as 60–75% whilst improving the growth of leafy vegetables (Yen et al. 2021).

Some of the problems of arid systems is soil salinity and sodicity, i.e. the increase of salts and particularly sodium in the soil (Ding et al. 2021) which generally occurs in arid and semi-arid regions of the world (Zaman et al. 2018). These phenomenon are considered serious environmental problems known to reduce crop yield by between 20 and 50% (Shrivastava and Kumar 2015). Salts are known to increase the osmotic pressure of water against absorption by the plant leading to plant death (Earth Observation System 2022). Vermicomposting can reduce soil salinity and sodicity [by as much as 37 and 34% (Ding et al. 2021)] which aids in the ability of plant to absorb water (Ding et al. 2021; Li et al. 2020a, b).

16.8 Problems Affecting Farmers in Adopting Vermicomposting Technology

The heavy use of and reliance on agrochemicals ever since the beginning of the "green revolution" boosted food productivity in the 1960s at the expense of the environment and society (Wako 2021). These chemicals destroyed the beneficial soil normal microflora, impaired the natural biological resistance in crops making them more susceptible to pests and diseases. On the other hand, these chemically grown foods have adverse effects on human health (Wako 2021). Vermicompost serves as an organic fertilizer which is nutrient rich and an excellent alternative soil conditioner. This impressive innovative technology offers a sustainable way to maintain both crop and soil health particularly in the context of the current and persistent climate changes (Prakash et al. 2021; Mubvuma 2013). Vermicomposting technology thus offers an economically viable, socially safe and environmentally sustainable alternative to the use of agrochemicals.

Although vermicomposting has been regarded as a clean, sustainable, and zero-waste approach to manage organic wastes, there are still some constraints in the popularization of this technology (Kavita and Vinod 2019). A study conducted by Padilla Carrillo et al. (2020) on the level of awareness on the benefits of the adoption of vermicomposting technology using dairy manures in selected cooperatives in Philippines showed that there were several issues and constrains in adopting this technology. Although the level of awareness and adoption were 3.2 and 3.6, respectively, indicating a relatively high utilization of vermicomposting using dairy cattle waste, the major concern amongst the farmers was difficulty in integrating vermicomposting technology (20%), lack of technical know-how (20%), added costs in the production process (30%), costly and limited source capital (20%) and insufficient financial support (10%) (Padilla Carrillo et al. 2020).

Capacity building as well as access to loans are some of the measures that can be taken to increase awareness and adoption and utilization of the technology. Given that knowledge is the prerequisite to any adoption of technology, capacity building helps to solidify the farmer's knowledge, skills, abilities and technical knowhow on how to utilize vermicomposting technology. This will further allow the farmers to adapt should challenges arise along the way. Loan windows are essential for poor resource farmers particularly in developing countries who face financial challenges that prevents them from investing and growing their income. Thus, access to loan windows becomes a powerful antidote to rural poverty alleviation. Shreedevi and Hanchinal (2017) conducted a study on the adoption of vermicomposting technology by farmers of Gulbarga district in Karnataka and found out that land holding, annual income and scientific orientation exhibited positive and significant relationship with their adoption level. If factors to do with land holding, income and training are not addressed, they could hinder adoption of vermicompost technology by the farmers.

Another study by Biswas and Islam (2019) to determine farmers' problem confrontation in organic farming at Magura Sadar Upazila of Bangladesh showed that all of the respondents were in medium to high problem confrontation category in organic farming practice. In this study, lack of knowledge ranked the number one challenge followed by lack of own poultry and livestock ranked. Other respondents indicated less production per unit area, low effectiveness of organic fertilizer, lack of fallow land for rearing poultry and livestock as some of their challenges in adopting organic farming (Biswas and Islam 2019).

High cost of production, inputs and lack of subsidies for organic farming was also shown to pose a great challenge to adopt organic farming (Barik 2017). It is therefore necessary to conduct in-depth empirical studies to identify the challenges, issues and problems faced by farmers so as to come up with appropriate interventions that would increase adoption of vermicomposting technology. Additionally, there is need for policies that facilitate access to resources, so that farmers would achieve higher production and increase income for the improvement of their livelihoods. There is evidence that those farmers having higher extension contact are likely to have lower level of problem confrontation in organic farming.

The economic challenges of a farmer to access resources can influence their ability and willingness to adopt technological innovations (Vedeld 1990). Farmers with a higher income may be less risk averse, have more access to information, and thus have greater capacity to mobilize resources including information hence a high level of innovativeness can be expected from them (Reij and Waters-Bayer 2001).

Farmers' knowledge of the usefulness of improving their soil fertility enhances their willingness to substitute inorganic fertilizer, which is expensive, with low cost technologies such as vermicomposting.

Another study in Ethiopia by Teka et al. (2019) showed that in addition to policy direction and financial constraints, farmers also experienced challenges with predators, parasites and pathogens of earthworms. Earthworms can be the target for a wide range of vertebrate predators, parasites and pathogens like of birds and ants, protozoa and nematodes bacteria, such as *Spirochaeta* sp. and *Bacillus* sp., and fungal have been reported to affect on the earthworms, respectively (Teka et al. 2019). Other farmers also reported actual earthworm handling as their major constraint. The worms are sensitive to moisture content as well as the feeding regimen.

Another factor which hindered investment on vermicomposting on a wide scale in Ethiopia was the sensitivity of the worms to different environmental factors. Earthworms have been shown to have well-defined limits of tolerance to environmental parameters, such as moisture and temperature and if these limits exceed, the earthworms may process the vermicompost very slowly (Teka et al. 2019).

16.9 Conclusion

Vermicomposting is a process where worms are used to transform organic waste products into fertilizer. This technology is organic, environmental friendly, sustainable and has been reported to improve seed germination, enhance seedling growth and development in horticultural crops, cereals and fruit trees. Vermicomposting technology has also been reported to culturally manage crop insect pests and weeds. Indeed, vermicompost technology is an innovative way of using available natural resources at farm level to improve soil fertility and crop yield. The liquid vermiwash from vermicompost has been reported to contain hormones (cytokines, auxin) that facilitate crop growth and development, decomposer bacteria that suppress pathogens, amylase and cellulose that stabilize physical, chemical and biological properties of soil. The technology has potential to improve agricultural production sustainably if further research is put on development of earthworm strains that are adaptable to the local environment. However, vermicompost have its challenges such as lack of knowledge amongst the farmers, economic problems, failures to access loan facilities. Thus, development of policies that promote farmer capacity building and climate-smart agriculture may reduce knowledge gap and promote vermicomposting. Although vermicomposting has several advantages, there are still some constraints in the popularization of this technology. The major concern amongst the farmers is difficulty in integrating vermicomposting technology into the existing conventional farming systems. However, with improvement in awareness which may be solved through conducting on farm research with farmers, conducting farmer training programmes on vermicompost and holding field days with farmers. Indeed, there is great potential of vermicomposting technology to play a major role in sustainable crop production systems.

References

Abou El Atta DA, Habash MG (2021) Evaluation the effectiveness of vermicompost and vermicompost tea on the tow-spotted spider mite, Tetranychus urticae Koch

Adhikary S (2012) Vermicompost, the story of organic gold: a review. Agric Sci 03(07):905–917. https://doi.org/10.4236/as.2012.37110

Ahirwar CS, Hussain A (2015) Effect of vermicompost on growth, yield and quality of vegetable crops. Int J Appl Pure Sci Agric 1(8):49–56

Ahmed R, Deka H (2022) Vermicomposting of patchouli bagasse—a byproduct of essential oil industries employing Eisenia fetida. Environ Technol Innov 25:102232

Arancon NQ, Owens JD, Converse C (2019) The effects of vermicompost tea on the growth and yield of lettuce and tomato in a non-circulating hydroponics system. J Plant Nutr 42(19): 2447–2458

Atiyeh RM, Arancon NQ, Edwards CA, Metzger JD (2002) The influence of earthworm-processed pig manure on the growth and productivity of marigolds. Bioresour Technol 81(2):103–108

Azarmi R, Giglou MT, Taleshmikail RD (2008) Influence of vermicompost on soil chemical and physical properties in tomato (*Lycopersicum esculentum*) field. Afr J Biotechnol 7(14): 2397–2401

Bachman GR, Metzger JD (2007) Physical and chemical characteristics of a commercial potting substrate amended with vermicompost produced from two different manure sources. HortTechnology 17(3):336–340. https://doi.org/10.21273/horttech.17.3.336

Balachandar R, Karmegam N, Subbaiya R (2018a) Extraction, separation and characterization of bioactive compounds produced by Streptomyces isolated from vermicast soil. Res J Pharm Technol 11(10):4569–4574

Balachandar R, Karmegam N, Saravanan M, Subbaiya R, Gurumoorthy P (2018b) Synthesis of bioactive compounds from vermicast isolated actinomycetes species and its antimicrobial activity against human pathogenic bacteria. Microb Pathog 121:155–165

Barik AK (2017) Organic farming in India: present status, challenges and technological breakthrough, The 3rd International conference on Bioresource and Management, 8–11 Nov 2017, Proceeding, pp 84–93

Bellitürk K, Adiloğlu S, Solmaz Y, Zahmacıoğlu A, Adiloğlu A (2017) Effects of increasing doses of vermicompost applications on P and K contents of pepper (*Capsicum annuum* L.) and eggplant (*Solanum melongena* L.). J Adv Agric Technol 4(4). https://doi.org/10.18178/joaat.4.4.372-375

Biri A, Kaba S, Taddesse F, Dechassa N, Jj S, Zewidie A, Chavhan D, A. (2016) Effect of vermicompost and nitrogen application on striga incidence, growth, and yield of sorghum [*Sorghum bicolor* (L.) Monech] in Fedis, eastern Ethiopia. Int J Life Sci 4:349–360

Biswas S, Islam MM (2019) Farmers' problem confrontation in organic farming at Magura Sadar Upazila of Bangladesh. South Asian J Agric 7:19–24

Blakemore RJ (2015) Eco-taxonomic profile of an iconic vermicomposter—the "African Nightcrawler" earthworm, Eudrilus eugeniae (Kinberg, 1867) etymology: named after Johan Gustaf Hjalmar Kinberg's Swedish survey ship, the. Afr Invertebr 56(November):527–548

Blouin M, Barrere J, Meyer N, Lartigue S, Barot S, Mathieu J (2019) Vermicompost significantly affects plant growth. A meta-analysis. Agron Sustain Dev 39(4):1–15

Ceritoglu M, Erman M, Ceritoglu F, Bektas H (2021) The response of grain legumes to vermicompost at germination and seedling stages. Legum Res 44(8):936–941

Coulibaly SS et al (2018) Vermicompost utilization: a way to food security in rural area. Heliyon 4(12). https://doi.org/10.1016/j.heliyon.2018.e01104

Das SK, Avasthe RK, Gopi R (2014) Vermiwash: use in organic agriculture for improved crop production. Popular kheti 2(4):45–46

Ding Z, Kheir AMS, Ali OAM, Hafez EM, ElShamey EA, Zhou Z, Wang B, Lin X, Ge Y, Fahmy AE, Seleiman MF (2021) A vermicompost and deep tillage system to improve saline-sodic soil quality and wheat productivity. J Environ Manag 277(August 2020). https://doi.org/10.1016/j.jenvman.2020.111388

Earth Observation System (2022) Soil salinization causes & how to prevent and manage it, earth observation system. Available at: https://eos.com/blog/soil-salinization/. Accessed 16 Feb 2022

Ebrahimi E, Ghorbani R, von Fragstein und Niemsdorff P (2020) Effects of vermicompost placement on nutrient use efficiency and yield of tomato (*Lycopersicum esculentum*). Biol Agric Hortic 36(1):44–52

Ebrahimi M, Souri MK, Mousavi A, Sahebani N (2021) Biochar and vermicompost improve growth and physiological traits of eggplant (*Solanum melongena* L.) under deficit irrigation. Chem Biol Technol Agricu 8(1):1–14

Edwards CA, Dominguez J, Arancon NQ (2002) The influence of vermicompost on plant growth and pest incidence. In: Shakir SH, Mickail WZA (eds) Soil zoology for sustainable development in the 21st century. Self-Publisher Cairo, Egypt, pp 397–420

Edwards CA, Arancon NQ, Emerson E, Pulliam R (2007) Suppressing plant parasitic nematodes and arthropod pests with vermicompost teas. Biocycle 48(12):38–39

Eo J, Park KC (2019) Effect of vermicompost application on root growth and ginsenoside content of Panax ginseng. J Environ Manag 234:458–463

Furlong C, Innovation S (2016) Worm Farming in South Africa Assessing the potential to supply composting worms for the humanitarian sector in Africa. International Federation of Red Cross and Red Crescent Society. https://doi.org/10.13140/RG.2.1.4176.6003

Gashaw B, Woldetsadik K (2020) Growth and yield performance of garlic (*Allium sativum* L.) varieties to application of vermicompost at Koga Irrigation Site, Northwestern Ethiopia. Growth 10(23). https://doi.org/10.7176/JBAH/10-23-01

Ghosh S (2018) Environmental pollutants, pathogens and immune system in earthworms. Environ Sci Pollut Res 25(7):6196–6208

Hashimoto S, Furuya M, You X, Wanibuchi G, Tokumoto H, Tojo M, Shiragaki K (2021) Chemical and microbiological evaluation of vermicompost made from school food waste in Japan. Jpn Agric Res Q 55(3):225–232. https://doi.org/10.6090/jarq.55.225

Hussain N, Abbasi SA (2018) Efficacy of the vermicomposts of different organic wastes as "clean" fertilizers: state-of-the-art. Sustainability 10(4):1205

Ievinsh G (2011) Vermicompost treatment differentially affects seed germination, seedling growth and physiological status of vegetable crop species. Plant Growth Regul 65(1):169–181

Ilker UZ, Sonmez S, Tavali IE, Citak S, Uras DS, Citak S (2016) Effect of vermicompost on chemical and biological properties of an alkaline soil with high lime content during celery (Apium graveolens L. var. dulce Mill.) production. Notulae Botanicae Horti Agrobotanici Cluj-Napoca 44(1):280–290

Jurik TW, Zhang S (1999) Tractor wheel traffic effects on weed emergence in Central Iowa. Weed Technol 13(4):741–746. JSTOR

Kakati LN, Thyug L, Mozhui L (2022) Vermicomposting: an eco-friendly approach for waste management and nutrient enhancement. Trop Ecol 63:1–13

Kamatchi KB, Petchiammal M, Poomari M, Priyanka S, Raja AA (2018) Effect of vermicompost extract on seed germination and seedling growth of some leafy vegetables. Int J Recent Res Aspects 65:898–901

Kaninga B, Chishala BH, Maseka KK, Sakala GM, Young SD, Lark RM, Tye A, Hamilton EM, Gardner A, Watts MJ (2020) Do soil amendments used to improve agricultural productivity have consequences for soils contaminated with heavy metals? Heliyon 6(11):e05502. https://doi.org/10.1016/j.heliyon.2020.e05502

Kavita S, Vinod KG (2019) Vermicomposting of waste: a zero-waste approach for waste management. In: Sustainable resource recovery and zero waste approaches. Elsevier, Amsterdam, pp 133–164. https://doi.org/10.1016/B978-0-444-64200-4.00010-4

Kiyasudeen K, Ibrahim MH, Ismail SA (2020) Vermicomposting of organic wastes and the production of vermicompost. In: Biovalorisation of wastes to renewable chemicals and biofuels. Elsevier, Amsterdam, pp 277–285

Li P et al (2020a) Soil quality response to organic amendments on dryland red soil in subtropical China. Geoderma 373(1):114416. https://doi.org/10.1016/j.geoderma.2020.114416

Li Y et al (2020b) Effects of earthworm cast application on water evaporation and storage in loess soil column experiments. Sustainability (Switzerland) 12(8):3112. https://doi.org/10.3390/SU12083112

Lin B, Li B, Huang ZIGUANG, Zheng X, Mu Y, Ye A, Zhang C, Ai SUILONG, Mustafad NS, Zhang L (2021) Effects of vermicompost application on quality of flue-cured tobacco. Pak J Bot 53(2):701–706

Little NG, DiTommaso A, Westbrook AS, Ketterings QM, Mohler CL (2021) Effects of fertility amendments on weed growth and weed–crop competition: a review. Weed Sci 69(2):132–146. https://doi.org/10.1017/wsc.2021.1

Liu YQ, Sun ZJ, Wang C, Li SJ, Liu YZ (2004) Purification of a novel antibacterial short peptide in earthworm *Eisenia foetida*. Acta Biochim Biophys Sin 36(4):297–302

McInga S et al (2020) Conservation agriculture practices can improve earthworm species richness and abundance in the semi-arid climate of Eastern Cape, South Africa. Agriculture (Switzerland) 10(12):1–12. https://doi.org/10.3390/agriculture10120576

Moles TM, Guglielminetti L, Reyes TH (2019) Differential effects of sodium chloride on germination and post-germination stages of two tomato genotypes. Sci Hortic 257:108730

Mondal T, Datta JK, Mondal NK (2015) An alternative eco-friendly approach for sustainable crop production with the use of indigenous inputs under old alluvial soil zone of Burdwan, West Bengal, India. Arch Agron Soil Sci 61(1):55–72

Mubvuma MT (2013) Climate change: matching growing season length with maize crop varietal life cycles in semi-arid regions of Zimbabwe. Greener J Agric Sci 3(12):809–816

Mupambwa HA (2015) Optimization of bio-conversion and nutrient release from coal fly ash amended cow dung—waste . . . . University of Fort Hare. https://doi.org/10.13140/RG.2.1.1324.3761

Mupambwa HA, Mnkeni PNS (2016) Eisenia fetida stocking density optimization for enhanced bioconversion of fly ash enriched vermicompost. J Environ Qual 45(3):1087–1095. https://doi.org/10.2134/jeq2015.07.0357

Mupambwa HA, Ncoyi K, Mnkeni PNS (2017) Potential of chicken manure vermicompost as a substitute for pine bark based growing media for vegetables. Int J Agric Biol 19(5):1007–1011. https://doi.org/10.17957/IJAB/15.0375

Mupambwa HA et al (2019) The unique Namib desert-coastal region and its opportunities for climate smart agriculture: a review. Cogent Food Agric 5(1). https://doi.org/10.1080/23311932.2019.1645258

Mupondi LT et al (2018) Vermicomposting manure-paper mixture with igneous rock phosphate enhances biodegradation, phosphorus bioavailability and reduces heavy metal concentrations. Heliyon 4(8):e00749. https://doi.org/10.1016/j.heliyon.2018.e00749

Murungweni C et al (2016) Climate-smart crop production in semi-arid areas through increased knowledge of varieties, environment and management factors. Nutr Cycl Agroecosyst 105(3):183–197. https://doi.org/10.1007/s10705-015-9695-4

Nadana GRV, Rajesh C, Kavitha A, Sivakumar P, Sridevi G, Palanichelvam K (2020) Induction of growth and defense mechanism in rice plants towards fungal pathogen by eco-friendly coelomic fluid of earthworm. Environ Technol Innov 19:101011

Ndegwa PM, Thompson SA (2000) Effects of C-to-N ratio on vermicomposting of biosolids. Bioresour Technol 75(1):7–12. https://doi.org/10.1016/S0960-8524(00)00038-9

Padilla Carrillo JM (2020) Estrategias agroecológicas urbanas para mitigar la disrupción de los sistemas agroalimentarios convencionales en el barrio Nueva Esperanza, Cantón Ambato, Provincia de Tungurahua en el periodo 2019–2020 (Bachelor's thesis, Ecuador, Latacunga: Universidad Técnica de Cotopaxi UTC.)

Pathma J, Sakthivel N (2012) Microbial diversity of vermicompost bacteria that exhibit useful agricultural traits and waste management potential. SpringerPlus 1(1):26. https://doi.org/10.1186/2193-1801-1-26icompostbacte. http://www.springerplus.com/content/1/1/26

Patwa A, Parde D, Dohare D, Vijay R, Kumar R (2020) Solid waste characterization and treatment technologies in rural areas: an Indian and international review. Environ Technol Innov 20: 101066

Prakash S, Sikdar S, Singh AK (2021) Level of farmers' knowledge & skills on vermi-compost: a post training behavioural exploration of farmers in Bihar state, India. Indian J Exten Educ 57(1): 45–48

Rahimian MH, Dadivar M, Zabihi HR (2021) The effect of three types of compost on grass water use efficiency. Agrotechniques Ind Crops 1(4):197–203

Reij C, Waters-Bayer A (2001) An initial analysis of farmer innovators and their innovations. Farmer innovation in Africa: a source of inspiration for agricultural development

Rupani PF, Embrandiri A, Rezania S, Shao W, Domínguez J, Appels L (2022) Changes in the microbiota during biological treatment of palm oil mill waste: a critical review. J Environ Manag 320:115772

Sharma K, Garg VK (2019) Vermicomposting of waste: a zero-waste approach for waste management. In: Sustainable resource recovery and zero waste approaches. Elsevier, St. Louis, pp 133–164

Sherathiya Pinalben H, Patel HM (2022) Vermicompost: an effective method of converting crop waste to gold

Shreedevi AS, Hanchinal SN (2017) Adoption of vermicomposting technology by farmers of Gulbarga district in Karnataka. Agric Update 12(4):639–642. https://doi.org/10.15740/HAS/AU/12.4/639-642

Shrivastava P, Kumar R (2015) Soil salinity: a serious environmental issue and plant growth promoting bacteria as one of the tools for its alleviation. Saudi J Biol Sci 22(2):123–131. https://doi.org/10.1016/j.sjbs.2014.12.001

Soto MDS, Zorpas AA, Pedreño JN, Lucas IG (2022) Vermicomposting of tomato wastes. In: Tomato processing by-products. Academic Press, Cambridge, pp 201–230

Tanase C, Bujor OC, Popa VI (2019) Phenolic natural compounds and their influence on physiological processes in plants. In: Polyphenols in plants. Academic Press, Cambridge, pp 45–58

Teka K, Githae E, Welday Y, Gidey E (2019) Vermi-composting for increased agricultural productivity, women empowerment and environmental sanitation in northern Ethiopia. AgriFoSe2030 Report 2019 (35), p 45

Thakur ANJANA, Kumar ADESH, Kumar C, Kiran B, Kumar SUSHANT, Athokpam VARUN (2021) A review on vermicomposting: by-products and its importance. Plant Cell Biotechnol Mol Biol 22:156–164

Tripathi YC, Hazarika P, Pandey BK (2005) Vermicomposting: an ecofriendly approach to sustainable agriculture. Verms and vermitechnology, SB Nangia. APH, New Delhi, pp 22–39

Ullah MI, Riaz M, Arshad M, Khan AH, Afzal M, Khalid S, Mehmood N, Ali S, Khan AM, Zahid SMA, Riaz M (2019) Application of organic fertilizers affect the Citrus Leafminer, Phyllocnistis citrella (Lepidoptera: Gracillariidae) infestation and citrus canker disease in nursery plantations. Int J Insect Sci 11:1179543319858634

Valembois P, Roch P, Lassegues M (1984) Simultaneous existence of hemolysins and hemagglutinins in the coelomic fluid and in the cocoon albumen of the earthworm *Eisenia fetida* andrei. Comp Biochem Phys A 78(1):141–145

Vedeld P (1990) Household viability and change among the Tugens: a case study of household resource allocation in the semi-arid Baringo district. Nomad People:133–152

Viljoen SA, Reinecke AJ (1990) Moisture preferences, growth and reproduction of the African nightcrawler, *Eudrilus eugeniae* (Oligochaeta). S Afr J Zool 25(3):155–160. https://doi.org/10.1080/02541858.1990.11448205

Viljoen SA, Reinecke AJ, Hartman L (1991) 'Life-cycle of the European compost worm' *Dendrobaena veneta* (Oligochaeta). S Afr J Zool 26(1):43–48

Vujoen SA, Reinecke AJ (1989) Life-cycle of the African nightcrawler, *Eudrilus eugeniae* (Oligochaeta). S Afr J Zool 24(1):27–32

Vuković A, Velki M, Ečimović S, Vuković R, Štolfa Čamagajevac I, Lončarić Z (2021) Vermicomposting—facts, benefits and knowledge gaps. Agronomy 11(10):1952

Wako RE (2021) Preparation and characterization of vermicompost made from different sources of materials. Open J Plant Sci 6(1):042–048

Wang F, Wang X, Song N (2021) Biochar and vermicompost improve the soil properties and the yield and quality of cucumber (*Cucumis sativus* L.) grown in plastic shed soil continuously cropped for different years. Agric Ecosyst Environ 315:107425

Yadav A, Gupta R, Garg VK (2013) Organic manure production from cow dung and biogas plant slurry by vermicomposting under field conditions. Int J Recycl Org Waste Agric 2(1):1–7. https://doi.org/10.1186/2251-7715-2-21

Yen YS, Chen KS, Yang HY, Lai HY (2021) Effect of vermicompost amendment on the accumulation and chemical forms of trace metals in leafy vegetables grown in contaminated soils. Int J Environ Res Public Health 18(12). https://doi.org/10.3390/ijerph18126619

Younas M, Zou H, Laraib T, Abbas W, Akhtar MW, Aslam MN, Amrao L, Hayat S, Abdul Hamid T, Hameed A, Ayaz Kachelo G (2021) The influence of vermicomposting on photosynthetic activity and productivity of maize (*Zea mays* L.) crop under semi-arid climate. PLoS One 16(8):e0256450

Zaman M, Shahid SA, Heng L (2018) Guideline for salinity assessment, mitigation and adaptation using nuclear and related techniques. Springer Nature, Cham, p 164

Chapter 17
Vermicompost and Vermi-leachate in Pest and Disease Management

K. Sivasabari, S. Parthasarathy, Deepak Chandran, S. Sankaralingam, and R. Ajaykumar

Abstract Vermiculture approach the synthetic rearing of worms to decompose natural meals wastes right into a nutrient-wealthy material. The output of vermiculture is known as vermicompost and the leachate of the vermicomposting process, referred to as vermi-leachate, is a brownish-colored substance created via way of means of the passage of water withinside the vermicomposting devices via the earthworm burrows. The produced vermicompost and vermi-leachate are wealthy in phrases of vitamins and different plant boom-selling substances that are able to present essential mineral vitamins to assist maintain plant boom and suppress plant root disease. Plants are subjected to biotic stress while they are attacked with the aid of using different residing organisms including bacteria, fungi, nematodes, protists, insects, viruses, and viroid. Two feasible mechanisms of pathogen suppression rely upon systemic plant resistance and the alternative is mediated with the aid of using microbial opposition, antibiosis, and hyper-parasitism. Antibiosis and opposition are extra factors taken into consideration inside the broader framework of ailment suppression. Competitiveness is an detail of the overall ailment control technique related with vermicomposting. It has been located that the use of vermicompost

---

K. Sivasabari (✉)
Department of Soil Science and Agricultural Chemistry, Amrita School of Agricultural Science, Coimbatore, Tamil Nadu, India
e-mail: k_sivasabari@cb.amrita.edu

S. Parthasarathy
Department of Plant Pathology, Amrita School of Agricultural Science, Coimbatore, Tamil Nadu, India

D. Chandran
Department of Veterinary Sciences and Animal Husbandry, Amrita School of Agricultural Sciences, Amrita Vishwa Vidyapeetham University, Coimbatore, Tamil Nadu, India

S. Sankaralingam
Department of Botany, Saraswathi Narayanan College, Madurai, Tamil Nadu, India

R. Ajaykumar
Department of Agronomy, Vanavarayar Institute of Agriculture, Coimbatore, Tamil Nadu, India

© The Author(s), under exclusive license to Springer Nature Singapore Pte Ltd. 2023
H. A. Mupambwa et al. (eds.), *Vermicomposting for Sustainable Food Systems in Africa*, Sustainable Agriculture and Food Security,
https://doi.org/10.1007/978-981-19-8080-0_17

crafted from meals waste drastically reduces the populace of aphids, bugs, beetles, mites, hornworms, pinworms, and mealybugs on positive vegetation inclusive of cabbage, cucumber, pepper, potato, and tomato. Vermicompost and vermi-leachate have been visible in an excessive capacity for suppressing sickness occurrences resulting from pathogens inclusive of *Aphanomyces*, *Aspergillus*, *Alternaria*, *Fusarium*, etc., withinside the experiments. It is suggested that earthworm ought to be inoculated in an agricultural field, or organized and implemented its vermi-leachate/vermicompost as spraying at the floor of plant life extensively suppresses a pathogen and pest.

Keywords Vermicompost · Vermi-leachate · Biotic stress · Pest control and disease suppression

17.1 Introduction

Agrochemicals like fertilizers, growth promoters, pesticides, and improved seed varieties have had a negative impact on soil health and agricultural production in India during the first green revolution, but they have also increased the rate of pests and disease in crops, which has resulted in increased pollution and environmental degradation (Chaoui et al. 2002). Annually, agricultural production suffers substantial crop because of pest and diseases. Worldwide annual production yield losses in the twenty-first century have been attributed to animal pests (18–24%), bacterial pathogens (16–20%), and weeds (34%), resulting in an average annual crop production loss of 68–78%. Across the globe, scientists are scrambling to find an agrochemical alternative that is economically viable, socially acceptable, and environmentally sound. The use of vermicompost as a preferable organic fertilizer replacement is currently the subject of numerous investigations (Sinha et al. 2010). Earthworm excrement, also known as vermicompost, is capable of enhancing soil health and nutrient status.

Earthworms are able to remodeling garbage into "gold." Charles Darwin known as them the "unrecognized soldiers of humanity." Aristotle referred to them as the "intestines of the earth" because of their ability to digest organic matter. Vermicompost, commonly referred to as "Black Gold," "farmer's friend," or "nature's plowman," can be an excellent soil ameliorant and highly effective tool for a sustainable future (Ganguly and Chakraborty 2020). They decompose large soil and plant debris, increasing the organic materials accessible for microbial breakdown. Using a combination of aerobic and anaerobic microorganisms, they turn organic waste into a valuable vermicompost. While passing through the worm's gut, a variety of biodegradable wastes, such as agricultural and food-based bio-wastes, kitchen and market wastes, and other bio-wastes, are transformed into nutrient-rich vermicompost. The content and quality of vermicompost/vermi-leachate change depending on the raw organic matter used in vermicomposting. Vermi-leachate is a brownish-colored liquid produced by the passage of water through the earthworm tunnels in the vermicomposting units (Aghamohammadi et al. 2016). Vermicompost and vermi-leachate generated from the same organic matter are almost identical in composition. Both products are made up of different combinations of different macro- and micronutrients as well as a wide range of different bacteria, hormones, mucus, enzymes and vitamins (Das et al. 2014; Nadana et al. 2020). As a fertilizer, it can be used to boost crop yield as well as for disease suppression and pest control because of its antibacterial and anti-pest properties.

An abundance of nutrient-rich vermicompost has been shown to improve soil fertility and plant growth. Vermicomposting is a very efficient method of recycling nutrients that utilizes earthworms as indigenous bioreactors to decompose organic materials through the combined efforts of soil microorganisms. Despite the fact that vermicomposting has been proved to be an effective strategy for reducing organic biomass and producing high-quality nourishment for plants, however little is understood about the microbial flora that is engaged in this breakdown activity. The activity of earthworms has been shown to boost beneficial microflora, enhance soil health, promote plant growth, and control the abundance of plant diseases and pests due to their high nutritional content and microbial activity. Additionally, vermicompost contains antifungal and insecticidal qualities derived from earthworm coelomic fluid and certain other bioactive chemicals, rendering it similarly efficient in reducing pests and diseases mainly in soil and rhizoplane. Many plant illnesses and insect issues can be prevented by using vermicompost materials, both solid and liquid, as described in the current chapter. The use of inhibitory systems as a natural or biological control technique has enormous promise, despite the fact that the mechanisms are not yet fully characterized. An important role in disease prevention is played by a diverse microflora, particularly by the increased diversity and number of antagonists seen in granular and fluid vermicompost preparations. The use of vermiproducts in conjunction with other bioproducts in soil and plants is also discussed in this chapter.

17.2 Potential of Vermicompost and Vermi-leachate in Plant and Soil Nutrient Supply

Vermicompost is wealthy in useful soil microbes such as nitrogen-fixing bacteria, mycorrhizal fungi, phosphate solubilizing bacteria, and actinomycetes, with an average of 1.5–2.2% N, 1.8–2.2% P, and 1.0–1.5% K. In the organic carbon, which ranges from 9.15 to 17.98%, are micronutrients such as sodium, calcium, zinc, sulfur, and magnesium. Soluble potassium, nitrates, calcium, and magnesium are also more easily accessible after vermicomposting (Manyuchi et al. 2013). Numerous studies have shown that vermicompost improves plant nutrient uptake and provides all nutrients in a form that is easily available to plants.

As the amount of vermicompost increased, so did the soil's ability to absorb nitrogen, with the maximum nitrogen uptake occurring at 50% of the recommended fertilizer rate plus 10 t ha^{-1} vermicompost. Fertilizer absorption of N, P, K, and Mg by rice (*Oryza sativa* L.) was maximized when administered in conjunction with vermicompost (Sharma and Banik 2012). As a result, soil micronutrients increased significantly, and soil physical and chemical qualities were enhanced to a higher quality due to the use of vermicompost and vermi-leachate in combination (Ansari and Sukhraj 2010). Liquid vermi-leachate is a high-nutrient biofertilizer that is produced by the decomposition of vermicompost in the soil (Nath et al. 2009; Palanichamy et al. 2011). Vermi-leachate application has been reported to improve soil quality and it restores decreased soil fertility by enriching the available pool of nutrients and conserving moisture as well as natural and biological resources. According to research, boosting the soil's organic carbon content and the populations and activities of soil microorganisms, particularly plant beneficial ones, would have enabled plants to better absorb nutrients, resulting in enhanced growth and yield when utilizing coconut leaf vermi-leachate. Mineral particles such as Ca, Mg, and K can be bound together in colloids of humus and clay by organic matter to achieve appropriate porosity in soil. Sundararasu and Jeyasankar (2014) found that vermi-leachate spray increased growth (plant height and leaf count) and yield (number of flowers and fruits per plant) characteristics in brinjal/eggplant (*Solanum melongena* L). Kumar (2019) discovered that using vermi-leachate increased plant height and number of leaves (56.29 cm and 6.14 days at 45 days after bud emergence), spike length and rachis (90.68 cm and 47.07 days), number of florets (15.08), vase-life (10.02 days), number of corms m-2 (28.66), corm weight (50.68 g), and cormels plant^{-1} (56.66). The treatment was also effective in shortening the time it took for spikes to emerge (81.73 day). When vermi-leachate and vermicompost were used together, plant growth parameters such as the number of leaves, plant height, and root length were elevated to a higher level (Tharmaraj et al. 2011).

Nonetheless, there are reasons to proceed with caution when using vermicompost (Lazcano et al. 2011). Plants and soil have been found to be negatively affected by the amount of vermicompost applied in some experiments. Higher vermicompost concentrations may adversely affect root development, resulting in phytotoxicity (Sangwan et al. 2010). A salt build-up in the soil can also be a problem when using a lot of vermicompost, which can swiftly kill a plant.

17.3   Mechanism of Vermicompost and Vermi-leachate in Crop Biotic Stress Management

When plants are attacked by other living species, such as bacteria, fungus, nematodes, protists, insects, viruses, and viroids, they are subjected to biotic stress. Diseases that thrive in harvested fruit and are often caused by bacteria, fungus, or yeasts are the most common cause of biotic stress in livestock operations. Biotic stress triggers a defense response in plants. Defending oneself is a response that is both innate and widespread. ROS are created following a pathogen infection and oxidative bursts inhibit the spread of the microorganisms (Lazcano et al. 2011). Pathogen attacks also cause plants to increase cell lignification. This technique serves to keep parasites at bay while also lowering the vulnerability of the host.

Vermicompost's microbiome has been shown to aid in biological control. As a result of sterilization, vermicomposts were no longer effective in preventing the spread of disease. Systemic plant resistance and microbial competition and hyperparasitism have been identified as two putative mechanisms of pathogen suppression (Sangwan et al. 2010). Microorganisms can suppress pathogens like Pythium and Phytophthora in two ways: "general suppression," where a wide range of microbes work together to do so, and "specific suppression," where a small range of organisms work together to do so, as in the case of Rhizoctonia-induced illness.

17.3.1 Plant Growth Regulation

Because vermicompost is rich in growth regulators like hormones, it is one of the best organic manures available (Thiruneelakandan and Subbulakshmi 2014). Plant growth responses in vermicompost have been compared to "hormone-induced activity" according to some research, because of the high concentrations of nutrients, humic acids, and humates in the compost (Sinha et al. 2009). Vermicompost has been shown to increase plant growth even when plants are receiving "optimal nutrition" according to research. Plant growth and germination were improved much more than if mineral nutrients were simply converted into plant-available forms. Also discovered in vermicompost were growth-promoting "auxins," "cytokinins," and flowering-inducing "gibberellins" hormones released by earthworm. Additionally, vermicompost encourages quick and strong plant growth. Seasonal crops such as okra, cucumber, pepper, eggplant, strawberry, and Amaranthus species were also grown in the same way (Chaoui et al. 2002).

17.3.2 Suppression of Pests and Diseases

Both vermicompost and vermi-leachates have been shown to control plant diseases in studies. Microorganisms in vermicompost and vermi-leachates can function as antagonistic agents of pathogens by competing for resources and space and creating an antibacterial chemical, eliminating parasitizing pathogens and systematically diminishing their resistance (Mehta et al. 2014; Naseri 2019).

Vermicompost has been studied extensively for its positive effects on pathogens, plant diseases, and physiologic abnormalities. *Phytophthora lycopersici* (*F. oxysporum*) wilt was found in tomato monocropping soil in a study. *Fusarium oxysporum* sp. lycopersici was successfully suppressed by rice straw, chicken dung compost, and vermicompost, respectively. Using vermicompost treatments reduced *F. oxysporum* f. sp. lycopersici most effectively, and there was an association between bacterial relative abundance and this pathogen (including the genera *Nocardioides, Ilumatobacter,* and *Gaiella*). Vermicompost included the same microbiotas found in soil, according to the study (Zhao et al. 2019).

It was also found that vermicompost application reduced the incidence of common scab disease in potatoes by Singhai et al. (2011). In the fight against root-knot nematodes, vermicompost has proven to be a useful tool for enhancing the health of host plants. First, Xiao et al. (2016) carried out a second investigation on tomato varieties' resistance to root-knot nematodes (*Meloidogyne incognita*). The tomatoes were inoculated with *M. incognita* to mimic root-knot nematode disease in field circumstances, and three amendments were compared: organic fertilizer, conventional compost, and vermicompost. On susceptible and resistant cultivar roots, vermicompost dramatically reduced nematode-induced galls by 77% and 42%, while inorganic fertilizer increased galls by 59% and 46%, respectively. Plant diseases such as *Pythium, Verticillium, Rhizoctonia,* and *Fusarium* were mitigated by inoculation of vermicompost in a study by Sundararasu and Jeyasankar (2014). Reddy et al. (2015) discovered that a combination of vermi-leachate and biopesticide is a superior alternative to chemical fertilizer and pesticides for *Leptocorisa varicornis* management and rice crop productivity. The extraordinary combos of vermi-leachate received from buffalo dung, gram bran with neem oil and aqueous extract of garlic is powerful for the management of pod borer (*Helicoverpa armigera*) infestation on gram plant. Additionally it will increase the growth, commence early flowering and improved the productiveness of gram up to a few instances.

17.4 General Suppression Mechanism

There are a variety of ways in which vermicomposting can boost the diversity and abundance of beneficial bacteria in a garden. The term "general suppression" refers to nutritional competition, antibiosis, in which helpful organisms emit drugs that directly inhibit the pathogen, hyper-parasitism/direct parasitism (one organism feeding on another), and possibly generated systemic plant resistance (Fig. 17.1). Suppression mechanisms have been proposed using the idea that large diversity of microbes functions as biocontrol agents, which prevent plant diseases from thriving in situations such as fungistasis. *Pythium* and *Phytophthora* plant pathogens were suppressed by a broad suppression mechanism.

One of the factors defined within "general disease suppression" by organic amendments is the concept of induced systemic resistance (ISR). Induced systemic resistance in plants triggered by compost applications is suggested to be possibly derived from the presence of some antibiotics and actinomycetes in vermicomposts which increases the "power of biological resistance" (Chaoui et al. 2002). However, this rationalization is the least jointly accepted claim among all those proposed for general disease suppression mechanisms. The idea of ISR is primarily based totally at the concept that the reaction of plants growing in the soil contributes to the

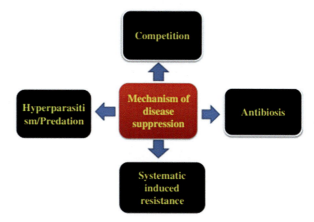

Fig. 17.1 Different mechanisms of disease suppression by vermicompost and vermi-leachate (Source: Adapted from Yatoo et al. (2021))

suppressiveness to disease incidences. That occurs, in case of soil-borne pathogens, while the rhizosphere is inoculated with a weakly virulent pathogen. After being challenged via way of means of a vulnerable pathogen, the plant develops the potential for future effective response to a more virulent pathogen. In maximum cases, adding mature high-qualified compost products to soil induces disease resistance in many plants (Chaoui et al. 2002; Sinha et al. 2010). Soil biodiversity is increased by vermicompost by accelerating the development of beneficial microorganisms, directly and indirectly boosting plant growth by suppressing plant diseases, nematodes, and other pests, improving plant health, and decreasing yield reduction, therefore it is considered as an effective biocontrol agent. Several studies discovered that vermicompost and its leachate reduced plant disease incidence by involving various antagonistic mechanisms, but to a smaller extent than chemical pesticides. After amendment with vermicompost, the rhizosphere population undergoes many characterized modifications establishing into a growing community for survival and fitness. This increased rhizosphere population of antagonistic fungi and bacteria might be involved in following disease-suppressive activities. General suppression characterized by nutrient competition, antibiosis, hyper-parasitism or predation, and systemic-induced resistance are discussed further below.

17.4.1 Competition

Vermicompost produced by vermicomposting competes with other vermicomposts in the general disease control process. Consequently, the level of disease suppression in the soil is inversely related to the total microbial activity. The soil's ability to use nutrients, energy, carbon, and space increases as the number of active bacteria in the soil increases, limiting the nutrients available to diseases. Disease suppression is possible when a helpful microbe outcompetes a phytopathogen for a resource. For instance, some microorganisms create low molecular weight siderophores under iron-limiting conditions, restricting the availability of iron to pathogens like Pythium species, reducing disease prevalence (Srivastava et al. 2010). Soil microbe activity can be improved by enhancing nutrient competition through any treatment that promotes soil microbial activity in general. It is possible that the mycorrhizal fungi transported and multiplied in the soil by worms may protect plant roots against illness by means such as antibiotic production, enhanced nutrient absorption by plant roots, and the creation of physical barriers. Some species of Trichoderma, such as *Trichoderma virens*, have been accepted as destructive mycoparasites capable of competing with pathogens at the infection site (Jogaiah et al. 2013), and Trichoderma species are the most commonly used organisms for agricultural yield enhancement (Sinha et al. 2010).

17.4.2 Antibiosis

Earthworms consume antagonistically and plant growth-promoting rhizobacteria such as *Bacillus*, *Pseudomonas*, and *Streptomyces*, among others, along with rhizosphere soil. These microorganisms may be activated or amplified by the favorable microenvironment of the stomach and released soil. These antagonists indirectly promote plant growth by inhibiting fungal and bacterial diseases through the secretion of antibiotics, fluorescent pigments, siderophores, chitinases, and glucanases that degrade the cell walls of pathogenic fungi and bacteria. Antibiosis is defined as a two-way interaction in which one organism directly affects another by producing non-toxic and/or unique metabolites or antibiotics. Enzymes such as amylase, cellulase, invertase, peroxidase, phosphatase, protease, urease, and dehydrogenase activities are elevated in wormcasts (Ananthakrishnasamy 2019). Dehydrogenase is an intestinal enzyme associated with the oxidative phosphorylation process and a marker of bacterial metabolism in the soil and other biological habitats. During 21–35 days of vermicomposting, the maximal enzymatic activity (cellulase, amylase, invertase, protease, and urease) was recorded, compared to 42–49 days for traditional decomposition. Due to the antimicrobial activities through antibiotics and chitinase-producing gene diversity in its microbiota, vermicompost is efficient against fungus like *Colletotrichum coccodes, Fusarium moniliforme, Rhizoctonia solani, Phytophthora capsici*, and *Pythium ultimum* (Yasir et al. 2009). Greater population levels of cellulolytic and oligotrophic actinobacteria and enterobacterial organisms on combinations containing mature compost were antagonistic to a range of pathogens, notably *Rhizoctonia solani*, by secretion of antibiotics and chitinolytic enzymes (Tuitert et al. 1998). The antibiotic "gliotoxin," isolated from *Gliocladium virens* from vermicompost, was discovered to be antagonistic to *Pythium ultimum*. According to Sinha et al. (2010), bacteria and fungi isolated from fruit-based compost have antagonistic activity against a variety of phytopathogens, including *Fusarium oxysporum*. Compost bacteria such as *Bacillus subtilis, Pseudomonas, Gliocladium virens*, and *Trichoderma harzianum* are antibiotics that can inhibit the growth, development, and replication of several plant pathogens. According to

Vinale et al. (2009), *Trichoderma harzianum* produces harzianic acid in vermicomposts, which has antibiotic activity against pathogenic organisms such as *Pythium irregulare*, *Rhizoctonia solani*, and *Sclerotinia sclerotiorum*. In a conducted experiment by Aira et al. (2006), pig slurry, a substrate rich in microorganisms, was treated in miniature vermireactors both with and without earthworms. The inclusion of earthworms in vermireactors greatly accelerated the rate of cellulose hydrolysis by increasing the microbial population and cellulase and glucosidase activity. During vermiculture, the activities of *E. fetida* also promoted the growth of fungi. The influence of four commercial composts on plant growth, total rhizobial microbiota, and the incidence of plant growth-promoting rhizobacteria in the root system of tomato plants was also observed. The results demonstrated that the addition of compost to the soil increased the frequency of siderophore-producing bacteria in the tomato rhizosphere that exhibited antagonistic activity against *Fusarium oxysporum* f. sp. *radicis-lycopersici*, *Pyrenochaeta lycopersici*, *Pythium ultimum*, and *Rhizoctonia solani*. The results (de Brito Alvarez et al. 1995) show that compost may help the root system make more of its own defenses against root diseases that come from the soil.

17.4.3 Hyper-parasitism/Predation

Phylogenetically unrelated microbes colonize infections and eventually destroy them through hyper-parasitism (Hoitink et al. 1997). Microbial predation, on the other hand, is the term given to the process of using phagocytosis to destroy infections. Carbon and other readily available nutrients have a substantial impact on parasitism because they inhibit the development and effect of lytic enzymes, which are employed to eliminate parasites. The degree of organic matter decomposition and their existence. The plant pathogen *Rhizoctonia solani* is not immediately attacked by several microorganisms, such as *Trichoderma hamatum* and *Trichoderma harzianum*, which produce lytic enzymes in fresh bark-based compost. Trichoderma species' chitinase genes are activated by readily available glucose and cellulose during the composting process, parasitizing *Rhizoctonia solani* and secreting the enzyme chitinase (Kwok et al. 1987). For instance, *Trichoderma harzianum* is an example of parasitism since it grows around the *Pythium ultimum* and punctures its cell wall before devouring the plant pathogen (Kumar 2019).

17.4.4 Induced Systemic Resistance

To stimulate disease resistance via induced systemic resistance (ISR) and to sell plant improvement for sustainable crop production, beneficial microorganisms have received a whole lot of hobby as an ecologically benign and economically efficient platform (Harel et al. 2014). The pathogen Pythium root rot, for example, was prevented when compost was fed to a section of cucumber root utilizing a split root system (Lievens et al. 2001). Other researchers have found that crop resistance to Phytophthora and *Botrytis cinerea* is boosted by Trichoderma species found in compost (Horst et al. 2005; Hoitink and Fahy 1986). Non-sterile compost tea was found to increase resistance when given to host okra plants, based on the identification of resistance-inducing chemicals (Siddiqui et al. 2008). According to Sang and Kim (2011), compost extract-mediated ISR gave pepper and cucumber protection from the disease anthracnose.

17.5 Specific Suppression Mechanism

An organism inhibits a known pathogen or one that can only be suppressed by a small number of microorganisms via this technique. a biocontrol agent extracted from compost/vermicompost is put to the soil in order to reduce disease occurrence (Hoitink et al. 1997). According to current thinking, this mechanism is what kills pathogens like *R. solani* and *S. rolfsii*. Plant roots can be colonized by some strains of non-pathogenic bacteria, such as the genus Bacillus, which can then induce particular resistance in the plant. Under order to control fruit post-harvest infection, Punja et al. (2016) used a *Bacillus subtilis* strain in greenhouse settings. Botrytis cinerea, the most common cause of tomato fruit rot, was likewise found to be inhibited by these strains of bacteria. Only two rhizobacterial strains, *Bacillus amyloliquefaciens* and *Ochrobactrum intermedium*, considerably reduced tomato wilt incidence and greatly enhanced seedling vigor index (i.e., percentage of seed germination and mean height of seedlings) in tomato wilt control (Kumar 2019).

17.6 Microbial Fauna in Vermicompost and Vermi-leachates

Earthworms have already been identified as a significant population of earth engineers in environments due to their ability to alter the physical properties of soil and hence regulate soil microbes, soil health properties, soil organic matter regulation, nutritive qualities, and thereby enhancing crop production (Eijsackers 2011). Because earthworms live closer to the rhizosphere, where organic matter is abundant, the drilosphere (the earthworm-influenced soil content and microorganisms) are inextricably linked. Various studies on soil earthworms have revealed that they enhance the earth's physiological, biochemical, and ecological aspects and promote its oxygenation and drain through their necessary feeding, tunneling, and castings (Pathma and Sakthivel 2012). Additionally, these processes contribute significantly to converting elements and nutrient status into forms that are available and accessible to plants and soil microbes, increasing the fertility of the soil and efficiency.

Earthworm-associated microflora contains essential plant growth-promoting microorganisms (PGPM) for sustained crop development (Raimi et al. 2022). In this regard, relationships between soil organisms' behaviors, particularly the microbiological features of earthworm casts and soil enzymes generated by soil microorganisms and serve as important markers of soil organic matter, have garnered increased interest in soil research.

Organic matter in the soil is the primary food source for most earthworms. The ecological condition of a product largely dictates its consumption characteristics (epigeic, endogeic, and anecic). In contrast to epigeic species (*Lumbricus rubellus*), endogeic and anecic species (*Aporrectodea caliginosa* and *Aporrectodea longa*) favor soil composition and organic matter in the soil, respectively. (Curry 2004). Additionally, they can consume soil microbes (Drake and Horn 2007). This changes the microbial population in the soil and the earthworm's digestive tract or feces. Sheehy et al. (2019) observed that an anecic type, *Lumbricus terrestris*, feeds on microorganisms in rich soil, and their decomposition was significantly slower in the earthworm gut system. However, the involvement of soil microbial communities in the ecology of earthworm feeding is not known precisely (Hoeffner et al. 2018). The mucus and fluid secreted from the dorsal pores contain bioactive compounds generated by diverse chains of amino acids to make them unique for fighting certain infections. These compounds were produced through an evolved process to allow earthworms to defend themselves against soil-inhabiting larvae, and soil-borne pathogens (Gudeta et al. 2021). Earthworm gut mucus includes proteins and sugars, organic and minerals content, amino acids, and microbial symbiotic relationships, such as bacteria and fungi. The higher organic carbon, organic matter content, nitrate, and moisture levels of the earthworm intestine develop a positive environment for hibernating microorganisms to be activated and endospores to germinate, among other things. Numerous proteolytic enzymes such as amylase, cellulase, chitinase, lipase, protease, and urease have been isolated from earthworm guts. These enzymes have been shown to digest a variety of organic substrates and diverse constituents of soil microflora (Parthasarathi et al. 2007). Cellulase and mannose activities were mediated by gut bacteria (Li et al. 2011). Comminuting the substrate by earthworms increases the surface area available for microbial breakdown, increasing the pressure of vermicomposting. The digestion of microbial populations in the earthworm gut system varies according to microbial community structure and abundance, earthworm ecology factors, and peripheral habitats.

Earthworms have been shown to increase microflora and biodiversity in organic wastes to levels comparable to thermophilic organic manure (Jack and Thies 2006). Thus, the general suppressive effect of vermicomposts against plant diseases is primarily a result of its microflora, as suppressiveness largely vanished following autoclaving, as documented (Jayaraman et al. 2021). Antibiosis and competition are two additional elements considered within the broader framework of disease suppression. Competitiveness is an element of the total disease management process linked with vermicomposting. As a response, disease suppression is directly dependent on the total microorganisms in the soil. More beneficial soil microorganisms lead to more nutrition, activity, carbon, and space in the soil, which may reduce the nutrient availability to pathogens. It happens when a positive microorganism beats a plant pathogen in a fight for a resource, which reduces disease. Antibiosis is when some beneficial organisms secrete antibiotics, and other beneficial microbes parasitize pathogens directly. The deployment of vermicompost materials to manage soil-borne plant pests and diseases needs to generate soil conditions that contain these characteristics. Thus, adding vermicompost to soil significantly increases the quantity and diversity of disease organism competitors, inhibitors, predators, and food supplies for these beneficial species. Compared to aerobic composting and other organic wastes, vermicomposts provide a more readily available food source for beneficial organisms. The relationship between earthworms and microorganisms has been discovered to be complex. Certain microorganisms have been a part of the earthworm's diet, as indicated through their destruction as they flow via the earthworm's digestive system.

The earthworms, *Drawida calebi, Lumbricus terrestris*, and *Eisenia fetida* digest a few more yeasts, protozoa, and certain species of fungi such as *Alternaria solani* and *Fusarium oxysporum* (Parthasarathi et al. 2007; Pathma and Sakthivel 2012). A review documented a diversity of bacterial species in vermicomposts produced by various earthworm species, including *Azospirillum, Azobacter, Nitrobacter, Nitrosomonas*, ammonifying bacteria, and phosphate solubilizers in *Eudrilus* sp. *Pseudomonas oxalaticus* in *Pheretima* sp., *Rhizobium japonicum*, and *Pseudomonas putida* in Lumbricus rubellus (Joshi et al. 2015). Vermicompost and related bacteria of the genera *Bacillus, Burkholderia, Pseudomonas*, and *Streptomyces* are capable of producing important compounds that inhibit plant pathogenic fungi in soil. Especially, *Bacillus* species that live in soil and produce spores are known to build mutualistic interactions with earthworms and produce a variety of bioactive compounds that are effective against plant diseases in composts and their derivative products. The spore-forming process enables bacteria to live and spread not just in the soil, but also in the stomach of earthworms (Soltan et al. 2022). They described all of the beneficial bacteria in vermicompost in their investigation, a total of 10 bacterial isolates using 16S rRNA gene sequencing. According to the similarity of their sequences, they all belong to *Bacillus* species, like *Bacillus amyloliquefaciens, Bacillus hemicentroti, Bacillus nakamurai, Bacillus subtilis,* and *Bacillus velezensis*. In another study, *Bacillus cereus* var. *mycoides* declined during an intestinal passage, although *Escherichia coli* and *Serratia marcescens* were entirely eradicated (Edwards and Fletcher 1998). Medina-Sauza et al. (2019) discussed that soil bacteria and fungi populations present in the feed intake surrounding *Pontoscolex corethrurus* were also found in its excrement without even any process of digestion, while fungal spore structures in the fecal matter of *Ulocladium botrytis* had a relatively low germination percentage. Arbuscular mycorrhizal fungi produce more cast when passed through the digestive walls of earthworm *Pontoscolex corethrurus,* they observed cast production, spore number, spore germination, infective propagules of *Rhizophagus clarus*, and *Claroideoglomus etunicatum* fungi were greatly enhanced in soil (de Novais et al. 2019).

Vermicomposting of *Cytisus scoparius*, a leguminous shrub, in a pilot-scale vermireactor for 3 months using earthworm species *Eisenia andrei*, increase in taxonomic and functional diversity of the bacterial community, and that included metabolic capacity, streptomycin, and salicylic acid production, and nitrogen fixation. These findings emphasize the importance of bacterial recurrence in the composting process which gives support for microbial activities that might also account for the positive effects of vermicompost on soil, environment, and vegetation (Domínguez et al. 2019). Another meta-study revealed, the microbial diversity analysis of vermicompost using high-throughput DNA metabarcoding of specific gene markers revealed the presence of the majority of dominant phyla such as Firmicutes, Proteobacteria, Planctomycetes, and Bacteroidetes. While *Aeromonas*, *Bacillus*, *Clostridium* sensu stricto 1, *Escherichia-Shigella*, *Morganella*, *Lysinibacillus*, *Pseudomonas*, and *Vibrio* were the most prevalent genera across the samples, they can contain potentially harmful organisms, demonstrating the importance of increasing vermicomposting safety and reliability. Functional profiling predicted that nitrogenases, phosphatases, and sulfatases were abundant in the bacterial populations. Additionally, the ability to synthesize a siderophore, indole acetic acids (IAA), and 1-aminocyclopropane-1-carboxylate (ACC) was envisaged. *Bacillus*, *Lysinibacillus*, *Paenibacillus*, and *Pseudomonas* were important bacterial communities that could be used to make bacterial formulations used on farms (Raimi et al. 2022).

Remarkably, the microbiome of decomposing worms is known to stimulate rapid substrate breakdown, which results in the release of nutrient elements. This is comparable to the methods that bacterial biofertilizers use to fertilize plants. Additionally, earthworm intestinal bacteria may perish due to toxin- and antibiotic-producing fungi in the earthworms' eating environment, such as *Aspergillus* spp., *Fusarium* spp., and *Penicillium* spp. (Ibrahim et al. 2016). Similarly, algae, bacteria, and protozoa have been observed (Kizilkaya et al. 2011). As a result, microorganisms not digested in the earthworm's intestinal system have access to organic matter-rich feces. Additionally, earthworm feces and burrows include mucus created by intestinal fluids with a higher carbon and nitrogen concentration and a lower carbon/nitrogen ratio (Kizilkaya et al. 2011). As a result, microbial populations and their activities are typically more significant in earthworm feces and burrow walls than in the soil. Along with microbial enumeration approaches, microbial biomass and respiration data can assess microbiota indirectly residing in earthworm feces. The total microbial biomass and activity of earthworm excrement were found to be greater than the feeding stuff (Van Groenigen et al. 2019).

The potency of vermicompost isolates as biofertilizer and biological antagonists has been recognized after the comprehensive investigations of thermophilic compost preparations. Accordingly, the preparation and utilization of vermi tea is a significantly newly emerging field; few papers are offering good scientific evidence on the issue. The application of vermi-leachate/teas as bio-control agents has increased during the previous decade (Becagli et al. 2022). There seems to be a considerable gap in the approach for manufacture and use of vermi teas in terms of appropriate dilution and rate of application concerning specific crop pests or diseases.

17.7 Vermicompost and Vermi-leachates in Pest Management

The interactions between the root system and vermicompost may alter the composition of secondary metabolites, the volatile organic chemicals released by the roots, induce host resistance, and/or the nutritional properties of plant parts. These impacts may cause the plant inhospitable to a variety of insect pests. Indeed, it has been shown that plants grown in organically changed soils that have been treated with vermicompost are less hospitable to agricultural pests than plants grown in regular soils that have been fed with agrochemicals (Cardoza and Buhler 2012). Numerous but speculative processes exist as to how vermicompost and vermicompost tea reduce or manage insect infestations. On the basis of the given research findings, four plausible methods for pest prevention can be proposed (Edwards and Arancon 2004). These include the production of phenolic compounds, a boost in the abundance and variety of beneficial bacteria and harmful parasites, the liberation of noxious chemicals, and increased nutritional availability to plants. Additional research is needed to substantiate these assumptions and to identify the appropriate processes by which these composts suppress pests of crop plants. In addition to these beneficial properties, it has been proposed that certain forms of vermicompost boost plant resistance to nematodes, and insect pests. For instance, it has been observed that using vermicompost made from food waste considerably reduces the population of aphids, bugs, beetles, mites, hornworms, pinworms, and mealybugs on certain crops such as cabbage, cucumber, pepper, potato, and tomato (Sedaghatbaf et al. 2018) (Fig. 17.2).

Numerous studies have demonstrated that vermicompost or vermicompost tea dramatically reduces insect and nematode infestations (Table 17.1). Tomato hornworms, cabbage caterpillars, cucumber beetles, and other sucking insects, such as aphids, whiteflies, red spider mites, and mealybugs, are all controlled effectively with vermicasts (Yatoo et al. 2021). Implementing vermicompost tea dramatically reduced the viability and fertility potency of Acalymma vittatum on cucumber and Manduca sexta on tomatoes, as found by Edwards et al. (2010) in their research. Due to the development of phenolic compounds that make plant tissues unpleasant to pest species, they concluded that this was the reason. Cardoza and Buhler (2012) found that vermicompost-mediated resistance in maize was effective at making maize resistant to *Helicoverpa zea* at both the adult and juvenile stages. They used vermicompost from a variety of sources to make maize resistant to the corn earworm. The effect of solid and aqueous vermicompost application in tomato potting media on host preference, survival, and fertility, life cycle parameters of tomato leaf miner, *Tuta absoluta*, was investigated, and the results indicated that 60% solid vermicompost treatment resulted in the lowest number of tomato leaf miner population eggs per leaf and the highest larval mortality (PeimaniForoushani and Poorjavad 2017). The study assessed the impact of vermicomposts from four distinct sources (pistachio waste, date waste, cattle manure waste, and food waste) on several life cycle parameters of the sweet potato whitefly *Bemisia tabaci* (Gennadius)

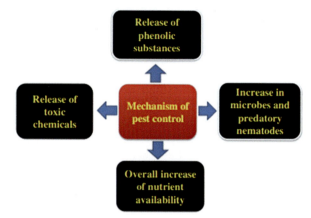

Fig. 17.2 Different mechanisms of pest control by vermicompost and vermi-leachate (Source: Adapted from Yatoo et al. (2021))

Table 17.1 Insect pest and nematodes suppression using vermicompost and vermi-leachates

Vermiworm species	Substrate	Pests	References
Eisenia fetida	Animal wastes and agro wastes	Lesion nematode	Nath and Singh (2011)
Eisenia fetida	Ipomoea leaves	Spiny bollworm	Hussain et al. (2020)
Eisenia fetida	Platanus leaf and cow manure	Red spider mite	Aghamohammadi et al. (2016)
Eisenia fetida	Animal dung and agro-kitchen wastes	Mustard aphid	Nath and Singh (2012)
Eisenia fetida	Food wastes	Northern root-knot nematode, root-knot nematode, and two-spotted spider mite	Chattopadhyay (2015)
Eisenia fetida	Food wastes	Stripped cucumber beetle, spotted cucumber beetle, and five-spotted hawkmoth	Yardim et al. (2006)
Eisenia fetida	Food wastes	Green peach aphid, mealy bug, and cabbage moth	Arancon et al. (2005)
Eisenia fetida	Food wastes	Green peach aphid, mealy bug, and red spider mite	Arancon et al. (2007)
Eudrilus eugeniae	Leaf litter	Cotton bollworm	Haralu et al. (2018)

(Hemiptera: Aleyrodidae). Whiteflies showed a substantially reduced preference for settling and oviposition on tomato plants treated with vermicomposts than on control plants, and chemical fertilizers performed better on pistachio waste plants (Sedaghatbaf et al. 2018). Jangra et al. (2019) conducted a study to determine the pesticide activity of vermicompost on the chili broad mite *Polyphagotarsonemus latus*. When plants were treated with vermicompost at a rate of 5 t/ha, it was revealed that both the population and egg production of *Polyphagotarsonemus latus* was dramatically reduced. They concluded that this is because vermicompost contains both major and minor nutrients, which strengthen plants' defensive systems against mites.

On brinjal crops, mixed vermicomposts and biocontrol agents were efficient against root-knot nematodes in rhizoplane and soil by antibiosis and enzymatic actions. Compared to other compositions of vermicomposts and biopesticides, the nematode *Meloidogyne incognita* was most effectively suppressed by vermicompost made from a mixture of buffalo manure and legume bran, together with aqueous extracts of garlic. In addition, brinjal yields increased due to the use of vermicompost and biopesticides in combination (Nath et al. 2011). Renčo and Kováčik (2015) observed that vermi tea significantly lowered the number of eggs and infectious juveniles of the potato cyst nematodes *Globodera rostochiensis* and *Globodera pallida*. Additionally, it suggests these vermicompost and vermicompost teas improved the young stem mass and stem length of potato plants. The use of vermicompost tea rather than dry vermicompost significantly lowered the volume of inputs required. These supplements, on the other hand, seem to be good for controlling potato cysts in conservation agriculture systems. Vermicompost water extract from vermicompost with moso-bamboo and kudzu as feedstock was observed to reduce *Meloidogyne incognita* and *Rotylenchulus reniformis* eggs and infective juveniles when compared to vermicompost water extract generated from vermicompost prepared from vegetable food waste (You et al. 2018). Awad-Allah and Khalil (2019) demonstrated that vermicompost, vermicompost tea, and Nemaless[®] were combinedly found to be nematicidal against the root-knot nematode *Meloidogyne incognita* infecting banana cv. "Grand Naine" in laboratory and field circumstances. Additionally, these applications aided in the growth of plants. Vermi-leachates used in combination with other plant-based pest management strategies, vermi-leachates has a synergistic effect on sap-feeding parasitic insects such as thrips and mites while also increasing the number of healthy pods to boost yield (Kanchan et al. 2013). Researchers have found that when bean leaves were treated with vermi-leachates in combination with acaricide azocyclotin, exhibited a strong repellent effect against two-spotted spider mite (*Tetranychus urticae*) on leaves surface in laboratory and glasshouse conditions (Aghamohammadi et al. 2016).

17.8 Vermicompost and Vermi-leachate in Disease Management

Vermicompost have been shown to be more beneficial as organic manures and inoculants than their thermophilic aerobic compost counterparts. It is widely considered to inhibit plant disease, increase soil availability of nutrients, and promote plant development through biotic and abiotic processes and the activities of competitive or antagonistic microorganisms. It is in comparison to chemical fungicides and has good potential for reducing plant diseases due to its efficacy, mutualistic association with beneficial microorganisms, and environmental responsibility. As a good source of helpful microflora, little is known about how beneficial bacteria and earthworms interact with each other to control disease in plants. Within the last two decades, there seems to have been a significant growth in the body of knowledge demonstrating the efficiency of vermicompost products in protecting plants against numerous diseases. Vermicomposts and related products such as vermi-leachates, coelomic liquid, bioactive secondary metabolites, and skin secretion of worms from decomposer bacteria in vermicompost were seen in a high potential for suppressing disease occurrences caused by pathogenic genus such as *Aphanomyces, Aspergillus, Alternaria, Fusarium, Pythium, Phytophthora, Rhizoctonia*, and *Verticillium* in the experiments, indoors, and in field conditions. The envisaged disease inhibition processes for plant pathogen influence through the use of traditional composts are classified into two categories: general and specific suppression processes (Gudeta et al. 2022). The general suppression processes include microbial competition for nutrients and space, production of antibiotics, siderophores, and hydrolytic enzymes, and induced systemic resistance in the host for disease suppression. Similarly, the beneficial and decomposer microflora in the vermi-leachates break down organic debris and produce essential substances that help prevent the occurrence of plant diseases. Additionally, induced systemic response has been proposed as pathways for vermicompost-mediated plant disease suppression. The reduction in bacterial canker of greenhouse tomato induced by *Clavibacter michiganensis* subsp. *michiganensis* suggests an induced systemic response to prophylactic foliar application of vermicompost (Utkhede and Koch 2004).

Early treatments of dry vermicomposts for plant disease control were primarily in the form of pot supplements in which plant growth substrates were replaced with modest rates of dry vermicomposts in greenhouses to evaluate their ability to inhibit soil-borne fungal root infections. Vermicompost added to the planting pot mixture greatly reduced *Phytophthora nicotianae* and *Fusarium oxysporum* infections of tomato plants. When it came to protecting plants from soil-borne illnesses, the vermicomposts derived from cow dung were 52% more efficient than the control peat medium (Szczech and Smolińska 2001). As shown in the experiment, vermicomposts supplied in the soil-free plant growth medium Metro Mix 360 may effectively reduce *Pythium ultimum, Rhizoctonia solani*, and *Verticillium* sp. under the field settings, which are all plant diseases. The ability of several commercially produced vermicomposts to suppress disease was evaluated against attacks caused by *Phomopsis viticola* and *Sphaerotheca fuliginea* on grapevine in the field (Edwards and Arancon 2004), they also mentioned about the destruction of tomato diseases caused by *Phytophthora infestans*, and *Phytophthora nicotianae* var. *nicotianae*, and Fusarium wilt by *Fusarium lycopersici* through vermicompost applications. Vermicomposts have also been shown to reduce a variety of diseases in the field, albeit at a slower and less consistent rate than in container experiments. Vermicompost top-dressing on turf has been shown to be beneficial in suppressing numerous *Typhula ishikariensis* grass diseases such as Fusarium patch by *Microdochium nivale*, red thread by *Laetisaria fuciformis*, damping-off by *Pythium graminicola*, brown patch by *Rhizoctonia solani*, dollar spot by *Sclerotinia homoeocarpa*, and snow mold by *Sclerotinia homoeocarpa*. Often, the disease-suppressive efficacy increases with the rate of application. Generally, compost inclusion rates of at least 20% are required to ensure disease control, particularly in peat-based media (Noble and Coventry 2005). Significant disease control in soil has been achieved with lower levels of compost application, on the other hand. Pseudomonad populations were found to have a positive correlation with the overall number of culturable microorganisms in the composts, according to Bradley and Punja (2010). Especially, antagonistic *Pseudomonas aeruginosa* strains showed the maximum antagonism in tests between other compost-isolated bacteria against fusarium root and stem rot development on greenhouse cucumber. In another study, the combinatorial efficacy of vermicompost in combination with *Trichoderma harzianum* was proven as effective in soil application against black scurf or stem canker or stolon canker of potato caused by *Rhizoctonia solani* (Rahul et al. 2016). The deliberate addition of microbial inoculants to composts in order to create reinforced composts is a relatively new practice. To provide a food source for the microbial inoculants without encouraging infections, previous studies recommended adding compost and biocontrol agents to potting media. Initiated by thermo-composting research, the fortification of composts has since expanded to include vermicomposts as well. *Bacillus subtilis* and *Rhizoctonia solani* were shown to prevent soil-borne potato illnesses in a study by Larkin and Tavantzis (2013) alone or with compost.

Vermi-leachates and fluids have been demonstrated to diminish the pathogenic hazardous impact in crops afflicted by pathogen infection by lowering mycelial proliferation and bacterial population. According to Naidu et al. (2012), vermi-leachates can suppress the powdery mildew fungus *Golovinomyces cichoracearum* from growing on melons. Additionally, it was used to combat the black mold disease caused by *Aspergillus niger* on onions and other alliums. It was found that the vermicompost and vermi-leachates had the greatest impact on preventing plant disease development, according to Mehta et al. (2014). Microorganisms in vermicompost and vermi-leachates can function as antagonistic agents of pathogens by competing for resources and space and creating an antibacterial chemical, eliminating parasitizing pathogens and systematically diminishing their resistance.

The review article indicated that the application of vermicompost aqueous extracts inhibited the vegetative mycelial growth of fungal plant pathogens such as *Botrytis cinerea*, *Corticium rolfsii*, *Fusarium oxysporum*, *Rhizoctonia solani*, and *Sclerotinia sclerotiorum*. Extracts of ventilated vermicompost teas were found to be effective in suppressing *Fusarium moniliforme* and preventing rice foot rot disease. Vermicompost made from cow dung inhibited the growth of the Phytophthora nicotianae fungus in the nursery (Joshi et al. 2015). The vermicompost associated *Bacillus subtilis* strain M29 combined with volatile organic compounds (VOCs) such as 1-butanol, acetic acid butyl ester, 1-heptylene-4-alcohol, and 3-methyl-3-hexanol exhibited the broadest spectrum of antifungal activity against Botrytis cinerea. This necrotrophic fungus causes a gray mold on fruits (Mu et al. 2017). The extract (mucus, vermicompost tea, and vermi-leachate) of redworm (*Eisenia fetida*) has been proved to have a substantial biocontrol effect on *Fusarium graminearum* development in the roots of wheat during seed germination (Akinnuoye-Adelabu et al. 2019). After being applied to agricultural land, earthworm mucus has been shown to suppress mycelial growth in the soil and to have antifungal action, resulting in a 26% reduction in the incidence of fungal diseases. The researchers have studied the impact of coelomic fluid earthworm species such as *Allolobophora chlorotica*, *Dendrobaena veneta*, and *Eisenia andrei* on major plant pathogenic fungi such as *Berkeleyomyces basicola*, *Fusarium culmorum*, *Globisporangium irregulare*, *Macrophomina phaseolina*, *Rhizoctonia solani*, and *Sclerotinia sclerotiorum*. Coelomic fluid from *Eisenia andrei* was more effective at stopping *Rhizoctonia solani* than any other earthworm that was used in the study (Ečimović et al. 2021). Earthworms and worm casts have been advocated for their favorable effects on crop production and disease suppression. Rarely, earthworm casts have been found to carry soil-borne diseases in a few instances. Earthworm casts were found to be infected with soil-borne *Phytophthora capsici* and *Pythium attrantheridium* at many organic farms in North Carolina where they were employed as a supplement in soilless planting mixtures for vegetable seedling preparation (Liu et al. 2012). *Eisenia andrei* and *Eisenia fetida*, two closely related temperate worms, as well as *Eudrilus eugeniae*, are the most commonly utilized species in the management of plant disease. It is common to see the terms "*Eisenia andrei*" as well as "*Eisenia fetida*" used interchangeably in the literature; however, these are two distinct species (Dominguez and Edwards 2011) (Table 17.2).

17.9 Conclusion

Vermicompost has tremendous potential to be an effective substrate for colonization of beneficial microorganisms and as consortia, it is a potential alternative to conventional pesticidal applications in soil and crop. It also aids in the control of soil-borne pathogens and pests by means of hostile microbial populations present in vermicomposted organic matter and a variety of biological, biochemical, and physiological mechanisms. Vermicompost assists in the field efficiency of vermiproducts, such as nitrification, chelation, accumulation, and the biogeochemical cycling of micronutrients due to their functional and metabolic activities. The utilization of vermicompost or associated microbial populations in agricultural systems is therefore a more environmentally friendly alternative to chemical input. That vermicompost improves soil fertility, microbial invasion, crop productivity, biological features and yield output is clear from this study. While vermicompost is a byproduct of the process of composting vermi-leachates (compost, fluid, and vermi-leachate) are natural substances. Co-friendly means of protecting plants from fungal, bacterial, and insect infections by applying it straight to the soil and spraying it all over the plant's surface soil microorganism and pest-killing bioactive macromolecules are found in earthworms' coelomic fluid, mucus, and skin secretions. Plant growth is stimulated by vermi-leachate metabolites, which also produce an unfavorable environment for soil-borne pathogens. As a result, we looked at the role of

Table 17.2 Plant pathogen suppression using vermicompost

Earthworm species	Substrate	Pathogen	References
Eisenia fetida	Cattle manure	*Fusarium oxysporum* f. sp. *Lycopersici*	Szczech (2008)
Eisenia fetida	Dairy manure separated solids	*Fusarium oxysporum* f. sp. *radiciscucumerinum*	Kannangara et al. (2000)
Eisenia fetida	Cattle manure, tree bark, potato culls, apples	*Rhizoctonia solani* AG-4	Simsek Ersahin et al. (2009)
Eisenia fetida	Paper mill and dairy sludge	*Botrytis cinerea, Didymella bryoniae, Fusarium oxysporum* f.sp. *melonis, Lecanicillium fungicola, Pythium aphanidermatum, Phytophthora parasitica* var. *nicotianae, Sclerotinia sclerotiorum*, and *Verticillium dahliae*	Marín et al. (2013)
Eisenia fetida	Goat manure, straw	*Bipolaris oryzae, Colletotrichum gloeosporioides, Curvularia lunata, Cylindrocladium floridanum, Cy. Scoparium, Fusarium oxysporum, Macrophomina phaseolina, Pestalotia theae*, and *Sarocladium oryzae*	Pathma and Sakthivel (2013)
Eisenia fetida	Cow manure, bed leachate	*Colletotrichum gloeosporioides*	Contreras-Blancas et al. (2014)
Eisenia fetida	Thermocompost	*Rhizoctonia solani*	Neher et al. (2017)
Eisenia fetida	Peat	*Phytophthora nicotianae* var. *nicotianae*	Szczech and Smolińska (2001)
Eisenia fetida	Crop debris	*Colletotrichum coccodes, Rhizoctonia solani, Pythium ultimum, Pythium capsici,* and *Fusarium moniliforme*	Yasir et al. (2009)
Eudrilus eugeniae	Leaf litter	*Xanthomonas campestris* pv. *vesicatoria*	Reddy et al. (2015)
Eisenia fetida, Dendrobaena venata	Dairy manure solids mixed 7:1:1 with spoiled corn and hay silage, and cured hot compost	*Pythium aphanidermatum*	Jack and Nelson (2018)
Lumbricus terrestris	Worm addition	*Fusarium oxysporum* f. sp. *asparagi, Fusarium proliferatum, Verticillium dahliae* and *Fusarium*	Elmer (2009)

(continued)

Table 17.2 (continued)

Earthworm species	Substrate	Pathogen	References
		oxysporum f. sp. *lycopersici* Race 1	
Lumbricus terrestris	Worm addition	*Oculimaculayallundae* sp.	Bertrand et al. (2015)
Eisenia fetida and *Dendrobaena venata*	Sterile sand	*Pythium aphanidermatum*	Jack and Nelson (2018)

vermicompost and vermi-leachates in pest control and disease suppression as well as the components of vermi-leachates such as antimicrobial peptides, cellulolytic enzymes, immune modulators, and phagocytosis in the case of applying these proven findings to agriculture in order to increase crop yield.

References

Aghamohammadi Z, Etesami H, Alikhani HA (2016) Vermiwash allows reduced application rates of acaricide azocyclotin for the control of two-spotted spider mite, *Tetranychus urticae* Koch, on bean plant (*Phaseolus vulgaris* L.). Ecol Eng 93:234–241

Aira M, Monroy F, Domínguez J (2006) *Eisenia fetida* (Oligochaeta, Lumbricidae) activates fungal growth, triggering cellulose decomposition during vermicomposting. Microb Ecol 52:738–747

Akinnuoye, Adelabu DB, Hatting J, de Villiers C, Terefe T, Bredenhand E (2019) Effect of redworm extracts against Fusarium root rot during wheat seedling emergence. Agron J 111(5):2610–2618

Ananthakrishnasamy S (2019) Comparative study of municipal solid waste using by *Lampito mauritii* and *Eudrilus eugeniae* earthworms enhance for microbial enzyme activity. Int J Pharmac Biol Arch 10(4):281–292

Ansari AA, Sukhraj K (2010) Effect of vermiwash and vermicompost on soil parameters and productivity of okra (*Abelmoschus esculentus*) in Guyana. African J Agric Res 5(14):1794–1798

Arancon NQ, Galvis PA, Edwards CA (2005) Suppression of insect pest populations and damage to plants by vermicomposts. Bioresour Technol 96(10):1137–1142

Arancon NQ, Edwards CA, Yardim EN, Oliver TJ, Byrne RJ, Keeney G (2007) Suppression of two-spotted spider mite (*Tetranychus urticae*), mealy bug (Pseudococcus sp.) and aphid (*Myzus persicae*) populations and damage by vermicomposts. Crop Prot 26(1):29–39

Awad-Allah SF, Khalil MS (2019) Effects of vermicompost, vermicompost tea and a bacterial bioagent against *Meloidogyne incognita* on banana in Egypt. Pak J Nematol 37(1):25–33

Becagli M, Arduini I, Cardelli R (2022) Using biochar and vermiwash to improve biological activities of soil. Agriculture 12(2):178

Bertrand M, Blouin M, Barot S, Charlier A, Marchand D, Roger-Estrade J (2015) Biocontrol of eyespot disease on two winter wheat cultivars by an Anecic earthworm (*Lumbricus terrestris*). Appl Soil Ecol 96:33–41

Bradley GG, Punja ZK (2010) Composts containing fluorescent pseudomonads suppress fusarium root and stem rot development on greenhouse cucumber. Can J Microbiol 56(11):896–905

Cardoza YJ, Buhler WG (2012) Soil organic amendment impacts on corn resistance to *Helicoverpa zea*: constitutive or induced? Pedobiologia 55(6):343–347

Chaoui H, Edwards CA, Brickner A, Lee SS, Arancon NQ (2002) Suppression of the plant diseases, Pythium (damping-off), Rhizoctonia (root rot) and Verticillium (wilt) by vermicomposts. In: Brighton crop protection conference pests and diseases, vol 2, pp 711–716

Chattopadhyay (2015) Effect of vermiwash of *Eisenia foetida* produced by different methods on seed germination of green mung, *Vigna radiata*. Int J Recycling Organic Waste Agric 4(4): 233e237

Contreras-Blancas E, Ruíz-Valdiviezo VM, Santoyo-Tepole F, Luna-Guido M, Meza-Gordillo R, Dendooven L, Gutiérrez-Miceli FA (2014) Evaluation of worm-bed leachate as an antifungal agent against pathogenic fungus, *Colletotrichum gloeosporioides*. Compost Sci Utiliz 22(1): 23–32

Curry JP (2004) Factors affecting the abundance of earthworms in soils. Earthworm Ecol 9:113–113

Das SK, Avasthe RK, Gopi R (2014) Vermiwash: use in organic agriculture for improved crop production. Popular Kheti 2(4):45–46. www.popularkheti.info

de Brito Alvarez MA, Gagné S, Antoun H (1995) Effect of compost on rhizosphere microflora of the tomato and on the incidence of plant growth-promoting rhizobacteria. Appl Environ Microbiol 61:194–199

de Novais CB, de Oliveira JR, Siqueira JO, de Faria SM, da Silva EM, Aquino AM, Júnior OJ (2019) Trophic relationships between the earthworm *Pontoscolex corethrurus* and three tropical arbuscular mycorrhizal fungal species. Appl Soil Ecol 135:9–15

Dominguez J, Edwards CA (2011) Relationships between composting and vermicomposting. In: Edwards CA, Arancon NQ, Sherman R (eds) Vermiculture technology: earthworms, organic wastes, and environmental management. CRC Press Taylor & Francis Group, Boca Raton, FL, pp 12–25. https://doi.org/10.1201/b10453

Domínguez J, Aira M, Kolbe AR, Gómez-Brandón M, Pérez-Losada M (2019) Changes in the composition and function of bacterial communities during vermicomposting may explain beneficial properties of vermicompost. Sci Rep 9(1):1–11

Drake HL, Horn MA (2007) As the worm turns: the earthworm gut as a transient habitat for soil microbial biomes. Annu Rev Microbiol 61:169–189

Ečimović S, Vrandečić K, Kujavec M, Žulj M, Ćosić J, Velki M (2021) Antifungal activity of earthworm coelomic fluid obtained from *Eisenia andrei*, *Dendrobaena veneta* and *Allolobophora chlorotica* on six species of phytopathogenic fungi. Environments 8(10):102

Edwards CA, Arancon NQ (2004) The use of earthworms in the breakdown of organic wastes to produce vermicomposts and animal feed protein. Earthworm Ecol 2:345–380

Edwards CA, Fletcher KE (1998) Interaction between earthworms and microorganisms in organic matter breakdown. Agric Ecosyst Environ 20:235–249

Edwards CA, Arancon NQ, Vasko-Bennett M, Askar A, Keeney G (2010) Effect of aqueous extracts from vermicomposts on attacks by cucumber beetles (*Acalymna vittatum*) (Fabr.) on cucumbers and tobacco hornworm (*Manduca sexta*) (L.) on tomatoes. Pedobiologia 53(2): 141–148

Eijsackers H (2011) Earthworms as colonizers of natural and cultivated soil environments. Appl Soil Ecol 50:1–13

Elmer WH (2009) Influence of earthworm activity on soil microbes and soilborne diseases of vegetables. Plant Dis 93:175–179

Ganguly RK, Chakraborty SK (2020) Eco-management of industrial organic wastes through the modified innovative vermicomposting process: a sustainable approach in tropical countries. In Earthworm assisted remediation of effluents and wastes. Springer, Singapore, pp 161–177

Gudeta K, Julka JM, Kumar A, Bhagat A, Kumari A (2021) Vermiwash: an agent of disease and pest control in soil, a review. Heliyon 7(3):e06434

Gudeta K, Bhagat A, Julka JM, Sinha R, Verma R, Kumar A, Kumari S, Ameen F, Bhat SA, Amarowicz R, Sharma M (2022) Vermicompost and its derivatives against phytopathogenic fungi in the soil: a review. Horticulturae 8(4):311

Haralu S, Karabhantanal SS, Jagginavar SB, Naidu GK (2018) Utilization of vermiwash as biopesticide in the management of pod borer, *Helicoverpa armigera* (hubner), in chickpea (*Cicer arietinum* l.). Appl Biol Res 20(1):37–45

Harel YM, Mehari ZH, RavDavid D, Elad Y (2014) Systemic resistance to gray mold induced in tomato by benzothiadiazole and mesocosm experiment. Sci Total Environ 514:147–154

Hoeffner K, Monard C, Santonja M, Cluzeau D (2018) Feeding behaviour of epi-anecic earthworm species and their impacts on soil microbial communities. Soil Biol Biochem 125:1–9

Hoitink HAJ, Fahy PC (1986) Basis for the control of soil-borne plant pathogens with composts. Annu Rev Phytopathol 24:93–114

Hoitink HA, Stone AG, Han DY (1997) Suppression of plant diseases by composts. Hortic Sci 32: 184–187

Horst LE, Locke J, Krause CR, McMahon RW, Madden LV, Hoitink HA (2005) Suppression of botrytis blight of begonia by *Trichoderma hamatum* 382 in peat and compost-amended potting mixes. Plant Dis 89:1195–1200

Hussain N, Abbasi T, Abbasi SA (2020) Evaluating the fertilizer and pesticidal value of vermicompost generated from a toxic and allelopathic weed ipomoea. J Saudi Soc Agric Sci 19(1):43–50

Ibrahim MH, Quaik S, Ismail SA (2016) Microbial ecology associated with earthworm and its gut. In: Prospects of organic waste management and the significance of earthworms. Springer, Cham, pp 123–145

Jack ALH, Nelson EB (2018) A seed-recruited microbiome protects developing seedlings from disease by altering homing responses of *Pythium aphanidermatum* zoospores. Plant Soil 422(1–2):209–222

Jack AL, Thies JE (2006) Compost and vermicompost as amendments promoting soil health. Biol Appr Sustain Soil Syst:453–466

Jangra M, Sindhu S, Sonika RG, Batra VK (2019) Studies on efficacy of vermicompost for the management of *Polyphago tarsonemus* latus (banks) (Acari: Tarsonemidae) infesting chilli (*Capsicum annuum* L.) in Haryana. Pharma Innov J 8:86–89

Jayaraman S, Naorem AK, Lal R, Dalal RC, Sinha NK, Patra AK, Chaudhari SK (2021) Disease-suppressive soils—beyond food production: a critical review. J Soil Sci Plant Nutr 21(2): 1437–1465

Jogaiah S, Abdelrahman M, Tran LS (2013) Characterization of rhizosphere fungi that mediate resistance in tomato against bacterial wilt disease. J Exp Bot 64:3829–3842

Joshi R, Singh J, Vig AP (2015) Vermicompost as an effective organic fertilizer and biocontrol agent: effect on growth, yield, and quality of plants. Rev Environ Sci Biotechnol 14(1):137–159

Kanchan M, Keshav S, Tripathi CP (2013) Management of pod borer (*Helicoverpa armigera*) infestation and productivity enhancement of gram crop (*Cicer aritenium*) through vermiwash with biopesticides. World J Agric Sci 9(5):401–408

Kannangara T, Utkhede RS, Paul JW, Punja ZK (2000) Effects of mesophilic and thermophilic composts on suppression of fusarium root and stem rot of greenhouse cucumber. Can J Microbiol 46(11):1021–1028

Kizilkaya R, Karaca A, Turgay OC, Cetin SC (2011) Earthworm interactions with soil enzymes. In: Biology of earthworms. Springer, Berlin, pp 141–158

Kumar D (2019) Effect of vermiwash from different organic resources on growth, yield and quality of organic blackgram [*Vigna mungo* (L.) Hepper]. M.Sc. (Ag.) thesis, Maharana Pratap University of Agriculture and Technology, Udaipur, p 50

Kwok O, Fahy P, Hoitink HA, Kuter G (1987) Interactions between bacteria and *Trichoderma hamatum* in suppression of rhizoctonia damping-off in bark compost media. Phytopathology 77: 1206–1212

Larkin RP, Tavantzis S (2013) Use of biocontrol organisms and compost amendments for improved control of soilborne diseases and increased potato production. Am J Potato Res 90(3):261–270

Lazcano C, Revilla P, Malvar RA, Domínguez J (2011) Yield and fruit quality of four sweet corn hybrids (*Zea mays*) under conventional and integrated fertilization with vermicompost. J Sci Food Agric 91(7):1244–1253

Li W, Wang C, Sun Z (2011) Vermipharmaceuticals and active proteins isolated from earthworms. Pedobiologia 54:S49–S56

Lievens B, Vaes K, Coosemans J, Ryckeboer J (2001) Systemic resistance induced in cucumber against Pythium root rot by source separated household waste and yard trimmings composts. Commun Sci Util 9:221–229

Liu B, Roos D, Buttler S, Richter B, Louws FJ (2012) Vegetable seedling diseases associated with earthworm castings contaminated with *Phytophthora capsici* and *Pythium attrantheridium*. Plant Health Progress 13(1):15

Manyuchi MM, Phiri A, Muredzi P, Chirinda N (2013) Bioconversion of food wastes into vermicompost and vermiwash. Int J Sci Mod Eng 1:1–2

Marín F, Santos M, Diánez F, Carretero F, Gea FJ, Yau JA, Navarro MJ (2013) Characters of compost teas from different sources and their suppressive effect on fungal phytopathogens. World J Microbiol Biotechnol 29(8):1371–1382

Medina-Sauza RM, Álvarez-Jiménez M, Delhal A, Reverchon F, Blouin M, Guerrero-Analco JA, Cerdán CR, Guevara R, Villain L, Barois I (2019) Earthworms building up soil microbiota, a review. Front Environ Sci 7:81

Mehta CM, Palni U, Franke-Whittle IH, Sharma AK (2014) Compost: its role, mechanism and impact on reducing soil-borne plant diseases. Waste Manag 34(3):607–622

Mu J, Li X, Jiao J, Ji G, Wu J, Hu F, Li H (2017) Biocontrol potential of vermicompost through antifungal volatiles produced by indigenous bacteria. Biol Control 112:49–54

Nadana GRV, Rajesh C, Kavitha A, Sivakumar P, Sridevi G, Palanichelvam K (2020) Induction of growth and defense mechanism in rice plants towards fungal pathogen by eco-friendly coelomic fluid of earthworm. Environ Technol Innov 19:101011

Naidu Y, Meon S, Siddiqui Y (2012) In vitro and in vivo evaluation of microbial-enriched compost tea on the development of powdery mildew on melon. BioControl 57(6):827–836

Naseri B (2019) Legume root rot control through soil management for sustainable agriculture. In: Meena R, Kumar S, Bohra J, Jat M (eds) Sustainable management of soil and environment. Springer, Singapore

Nath G, Singh K (2011) Combination of vermicomposts and biopesticides against nematode (Pratylenchus sp.) and their effect on growth and yield of tomato (*Lycopersicon esculentum*). IIOAB J 2(5):27e35

Nath G, Singh K (2012) Combination of vermiwash and biopesticides against aphid (*Lipaphis erysimi*) infestation and their effect on growth and yield of mustard (*Brassica campestris*). Dyn Soil Dyn Plant 1(1):96e102

Nath G, Singh K, Singh DK (2009) Chemical analysis of vermicompost/vermiwash of different combinations animal, agro and kitchen wastes. Austral J Basic Appl Sci 3(4):3672–3676

Nath G, Singh DK, Singh K (2011) Productivity enhancement and nematode management through vermicompost and biopesticides in brinjal (*Solanum melogena* L.). World Appl Sci J 12(4): 404–412

Neher DA, Fang L, Weicht TR (2017) Ecoenzymes as indicators of compost to suppress *Rhizoctonia solani*. Compost Sci Utiliz 25(4):251–261

Noble R, Coventry E (2005) Suppression of soil-borne plant diseases with composts: a review. Biocontrol Sci Tech 15(1):3–20

Palanichamy V, Mitra B, Reddy N, Katiyar M, Rajkumari RB, Ramalingam C, Arangantham (2011) Utilizing food waste by vermicomposting, extracting vermiwash, castings and increasing relative growth of plants. Int J Chem Anal Sci 2:1241–1246

Parthasarathi K, Ranganathan LS, Anandi V, Zeyer J (2007) Diversity of microflora in the gut and casts of tropical composting earthworms reared on different substrates. J Environ Biol 28(1):87–97

Pathma J, Sakthivel N (2012) Microbial diversity of vermicompost bacteria that exhibit useful agricultural traits and waste management potential. Springerplus 1(1):1–19

Pathma J, Sakthivel N (2013) Molecular and functional characterization of bacteria isolated from straw and goat manure-based vermicompost. Appl Soil Ecol 70:33–47

PeimaniForoushani A, Poorjavad N (2017) Effects of different application methods of vermicompost on tomato leaf miner, *Tuta absoluta* (Lep., Gelechiidae) population. Iran J Plant Prot Sci 48(1):59–67

Punja ZK, Rodriguez G, Tirajoh A (2016) Effects of Bacillus subtilis strain QST 713 and storage temperatures on post-harvest disease development on greenhouse tomatoes. Crop Prot 84:98–104

Rahul SN, Khilari K, Jain SK, Dohrey RK, Dwivedi A (2016) Management of black scurf of potato caused by *Rhizoctonia solani* with organic amendments and their effect on different parameter of potato crop. J Pure Appl Microbiol 10:2433–2438

Raimi AR, Atanda AC, Ezeokoli OT, Jooste PJ, Madoroba E, Adeleke RA (2022) Diversity and predicted functional roles of cultivable bacteria in vermicompost: bioprospecting for potential inoculum. Arch Microbiol 204(5):1–13

Reddy SA, Bagyaraj DJ, Kale RD (2015) Management of tomato bacterial spot caused by *Xanthomonas campestris* using vermicompost. J Biopest 5(1):10–13

Renčo M, Kováčik P (2015) Assessment of the nematicidal potential of vermicompost, vermicompost tea, and urea application on the potato cyst nematodes *Globodera rostochiensis* and *Globodera pallida*. J Plant Protect Res 55(2):187–192

Sang MK, Kim KD (2011) Biocontrol activity and primed systemic resistance by compost water extracts against anthracnoses of pepper and cucumber. Phytopathology 101:732–740

Sangwan P, Garg VK, Kaushik CP (2010) Growth and yield response of marigold to potting media containing vermicompost produced from different wastes. Environmentalist 30(2):123–130

Sedaghatbaf R, Samih MA, Zohdi H, Zarabi M (2018) Vermicomposts of different origins protect tomato plants against the sweet potato whitefly. J Econ Entomol 111(1):146–153

Sharma RC, Banik P (2012) Effect of integrated nutrient management on baby corn-rice cropping system; economic yield, system productivity, nutrient use efficiency and soil nutrient balance. Indian J Agric Sci 82(3):220–224

Sheehy J, Nuutinen V, Six J, Palojärvi A, Knuutila O, Kaseva J, Regina K (2019) Earthworm *Lumbricus terrestris* mediated redistribution of C and N into large macroaggregate-occluded soil fractions in fine-textured no-till soils. Appl Soil Ecol 140:26–34

Siddiqui Y, Meon S, Ismail R, Rahmani M, Ali A (2008) Bio-efficiency of compost extracts on the wet rot incidence, morphological and physiological growth of okra *Abelmoschus esculentus* [(L.) Moench]. Sci Horticult 117:9–14

Simsek Ersahin Y, Haktanir K, Yanar Y (2009) Vermicompost suppresses *Rhizoctonia solani* Kühn in cucumber seedlings. J Plant Dis Protect 116(4):182–188

Singhai PK, Sarma BK, Srivastava JS (2011) Biological management of common scab of potato through Pseudomonas species and vermicompost. Biol Control 57:150–157

Sinha RK, Herat S, Valani D, Chauhan K (2009) Vermiculture and sustainable agriculture. Am Eurasian J Agric Environ Sci IDOSI Publica 155

Sinha RK, Herat S, Valani D, Chauhan K (2010) Earthworms- the environmental engineers: review of vermiculture technologies for environmental management and resource development. Int J Glob Environ 10(3–4):265–292

Soltan HA, Dakhly OF, Mahmoud MA, Fayz YF (2022) Microbiological and genetical identification of some vermicompost beneficial associated bacteria. SVU-Int J Agric Sci 4(1):21–36

Srivastava R, Khalid A, Singh US, Sharma AK (2010) Evaluation of arbuscular mycorrhizal fungus, fluorescent Pseudomonas and *Trichoderma harzianum* formulation against *Fusarium oxysporum* f. sp. lycopersici for the management of tomato wilt. Biol Control 53:24–31

Sundararasu K, Jeyasankar A (2014) Effect of vermiwash on growth and yield of brinjal (*Solanum melongena*). Asian J Sci Technol 5(3):171–173

Szczech M (2008) Mixtures of microorganisms in biocontrol. In: Kim M-B (ed) Progress in environmental microbiology. Nova Science Publishers, Hauppauge, New York, pp 69–110

Szczech M, Smolińska U (2001) Comparison of suppressiveness of vermicomposts produced from animal manures and sewage sludge against *Phytophthora nicotianae* Breda de Haan var. nicotianae. J Phytopathol 149(2):77–82

Tharmaraj K, Ganesh P, Kolanjinathan K, Suresh Kumar R, Anandan A (2011) Influence of vermicompost and vermiwash on physico-chemical properties of rice cultivated soil. Curr Bot

Thiruneelakandan R, Subbulakshmi G (2014) Vermicomposting: a superlative for soil, plant, and environment. Int J Innov Res Sci Eng Technol 3(1):930–938

Tuitert G, Szczech M, Bollen GJ (1998) Suppression of *Rhizoctonia solani* in potting mixtures amended with compost made from organic household waste. Phytopathology 88:764–773

Utkhede R, Koch C (2004) Biological treatments to control bacterial canker of greenhouse tomatoes. BioControl 49(3):305–313

Van Groenigen JW, Van Groenigen KJ, Koopmans GF, Stokkermans L, Vos HM, Lubbers IM (2019) How fertile are earthworm casts? A meta-analysis. Geoderma 338:525–535

Vinale F, Flematti G, Sivasithamparam K, Lorito M, Marra R, Skelton BW, Ghisalberti EL (2009) Harzianic acid, an antifungal and plant growth promoting metabolite from *Trichoderma harzianum*. J Nat Prod 72:2032–2035

Xiao Z, Liu M, Jiang L, Chen X, Griffiths BS, Li H, Hu F (2016) Vermicompost increases defense against root-knot nematode (Meloidogyne incognita) in tomato plants. Appl Soil Ecol 105:177–186

Yardim EN, Arancon NQ, Edwards CA, Oliver TJ, Byrne RJ (2006) Suppression of tomato hornworm (*Manduca quinquemaculata*) and cucumber beetles (*Acalymma vittatum* and *Diabotrica undecimpunctata*) populations and damage by vermicomposts. Pedobiologia 50(1):23–29

Yasir M, Aslam Z, Kim SW, Lee SW, Jeon CO, Chung YR (2009) Bacterial community composition and chitinase gene diversity of vermicompost with antifungal activity. Bioresour Technol 100:4396–4403

Yatoo AM, Ali M, Baba ZA, Hassan B (2021) Sustainable management of diseases and pests in crops by vermicompost and vermicompost tea. A review. Agron Sustain Dev 41(1):1–26

You X, Tojo M, Ching S, Wang KH (2018) Effects of vermicompost water extract prepared from bamboo and kudzu against *Meloidogyne incognita* and *Rotylenchulus reniformis*. J Nematol 50(4):569

Zhao FY, Zhang Y, Dong W, Zhang Y, Yang L (2019) Vermicompost can suppress *Fusarium oxysporum* f. sp. lycopersici via generation of beneficial bacteria in a long-term tomato monoculture soil. Plant Soil 440:491–505

Part IV
Vermicomposting and Wastes

Chapter 18
Vermicompost: A Potential Reservoir of Antimicrobial Resistant Microbes (ARMs) and Genes (ARGs)

Zakio Makuvara, Jerikias Marumure, Rangarirayi Karidzagundi, Claudious Gufe, and Richwell Alufasi

Abstract Vermicompost is an agronomic humus-like material produced by an earthworm and microorganism-based process that offers decomposition, conversion, and detoxification of organic waste that harbors important pathogenic microbes and undesirable substances, such as AMR genes and antibiotic residues. This chapter is based on a literature survey in databases such as Web of Science, Google Scholar, Scopus, and PubMed. Though vermicomposting significantly reduces pathogenic microbes, some microbes develop resistance and persist within vermicomposting-derived organic fertilizers. Subsequent application of vermicompost (biofertilizers) into the soil inevitably transmits antimicrobial resistant microbes and associated antimicrobial resistance genes into the environment. Vermicompost has been implicated as essential sources of verotoxin-producing bacteria, some of which are rarely reduced by vermicomposting processes due to their spore-forming nature. Therefore, vermicompost can be reservoirs and vehicles of antimicrobial resistant microbes and genes. However, there is limited data on the potential of vermicomposts to harbor AMR microbes (ARMs) and AMR genes (ARGs), and there is paucity on mitigatory measures of preventing transmission of ARMs and ARGs to human beings. In this chapter, we explore (1) the nature and drivers of antimicrobial resistance in vermicomposting processes, (2) generation of AMR during vermicomposting, (3) fate of vermicomposting-derived antimicrobial resistant microbes (ARMs) and

---

Z. Makuvara (✉) · J. Marumure
School of Natural Sciences, Great Zimbabwe University, Masvingo, Zimbabwe
e-mail: zmakuvara@gzu.ac.zw

R. Karidzagundi
Materials Development Unit, Zimbabwe Open University, Harare, Zimbabwe

C. Gufe
Department of Veterinary Technical Services, Central Veterinary Laboratories, Harare, Zimbabwe

R. Alufasi
Biological Sciences Department, Bindura University of Science Education, Bindura, Zimbabwe

© The Author(s), under exclusive license to Springer Nature Singapore Pte Ltd. 2023
H. A. Mupambwa et al. (eds.), *Vermicomposting for Sustainable Food Systems in Africa*, Sustainable Agriculture and Food Security,
https://doi.org/10.1007/978-981-19-8080-0_18

genes (ARGs), (4) current techniques of monitoring antimicrobial resistance during and after vermicomposting, (5) removal of antimicrobial resistant microbes and genes in vermicompost, and (6) effects of ARGs and ARMs derived from vermicompost on public health.

Keywords Vermicomposting · Antimicrobial resistance · Vermicasts · Worms · Microorganisms · Public health · Biotransformation

18.1 Introduction

Vermicomposting is a mesophilic biodegradation and organic matter stabilization process that takes advantage of earthworms' coexistence and dual activity (Du Plessis 2010). Furthermore, vermicomposting is the biochemical breakdown process by which earthworms and their associated microbes transform vermicomposting substrates into earthworm biomass and vermicompost (Lee et al. 2018; Lim et al. 2015). Generally, vermicomposting has a higher stabilization capability and a faster physical and biochemical transformation rate than natural composting (Liew et al. 2022). The efficacy of vermicomposting lies in its ability to lower the C:N ratio and produce a homogeneous biofertilizer. Additionally, vermicomposting is a cost-effective and eco-friendly waste management intervention (Liew et al. 2022; Soobhany 2019; Yadav and Garg 2011). Vermicomposting, because of its inherent biological, biochemical, and physicochemical features, can be utilized to enhance sustainable organic fertilizer management (Kumar 2011). The main product of vermicomposting processes is vermicompost, a rich source of nutrients that improves agricultural soils' biological, physical, and chemical characteristics. In addition to vermicompost, earthworms, which are integral parts of vermicomposting processes, are important protein sources for fish and monogastric animals (Pierre-Louis et al. 2021; Ansari et al. 2020).

Earthworms are supplied with organic supplements such as cow dung, chicken droppings, and microbial inoculum to maximize their activity and survival in various wastes and vermicomposting processes (Munnoli et al. 2010). Organic additives help in decreasing the toxicity of some vermicomposting substrates and establish an appropriate C:N ratio (Liew et al. 2022; Soobhany 2019; Yadav and Garg 2011). The overall quality of the stages of vermicomposting is determined by the selection of an appropriate earthworm species (e.g., those with high adaptability and reproductive capacity) and organic waste to be degraded. This is so since there is a diversity of earthworms which are classified as anecic, endogeic, and epigeic based on their behavior in the environment (Singh et al. 2011; Domfnguez 2004). The most common earthworms used in vermicomposting are *Eisenia fetida* (used worldwide) and *Eudrilus eugeniae* (used in tropical and subtropical regions). These earthworms are utilized in vermicomposting because of their ability to colonize, ingest, grind, digest (facilitated by earthworms' gut microflora which is both aerobic and anaerobic), and absorb organic waste (Ibrahim et al. 2016). Generally, the earthworms are involved in both physical (e.g., organic waste aeration and grinding) and biochemical (e.g., microbial degradation of substrates) (Ibrahim et al. 2016; Singh et al. 2011; Domfnguez 2004).

Although vermicomposting processes are beneficial, the same processes have been linked to pathogen tolerance, making it an essential reservoir of pathogenic bacteria, antimicrobial resistant microorganisms, and genes (Ezugworie et al. 2021; Qiu et al. 2021). This is correlated with the highest temperature in a vermicomposting process, which is about 35 °C. Notably, harmful microbes persist at this low temperature and lead to antibiotic resistance (Tripathi et al. 2005). Even though vermicomposting has been identified as an environmentally friendly process capable of reducing pathogenic microbes, it has been reported that the number of microorganisms that produce spores, such as *Bacillus and Clostridium* spp., as well as genes for antibiotic resistance, remained high during and after vermicomposting (Awasthi et al. 2019; Manyi-Loh et al. 2016). Temperature exceeding 70 °C is necessary to remove the pathogenic microorganisms and associated antimicrobial resistance genes. However, it has been observed that at such temperatures, earthworms, which are at the core of vermicomposting operations, are killed, and the process is halted (Gajalakshmi et al. 2002). However, it is noted that, though pathogenic microorganisms and their antimicrobial resistant gene persist, vermicomposting processes remove ARB and ARGs (Cui et al. 2018). Vermicomposts are biologically varied in that they contain a wide range of microbial and macroorganism populations that function on a variety of vermicompost substrates (Gómez-Brandón and Domínguez 2014a, b; Domínguez et al. 2010).

Vermicompost substrates typically include a variety of harmful compounds, antimicrobial resistant microorganisms, genes for antimicrobial resistant, and residues of antibiotics (Grantina-Ievina and Rodze 2020; Liu et al. 2019). In recent years, there has been intensive utilization of antibiotics in animal husbandry, but up to 30–90% of the administered antibiotics are excreted through urine and manures (Xu et al. 2020). This leads to accumulation of residual antibiotics in animal waste (which can be utilized as vermicompost substrates) and concentrations in the μg/kg to mg/kg level have been reported (Song et al. 2021). These high concentrations of residual antibiotics in manure and other vermicompost substrates are highly implicated in the acquisition of antibiotic resistance genes (ARGs) by bacteria via horizontal gene transfer or spontaneous mutation during vermicomposting, thereby causing the proliferation of antibiotic-resistant bacteria (Xu et al. 2020). Antibiotic-resistant human pathogenic taxa such as *Enterococcus, Escherichia coli*, and *Alcaligenes faecalis* have been discovered in chicken manure (a possible vermicomposting substrate) (Song et al. 2021), and *Salmonella* spp. and *Campylobacter*, in pig farm lagoons (Swati and Hait 2018). This suggests that applying these substrates to soil potentially disseminates ARGs into the environment. Pathogenic bacteria linked with substrates utilized in vermicomposting processes include bacteria capable of producing verotoxins and these bacteria include *Enterococcus faecalis, E. coli, Clostridium perfringens, S. enterica, Listeria monocytogenes*, and *Neisseria meningitides* (Grantina-Ievina and Rodze 2020).

Antimicrobial resistance (AMR) jeopardizes disease prevention and therapy for many illnesses (Zrnčić 2020; Thanner et al. 2016; Jasovský et al. 2016). Because

ARBs and ARGs are found in vermicomposting substrates, ARGs monitoring in vermicompost should be achievable using repeatable, predictable, and reliable diagnostic testing procedures (Jasovský et al. 2016). A range of approaches exists for detecting, monitoring, and profiling antimicrobial resistance in vermicompost. Antimicrobial resistance has traditionally been investigated through isolation of bacterial cultures and evaluating resistance using culture-dependent tests and anti-biotic breakpoints established by the European Committee on Antimicrobial Susceptibility Testing (EUCAST) or the Clinical and Laboratory Standards Institute (CLSI) (Eckstrom 2018). However, these methods are time-consuming, have low throughput, and are limited to microorganisms that can be cultivated. As a result, culture-independent AMR detection techniques including quantitative PCR, fluorescence in situ hybridization, and, most recent, rapid genomic sequencing are gaining popularity (Thanner et al. 2016). As more of these technologies are put in place, our capacity to track the spread of AMR in vermicompost becomes stronger. In line with paucity and inclusive evidence on the potential of vermicompost as a reservoir of AMR microbes (ARMs) and genes (ARGs) (emerging pollutants), this chapter explores (1) the nature and drivers of AMR during vermicomposting processes, (2) generation AMR during vermicomposting, (3) fate of vermicomposting-derived AMR microbes and genes, (4) current techniques of monitoring AMR during and after vermicomposting, (5) removal of AMR microbes and genes in vermicompost, and (6) effects of ARGs and ARMs derived from vermicompost on public health.

18.2 Nature and Drivers of Antimicrobial Resistance During Vermicomposting

18.2.1 Nature

Antimicrobial resistance can be characterized as (1) intrinsic (inherent) or (2) acquired, according to Cox and Wright (2013). Though there is insufficient evidence on the nature of antimicrobial resistance during and after vermicomposting, antimicrobial resistance is grouped into these two classifications when it exists. Intrinsic antimicrobial resistance occurs spontaneously; for example, all bacterial species are typically antibiotic resistant (Reygaert 2018). According to Tadesse et al. (2017), enterococci and anaerobic bacteria transmit inherent antibiotic resistance to drugs classed as cephalosporins and aminoglycosides. This type of resistance is mediated by many factors, including bacterial outer membranes (for example, Gram-negative bacteria are resistant to a variety of antimicrobials due to the presence of gram-negative membranes) (Alekshun and Levy 2007), naturally produced antibiotic inactivating compounds (e.g., extended-spectrum beta-lactamase, which hydrolyzes ceftriaxone) (Tadesse et al. 2017), and active efflux. Furthermore, inherent antimicrobial resistance is affected by inherent genes and exists regardless of antimicrobial selection pressure. Acquired antimicrobial resistance is a global concern (Amábile-Cuevas 2021) since susceptible microbes become resistant through various methods such as chromosomal and gene mutations, horizontal gene transfer, and the development of efflux pumps (Tenover 2006; Roberts 2005; Poole 2002). Resistance to antibiotics such as amoxicillin (MR 72.9%, QR9.1%–87.3%) and trimethoprim/sulfamethoxazole (MR 75.0%, IQR 49.5%–92.3%) were described as typical examples of documented acquired antimicrobial resistance attributed to antibiotic usage (Tadesse et al. 2017). Furthermore, it has been claimed that up to 90% of gram-negative bacteria have developed resistance to the widely used antibacterial such as chloramphenicol (Leopold et al. 2014).

18.2.2 Drivers of Antimicrobial Resistance During Vermicomposting

Antimicrobial resistance appears to be induced by three factors: (1) antimicrobial residues and heavy metals associated with vermicomposting substrates, (2) microorganisms in substrates, and (3) worm gut microbiota as well as vermicomposting processes. According to Heuer et al. (2011), antibiotics in agriculture, particularly in animal husbandry, have hastened antibiotic resistance's emergence and persistence in farm produce and the environment. Notably, only a small percentage of animal administered antimicrobials such as antibiotics are absorbed, and the majority are excreted in animal waste (Blaser 2016). This is consistent with reportedly high antibiotic residues in livestock dung (Deng et al. 2018). Manure, an animal husbandry product, has been identified as a reservoir of antimicrobial residues (e.g., antibiotics), pathogenic bacteria, antimicrobial resistant microbes (ARMs), and related ARGs. According to Heinonen-Tanski et al. (2006), manure is always linked with microorganisms originating from feces, feed leftovers, and bedding substrates. However, animal dung is an essential vermicompost substrate that is occasionally utilized to supplement vermicomposting operations (Heuer et al. 2011). Although there is limited data on the presence and abundance of antimicrobial resistant or pathogenic microbes and antimicrobial residues in other vermicompost substrates, these two (antimicrobial resistant or pathogenic microbes and antimicrobial residues) likely exist in significant quantities in vermicompost substrates. For example, sewage as a vermicompost substrate resulted in mature vermicompost containing substantial amounts of pathogenic *Escherichia coli* and pathogenic fungal species such as *Aspergillus fumigatus*, *Scedosporium prolificans*, and *Aphanoascus terreus* (Grantina-Ievina et al. 2013).

Thus, during vermicomposting, antibiotic residues in vermicompost substrates exert selection pressure on accessible bacteria (from vermicompost substrates and worm microbiota), boosting antimicrobial resistance, and the propagation of ARGs (Zarei-Baygi et al. 2019). In one study, for example, the addition of tetracycline to a vermicomposting process including sludge substrates elevated pathogenic bacteria in two genera (1) *Bacillus* and (2) *Mycobacterium*, as well as ARGs such as tetC and int-1 by 4.7–186.9 and 4.25 folds, respectively (Xia et al. 2019). The findings of this study demonstrate that substantial amounts of antibiotic residues, such as tetracycline, can transform microorganism ecosystems and promote ARGs transmission in vermicompost. Additionally, it is anticipated that during vermicomposting, the genomic loci of resistance genes in bacteria would shift towards mobile genetic elements such as transposable elements, integrons, and broad-host-range plasmids. During horizontal and vertical transmission, these mobile genetic elements can be transferred to other microorganisms in the vermi-reactor (Heuer et al. 2011).

Metals found in vermicomposting substrates significantly influence the spread of antimicrobial resistance during vermicomposting. Heavy metals (e.g., copper and zinc) have been found in vermicomposting substrates such as manure, and this has been linked to cross-selection and co-selection mechanisms in vermicomposting microbial populations such as bacteria (Zhang et al. 2022; Tian et al. 2021; Dai et al. 2021). Heavy metals generally improve microbial tolerance and expose antimicrobial resistant microorganisms, such as antibiotic-resistant bacteria, under selection pressure. Though not explicitly recorded, the presence of heavy metals in vermicomposting processes is anticipated to increase ARGs via horizontal and vertical gene transfer rapidly.

18.3 Vermicomposting Processes and Antimicrobial Resistance

The complicated symbiotic relationship between earthworms (such as *Eisenia fetida* and *Perionyx excavatus*) and microbes produces vermicompost (Sim and Wu 2010). The disintegration, bio-degradation, and stabilization of organic waste result in nutrient-rich biofertilizer as a result of these complex interactions (Domínguez et al. 2010). Although microorganisms are fundamental in the biochemical decomposition of organic wastes, earthworms promote metabolic breakdown by modifying the substrate and modulating biological activity. This is accomplished primarily by (1) the earthworm directly feeding on the microorganisms (Gómez-Brandón and Domínguez 2014a, b) and (2) making the vermicomposted material more granular (thus increasing surface area) as a result of the earthworms' digestion and fragmentation (Lim et al. 2014). Compared to a biodegradation system that does not include earthworms, these activities are reported to increase the rate of organic matter turnover and microbial community production, resulting in faster breakdown (Lim et al. 2015; Blouin et al. 2013). The produced vermicompost improves soil fertility on both a physical and biological level, as well as soil conditioning (Lim et al. 2015).

Apart from producing a nutrient-rich product, vermicomposting can significantly reduce ARGs, particularly in aerobic conditions, which is critical in preventing their release into the different ecosystems (Cui et al. 2019). Huang et al. (2020) discovered that earthworm activity during vermicomposting altered the microbial ecology and decreased mobile genetic elements, indicating that ARGs were reduced.

Additionally, significant decreases in the number of bacteria in the castings were detected, which coincide with the elimination of ARGs (Cui et al. 2018). It is unknown, however, how earthworm gut digestion decreases ARGs (Cui et al. 2019). This might be owing to the breaking, biosynthesis, and catalytic and microbiological enrichment that has been observed in the earthworm's gut (Cardoso et al. 2008).

Variations in the ARGs and ARBs eliminated during vermicomposting are linked to environmental conditions, microbiome composition, and earthworm species (Huang et al. 2018). Temperature, for example, has been linked to the fate of ARGs in vermicomposting. Cui et al. (2022) reported that 20 °C is an appropriate temperature for vermicomposting activity which achieves the highest ARGs removal efficiency while maintaining good bio-stability of the final biofertilizers. Bacteria from the genera *Aeromonas* and the *Chitinophagaceae* were found at 25 °C, which was problematic since they might influence the recurrence of genes such as qnrA, and qnrS in vermicompost. Another critical factor contributing to the spread of ARGs and ARBs is antibiotic residue contamination in vermicomposting processes. According to Chen et al. (2019a, b), higher concentrations of antibiotic residuals can alter microbial communities and ARG dissemination risk in vermicompost is increased. The same authors revealed that concentration of tetracycline above 100 mg/kg reduced the abundance and diversity of active microbial cells while increasing the abundance of ARGs including *tet*M, *tet*X, *tet*C (4.7–186.9 folds increase), and *int-1* (4.25 folds increase), in spiked sewage sludge. In this regard, while reducing the high levels of antibiotics in vermicomposts is difficult (Kui et al. 2020), mechanisms for removing them should be investigated. In situations where antibiotic residues free organic matter is unavoidable, antibiotic biodegradation mechanisms involving vermicomposting should be considered. For example, housefly larvae (*Musca domestica*) attenuate nine antibiotics, including tetracyclines, sulfonamides, and fluoroquinolones, during a 6-day vermicomposting process (Zhang et al. 2014). Similarly, adding corncob and rice husk biochars to sludge prior to vermicomposting and vermicomposting for 60 days using *Eisenia fetida* resulted in a significant reduction in antibiotics, with tetracycline being completely removed (Kui et al. 2020).

Quite a number of studies have found that while ARGs and ARB are reduced during vermicomposting, they are not entirely removed (Cui et al. 2018; Huang et al. 2018, 2020; Tian et al. 2021; Zhao et al. 2022). For example, Huang et al. (2020) demonstrated a 41.5% decrease in ARGs abundance in sludge vermicompost. Similarly, 90% of ARGs (particularly *qnr*S, *tet*M, and *tet*X) were found to be removed from earthworm casts when compared to unvermicomposted organic matter (Cui et al. 2019), and quinolone resistance genes abundance in excess sludge was reported to be reduced by earthworms, with a drop ratio of 85.6–100% and 92.3–95.35% for qnrA and qnrS, respectively (Cui et al. 2018). Interestingly, complete removal of parC was discovered in another study, despite the fact that vermicomposting lowered the abundances of tetracycline and int1 in the same study but could not completely remove them (Huang et al. 2018). To that end, earthworms' inability to completely remove ARGs and ARB reduces the agricultural value of vermicompost (Cui et al. 2018). This is a cause for concern as ARGs and ARB are released into the environment, mainly through agricultural purposes, posing a potential public health risk (Huang et al. 2020). However, vermicomposting with scarab larvae (*Protaetia brevitarsis*) in a recent pot experiment revealed significantly lower percentage of ARGs and MGEs in rhizosphere, and root endophytes. The findings indicate that utilizing P. brevitarsis in vermicomposting may lower the danger of ARGs propagating through the plant-soil system, offering an alternative option for lowering ARGs in organic waste (Zhao et al. 2022).

Pathogenic microbes, including the ARB from vermicompost substrates, are generally reduced as earthworms ingested them. The ingested microbes are mixed with hydrolytic enzymes within the gut, which degrade cell walls, inactivate proteins and interfere with nucleic acids, and suppress cellular functions, causing death (Alberts et al. 2002). Furthermore, earthworms secrete antibacterial factors and coelomic fluids that inactivates pathogens including *A. hydrophila* (Pedersen and Hendriksen 1993), contributing to the removal of microbes from waste. However, the removal of the pathogens is largely selective as the earthworm exerts a differential effect depending on the earthworm genera and pathogen species (Swati and Hait 2017).

18.4 Novel Techniques of Monitoring Antimicrobial Resistance in Vermicompost

Specific genes from vermicompost can be identified using PCR-based approaches such as quantitative PCR. These approaches use fluorescent short DNA sequences that anneal to the target gene. The need for existing sequences knowledge and the detection limit are two major limitations of this approach (Luby et al. 2016; Raith et al. 2013). As microarray chips become more automated, they may be able to identify multiple genes in a single hybridization experiment (Kralik and Ricchi 2017; Peed et al. 2011). Using sequencing data, microarray chips create oligonucleotide probes that are affixed to customized microscope slides (Luby et al. 2016). This supports the production of unique arrays as well as the quick identification of target sequences from a wide range of DNA molecules (Haben Fesseha 2020). Microarray analysis would be an excellent tool for detecting AMR genes in vermicompost because of its versatility and speed. However, the method provides just information on AMR genes and must be augmented with additional techniques to detect microorganisms or gene expression.

Shotgun metagenomic sequencing is gaining popularity for studying AMR in vermicompost. The capacity to determine the genetic composition of each vermicompost sample in a single sequencing session has resulted in a better understanding of the dynamics and variety of AMR transmission. Shotgun metagenomics does not utilize a target sequence but rather fragments "shotgunned" the entire extracted DNA and then sequences it (Sharpton 2014). Scientists may utilize this single technique to find AMR microbes and genes without amplicon's primer bias and limits (Sharpton 2014). Among the most prominent benefits of shotgun metagenomics, is its capacity to establish the composition of microbial communities without prejudice (Sharpton 2014). Shotgun metagenomic sequencing has been successfully utilized to detect AMR genes in vermicompost, manure, and soils (Durso et al. 2012; Wichmann et al. 2014). Scientists were able to determine the presence of AMR genes in these cases, as well as the most likely gene transfer routes and the microbiological background of each vermicompost sample (Durso et al. 2012; Wichmann et al. 2014). Shotgun sequencing is an excellent tool for detecting AMR microorganisms and genes in vermicompost because of its adaptability and data range.

While shotgun sequencing lacks the precision of qPCR, it can reveal new genes, co-resistance patterns, and the genomic background of AMR genes in vermicompost (Bengtsson-Palme et al. 2017). However, due of its complexity, it creates substantially more data and necessitates data processing issues (Bengtsson-Palme et al. 2017). Sample preparation, sequencing depth and analysis are all aspects that should be taken into account shotgun sequencing is utilized (Bengtsson-Palme et al. 2017). When all DNA is sequenced, any contaminants and errors introduced during collection of samples, processing, and DNA isolation will be transmitted along (Xavier et al. 2016). Following sequencing, users are presented with a plethora of options for evaluation and AMR gene annotation; there are many databases designated solely to AMR gene annotation, each with its own set of categorization algorithms (Xavier et al. 2016).

Finally, combining classical microbiology with sophisticated next-generation sequencing, a new class of approaches called functional metagenomic sequencing has emerged. The aforementioned culture-independent techniques have the limitation of just identifying genetic characteristics and not demonstrating functional AMR (Mullany 2014; Luby et al. 2016; Boolchandani et al. 2017). To remedy this, a functional metagenomic method can be utilized. This procedure entails fragmenting and inserting DNA into vectors, turning the plasmids into competent experimental bacteria, and growing them on media (Mullany 2014; Luby et al. 2016; Boolchandani et al. 2017). Metagenomics' investigative capability has led to its application in a wide range of sectors, as has the requirement for rapid, accessible, and reliable bioinformatics analytic tools. Shotgun sequence characterization is frequently separated and therefore include reference alignment, k-mer analysis, assembly, and phylogenetics (McIntyre et al. 2017).

18.5 ARGs and ARBs Derived from Vermicompost and Their Effects on Public Health

ARGs and ARB in livestock waste and other vermicomposting substrates are hard to get rid of using standard sludge treatment methods (Huang et al. 2018). The presence of ARGs in vermicomposting substrates such as high organic matter animal manure and various microorganisms enables ARGs to proliferate and spread across bacterial species via horizontal gene transfer, posing considerable public health threats (Xu et al. 2020). ARBs and ARGs have emerged as novel contaminants since they are often present in most vermicomposting products obtained from livestock manure (Ezugworie et al. 2021). These contaminants in agricultural soils receiving vermicompost have sparked major concern due to the potential environmental and human health risks to the whole ecosystem (Ezugworie et al. 2021; Xu et al. 2020). Table 18.1 highlights the various ARGs and ARBs found in vermicomposts, as well as the possible public health concerns.

Vermicomposting was seen to harbor bacterial species, some of which conferred antibacterial resistance. This is confirmed by the existence of diverse bacteria such as *Chloroflexi, Acidobacteria, Bacteroidetes,* and *Gemmatimonadetes* in the post-vermicompost (Huang et al. 2020). Additionally, the post-vermicompost is enriched with *Proteobacteria* (mostly *Ignatzschineria, Providencia*, and *Pseudomonas)*, *Bacteroidetes*, and *Alcaligenaceae* (Wang et al. 2017). Compared with raw manure (vermicompost substrate), vermicompost had a significantly higher operational taxonomic unit belonging to family *Alcaligenaceae*, genus *Pseudidiomarina*, *Erysipelothrix* and order *Bacillales*, implying that these operational taxonomic units increased during vermicomposting (Wang et al. 2017). These bacterial species from these families and order have been implicated for enriching vermicomposts with ARGs such as *mphA, aadA5–2,* and *tetG-01*. It was reported that ARGs, in this case, had 97% similarity with those from bacterial genera belonging to the family *Alcaligenaceae, Weeksellaceae* and the order *Bacillales* (Wang et al. 2017). *Pseudomonas* has also been linked to ARGs produced from vermicompost, such as ampC-07, tet-x, and tetL-02. After vermicomposting, the most prevalent taxon, Genus *Ignatzschineria,* coincided with the vermicompost discovered ARG, aacC1 (Wang et al. 2017). According to Wang et al. (2015), during vermicomposting, the relative abundance of *Proteobacteria* phylum and *Flavobacteriaceae* spp. increases considerably.

Vermicomposting help propagates and creates reservoirs of ARB and ARGs in the soil. ARG and ARB transfer from the surroundings to humans through food farm products poses a greater health concern. The advent of ARB in humans has the potential to disrupt the gut microbiota and encourage the spread of harmful microbes, resulting in bowel cancer, and mortality due to incurability (Anand et al. 2021). Pathogenic ARB is of most direct concern as they can lead directly to adverse health outcomes such as increased duration of hospitalization, increased failure rates of antimicrobials, the emergence of antimicrobial resistance, increasing mortality rates (Hong et al. 2018). *Pseudomonas aeruginosa* and *Legionella pneumophila,*

Table 18.1 ARGs and ARBs derived from vermicomposting processes, possible sources, and human health impacts

ARGs/ARBs derived from vermicomposting processes		Possible sources	Human health risk/impacts	Remarks	References
Antibiotic-resistant genes					
Tetracycline resistant genes	*tet*A, *tet*C, *tet*W, *tet*M, *tet*O, *tet*X, *tet*G, *tet*S	Vermicomposting substrates, soil, harvested vegetable roots and leaves of lettuce, tomato, carrot, reddish and endive vegetable, vermicompost	Increased microbial infections, mortality, incurable diseases due to antimicrobial resistance	It may be found free in the soil or bacterial cells	Smith et al. (2019), Sanganyado and Gwenzi (2019), Heuer et al. (2011), Marti et al. (2013)
Fluoroquinolone resistant genes	gryA, parC, qnrS	Soil, vermicomposting substrates, crop plants, vermicompost	Development of antimicrobial resistance, high infection rate, prolonged hospitalization	It may be transferred to other microbes by horizontal or vertical transfer	Cui et al. (2022), Amarasiri et al. (2020), Heuer et al. (2011)
Sulfonamide resistant genes	*sul*1, *sul*2, *dfr*A1, *dfr*A7	Soil, water, harvested vegetable roots, leaves of lettuce and endive vegetable, cucumber, vermicomposting substrates, vermicompost	High exposure when drinking, contaminated water, high rate of infection and transmission	May increase antimicrobial resistance	Liu et al. (2019), Li et al. (2015), Coleman et al. (2012), Amarasiri et al. (2020)
Vancomycin-resistant genes	vanB, vanC1	Soil, vermicomposting substrates, vermicompost	Morbidity and prolonged hospitalization	It may be transferred to other microbes by horizontal or vertical transfer May increase antimicrobial resistance	Sanganyado and Gwenzi (2019)
Aminoglycoside resistant gene	Aac3-VI, aacC4, aminoglycoside-acetyltransferase	Soil, river water, tomato, carrots, lettuce vegetables, vermicomposting substrates, vermicompost	High exposure when bathing and water sporting in the contaminated water	May increase antimicrobial resistance	Chen et al. (2019a, b), Li et al. (2015), Coleman et al. (2012), Amarasiri et al. (2020), Marti et al. (2013)

(continued)

Table 18.1 (continued)

ARGs/ARBs derived from vermicomposting processes	Possible sources	Human health risk/impacts	Remarks	References
Antibiotic-resistant bacteria				
Erythromycin resistant bacteria	*Bacillus cereus, Pseudomonas aeruginosa, Staphylococcus lugdunensis*	Soil Air (aerosol/bioaerosols). Vermicomposting substrates Vermicompost Worms	Toxic allergies, respiratory disorders, infectious diseases, and malignancies	Air pollution components are usually taken into the human body through inhalation	Anand et al. (2021), Heuer et al. (2011)
Methicillin-resistant bacteria	*Staphylococcus aureus*	Soil Vermicomposting substrates Vermicompost Worms	Enhanced pathogenicity and disease outbreaks	Presumably transferred to other environments, e.g., water, sediments	Pepper et al. (2018), Sanganyado and Gwenzi (2019)
Sulfonamide resistant bacteria	*Acinetobacter, Bacillus, Psychrobacter, Escherichia coli, Salmonella, Staphylococcus, Streptococcus, Streptomyces*	Soil Vermicomposting substrates Vermicompost Worms	Increased morbidity and hospitalization, enhanced virulence	Demonstrated the co-presence of *sul*1, *sul*2, and *sul*3 in a single cell	Sanganyado and Gwenzi (2019)
Vancomycin-resistant bacteria	*Enterococcus faecium Enterococcus durans Enterococcus faecalis Enterococcus hirae*	Meat Vermicomposting substrates Vermicompost Worms	Urinary tract, lower respiratory tract, and bloodstream infections	Infect humans through meat requiring minimum processing	Elmalı and Can (2018), Sanganyado and Gwenzi (2019)
Rifampicin resistant bacteria	*Enterococcus faecalis, Enterococcus hirae*	Meat Vermicomposting substrates	Enhanced infection and prolonged hospitalization	Transmitted by vectors such as house flies	Sanganyado and Gwenzi (2019), Elmalı and Can (2018)

| Ampicillin resistant bacteria | Clostridium, Escherichia coli | Vermicompost Worms Carrots, cucumber, red-dish Vermicomposting substrates Vermicompost Worms | Abdominal infection, disease outbreaks | Transferred through raw vegetables | Li et al. (2015), Marti et al. (2013) |

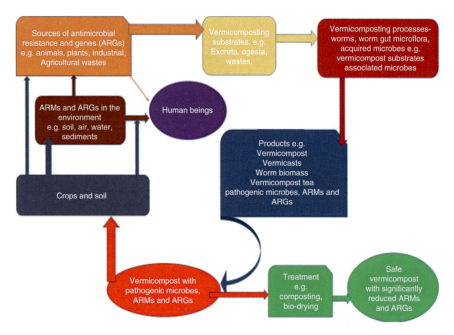

Fig. 18.1 A theoretical summary of transfer of ARGS and ARMs from the sources to the vermicompost and human beings

two opportunistic ARBs associated with vermicomposting, can cause pneumonia and skin infections, and are known to cause serious illnesses in burn and patients undergoing surgery (Hong et al. 2018; Mena and Gerba 2009). The World Health Organisation (2017) highlighted that ARB *Klebsiella pneumoniae* (associated with vermicomposting processes) is one of the leading global causes of urinary tract, lower respiratory tract, and bloodstream infections due to carbapenem-resistance. Sanganyado and Gwenzi (2019) noted that in African and Asian regions, ciprofloxacin-resistant *Shigella* species continues to cause diarrhea and dysentery, which are highly fatal in infants. ARGs and ARBs in the soil can be absorbed by crop plants, bio-accumulate inside plants, and then enter food webs, possibly having serious public health consequences to consumers. The enrichment of vermicompost-amended soils with ARGs and ARBs can potentially disseminate ARGs and ARBs to vegetables, particularly those eaten raw or minimally processed, representing an essential vehicle for ARGs transmission into humans and posing a potential threat to human health (summarized in Figs. 18.1 and 18.2).

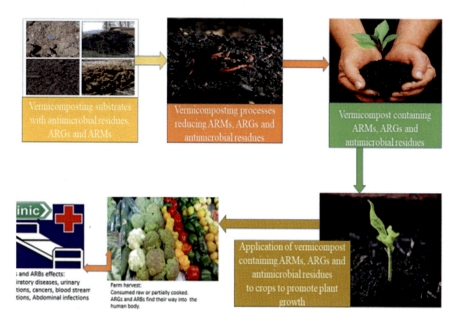

Fig. 18.2 Possible routes taken by ARGs and ARBs from substrates of vermicomposting to human beings and their effects

18.6 Fate of Antimicrobial Resistant Microbes and Genes from Vermicomposting Processes

The occurrence of ARGs and ARB in animal waste threatens the environment animal, and human health (Fu et al. 2021; Xu et al. 2020; Friedman et al. 2016). Waste stabilization methods that have higher efficiency in removal of the ARGs and ARB are critical to reduce their concentrations to safe levels to prevent health risks (Swati and Hait 2018). Composting has been found in several studies to minimize antibiotic resistance by modifying microbial structure and lowering microbial burden (Fu et al. 2021). However, the findings are not applicable to all ARGs. According to Fu et al. (2021), while most tetracycline resistance genes decreased after composting, the quantity of some remained same while others rose dramatically. Nevertheless, vermicomposting has shown great potential in inactivation of pathogenic microbes during decomposition of vermicomposting substrates (Swati and Hait 2018). The earthworm's body acts as a reactor in vermicomposting processes, purifying, disinfecting, and detoxifying organic waste (Kumar 2011). Disinfection of the waste is assumed to be able to remove most of the pathogenic microbes from the waste materials. Several researches investigated the fate and removal mechanisms of ARB and ARGs from many of organic wastes (Fu et al. 2021; Xu et al. 2020; Huang et al. 2018; Swati and Hait 2018; Wang et al. 2015). Swati and Hait (2018) reports that pathogen removal during vermistabilization is a result of the joint action of earthworms and endosymbiotic microbes within the earthworm's gut. Pathogens are removed from vermicomposting substrates through mechanisms such as gizzard churning action, microbial suppression due to intestinal catalytic hydrolysis, antibacterial activity of intestinal secretions, and cell lysis (Swati and Hait 2018). Pathogenic microbes including the ARB are ingested by earthworms as food. Furthermore, earthworms secrete antibacterial factors as well as coelomic fluids that inactivate pathogens (e.g., *E. coli, P. putida, E. cloacae,* and *A. hydrophila*) (Pedersen and Hendriksen 1993) contributing to removal of microbes from waste. For example, the earthworm *L. Lumbricus* secretes lumbricin, a compound that exhibits antimicrobial activity against a spectrum of microbes (Sinha et al. 2010). However, the removal of the pathogens is largely selective as the earthworm exerts a differential effect depending on the earthworm species and pathogen considered (Swati and Hait 2018).

Besides the microbial interactions that occur within earthworms, there is an establishment of microbial communities outside the earthworm within the vermi-reactor. These microbial communities establish antagonistic and competitive interactions that effectively eliminate pathogenic microbes from waste during decomposition (Pedersen and Hendriksen 1993). The mechanism mentioned above and other methods effectively remove bacteria from waste during vermicomposting (Figs. 18.1 and 18.2). For example, Kumar (2011) reported 100% sanitization of bacterial species for 60 days and stabilization of various agricultural wastes (e.g., rice straw, and sugarcane trash) in the presence of *E. fetida*.

Antibiotic-resistant genes are known to be harbored in microbial hosts (Li et al. 2015), and the removal of the host microbes can effectively reduce the concentrations of ARGs in vermicompost. For example, *Flavobacteriales, Campylobacterales,* and *Spirochaetales* take up diverse ARGs (*text, tetG,* and *tetO*) (Huang et al. 2018). The activity of earthworms and accompanying bacteria in vermi-reactors are therefore vital in determining the fate of the ARB as well as the ARGs. Promotion of multiplication of ARGs hosts can also cause a significant increase in the ARGs as Huang et al. (2018) reported a strong link between ARGs and their host bacteria. Composting conditions play a vital role in determining the fate of ARGs and ARB. For example, aerobic conditions created by burrowing action, secretion of mucus and aeration, by, earthworms have an impact on density, diversity, and activity of microbes (Suthar 2008). These composting conditions can effectively remove ARB from waste and studies have reported reductions of 7 and 4 logs for bacteria resistant to tetracycline and erythromycin, respectively, from livestock manure (Xu et al. 2020; Swati and Hait 2018). Abundance of *Actinomadura* and *Virgibacillus* in animal manure and organic fertilizers was investigated by Xu et al. (2020). Results of study showed that abundance of the two ARB was 200 to 300 times more in animal organic waste than in fertilizer derived from this waste. The study reported a similar trend for ARGs during composting. In a related study, Huang et al. (2018) investigated the effects vermicomposting processes on abundancy of tetracycline resistant genes and the relationship between the earthworms and possible hosts of ARGs in sludge. The study showed that tetracycline resistant genes in control experiment increased up to 27 times during the experiment, with *tetX* recording the highest increment.

However, the abundances of both tetracycline resistant genes decreased during vermicomposting with abundances of *tet*C, *tet*G, *tet*M, *tet*O, *tet*W, and *tet*X genes in the end vermicompost decreasing by 83.1%, 39.6%, 99%, 80.2%, 94.1%, and 60.9%, respectively (Huang et al. 2018). A similar trend was observed for fluoroquinolone resistant genes *qry*A, *par*C, and *qnr*S which recorded 57.2%, 100%, and 90% reductions, respectively.

Although studies have shown that earthworms can attenuate ARGs and ARB, there is a chance that earthworms may pick up these genes and the associated microbes within the vermin-reactor. The conditions are favorable for transferring ARGs from the substrate into their bodies (Wang et al. 2015). Also, high organic matter content and the diverse microorganisms within the vermi-reactor enable propagation and dissemination of ARGs among bacterial species through horizontal gene transfer (Guo et al. 2017a, b), leading to the spread of ARGs from free-living and commensal to pathogenic species (Huang et al. 2018). This, therefore, makes it very difficult to eliminate the ARGs from the environment. Although vermicomposting removes ARGs and ARB from waste, only a few kinds of research produced 100% efficiency, and antibiotic-resistant genes are always detected in the vermicompost. Therefore, there are higher chances that the ARGs and ARB are transferred to the environment as the compost is applied to land (Figs. 18.1 and 18.2). However, there is little evidence that the resistance genes residues will promote antimicrobial resistance in soil microbial communities (Negreanu et al. 2012), and this warranty further investigation.

Recently, a waste management technique, vermicomposting has been shown to effectively reduce the overall ARGs and ARB, but significant levels remain in earthworms and vermicompost (Huang et al. 2020; Wang et al. 2015). In a recent study, 94 out of 158 antibiotic resistance genes (ARGs) were greatly reduced (by 85%) following vermicomposting of swine dung employing housefly larvae, with 23 of the ARGs highly enriched (Wang et al. 2017). Although earthworm conversion significantly decreased certain ARGs in cow dung, a considerable quantity of ARGs remained in earthworms and vermicompost, according to Tian et al. (2021). This indicates that vermicomposting does not completely attenuate ARGs and ARBs and thus finds its way into cropland and water bodies (Tian et al. 2021). Research has confirmed the presence of ARGs and ARBs in the soil (Zhang et al. 2015; Zhou et al. 2019). ARGs and ARBs have been reported in aquatic environments, sludge, soil, and sediment (Philip et al. 2018).

According to recent study, compost manure treated soil increased the amount of ARGs and ARBs in the earthworm stomach, potentially increasing the likelihood of ARGs spreading down the food chain (Li et al. 2020). Amended soils have been observed to have ARGs and ARBs in higher abundances, a phenomenon that is common when soils are amended with vermicompost. In research to quantify selected antibiotic resistance genes in agricultural and non-agricultural soils genes such as blaCTX-M-1, and tet(X) were identified. The ARGs were detected in many of the soils, with the cropland amended soils having a higher relative abundance of ARGs compared with soils (Dungan et al. 2019). Water runoff may rapidly transport ARGs and ARBs into aquatic ecosystems, and eventually transferred to humans.

Additionally, ARGs and ARBs have been detected on harvested vegetables such as tomatoes, cucumbers, carrots, and reddish grown in manure-amended soil.

18.7 Removal of Antimicrobial Resistant Microbes and Genes During Vermicomposting

The occurrence and transmission of AMR are rapidly becoming the most severe issues in human health (Prestinaci et al. 2015; Samreen et al. 2021). A multi-sectoral strategy is required to control the transmission of AMR microorganisms and genes. Vermicomposting uses many organic wastes, including plants, humans, and animals (Grantina-Ievina and Rodze 2020), most of which are contaminated by AMR bacteria and genes. It is critical to monitor and identify solutions to minimize or eradicate AMR bacteria and genes during vermicomposting. It is well accepted that the establishment and spread of AMR from one bacterium to another occur via horizontal gene transfer via mobile components (Samreen et al. 2021), which significantly promote the proliferation of AMR genes or microorganisms. The elimination of AMR genes became more effective as the number and variety of bacterial hosts for AMR genes reduced. As a result, structural changes, bacterial abundance, and diversity caused by composting are principally responsible for the drop in AMR genes (Sun et al. 2016; Zhang et al. 2019; Zou et al. 2020; Zheng 2016). The number and variability of AMR bacteria and genes are shaped and modified by the physical and chemical parameters of a compost. A link has been observed between the physical properties of compost and the frequency and content of AMR bacteria and genes in several studies (Guo et al. 2017a, b; Sun et al. 2016; Zhang et al. 2019; Zou et al. 2020; Zheng 2016). Higher temperatures in composting can kill AMR bacteria, break down antibiotic residues in animal, plant, and human organic waste, and remove AMR genes (Fig. 18.1) (Guo et al. 2017a, b; Sun et al. 2016; Zhang et al. 2019; Zou et al. 2020; Zheng 2016).

Vermicomposting has also shown great potential in removing antibiotic-resistant genes from organic waste. Wang et al. (2015) investigated the fate of *tet* genes in swine-derived manure under vermicomposting process. After one day, except for tet (Q), the abundance of tet genes declined, and by day 6, concentrations were significantly reduced for example tet(M) was reduced to 54.2%, from its starting abundance. However, after 6 days the abundance of genes such as sul1 and sul2 significantly increased (Wang et al. 2015). Cui et al. (2018) examined the fate and stability of quinolone resistance genes in vermicomposting processes involving *E. fetida*. They recorded an 85.6–100% reduction ratio for *qnrA* and 92.3–95.3% for *qnrS*. The findings show that earthworms enhance quinoline resistance gene attenuation by reducing bacterial abundance and activity. However, Huang et al. (2018) recorded a significant increase of both *qnrA* and *qnrS* on day one, but the abundance rapidly decreased to pre-treatment levels throughout vermicomposting.

Water content, temperature, pH, carbon and nitrogen content are some of the most important environmental variables that influence the variability and abundance of ARGs in compost. Organic carbon has been identified as one of the most important elements regulating ARG abundance (Fu et al. 2021; Huang et al. 2018). Nevertheless, temperature also plays a role in the removal of ARGs, as high temperatures may cause damage to plasmids and effectively reduce transmission of the ARGs (Fu et al. 2021). The study by Huang et al. (2018) reported negative correlations between ARGs and changes in temperature and pH. Furthermore, ARG abundance is positively correlated to moisture and organic wastes in the vermicomposting substrate (Huang et al. 2018). Furthermore, high temperatures were found to decrease the quantity of genes which confer resistance to tetracycline during the composting of pig dung (Guo et al. 2017a, b). High-temperature composting efficiently eliminated antibiotic residues and AMR genes from vermicompost; nevertheless, it cannot prevent the formation of mobile genetic elements. The addition of mature compost to composting organic matter revealed that manure compost accelerated organic matter decomposition and prolonged the thermophilic phase as a result of increased proliferation and multiplication of thermophilic microbes such as those found in the *Bacillus*, *Thermobifida*, and *Thermobacillus* genera (Zou et al. 2020). By decreasing the number of bacteria, manure compost lowered AMR gene transmission. As a result, manure composting may be a viable technique for decreasing the transmission of AMR bacteria and genes.

Exogenous substrates added to vermicompost have been demonstrated to increase ARG elimination in recent tests (Zhang et al. 2016; Fu et al. 2021). Adding additives such as zeolite, bamboo, biochar, and charcoal to compost have reduced the total quantity of compost associated ARGs (Zhang et al. 2015; Fu et al. 2021). Biochar, for example, has a higher pore structure and is rich in functional groups, which promote microbial activity and humification process affecting the abundance of ARGs (Fu et al. 2021). Besides promoting microbial activity, biochar has inhibitory effects on ARGs such as *intl1* and therefore reduces the risk of diffusion of ARGs in compost (Li et al. 2017). Therefore, more exogenous additives could be identified and used in composting to improve ARGs elimination from wastes. Additionally, the modification of external composting variables such as temperature affects antibiotic breakdown and elimination of resistance genes, making composting an excellent strategy for eradicating AMR bacteria and genes in vermicompost. Temperature significantly influences the amount and type of AMR bacteria and genes during composting. Higher temperatures, except for thermophiles, inhibit the development of other microorganisms. These thermophiles are frequently beneficial organisms less likely to spread AMR genes. Zheng (2016) showed that high-temperature composting of pig manure successfully reduced the quantity of -lactam resistance genes and quinolone resistance genes. After vermicomposting, vermicompost may be further treated using bio-drying, composting at higher temperatures around 55 °C, anaerobic digestion in the presence of zerovalent iron. Though these techniques have been reported to reduce ARGs and ARMs (Liao et al. 2018), their application to vermicompost and impacts on vermicompost nutrient composition are still explored.

18.8 Future Research Directions

- Although antimicrobial resistance is widely understood, additional research into the relationship between intrinsic and acquired antimicrobial resistance and vermicomposting processes is required. There is little data on the drivers of antimicrobial resistance during vermicomposting. However, it is assumed to be due to (1) antimicrobial residues and heavy metals associated with vermicomposting substrates, (2) microorganisms in substrates, and (3) worm gut microbiota and vermicomposting processes. As a result, more research should be conducted to identify the causes of antimicrobial resistance in vermicomposting processes.
- Vermicomposting techniques have been shown to reduce ARGs and ARMs in vermicomposts significantly. Several studies, however, have found an increase in ARGs and ARMs during vermicomposting. This lack of reliable data suggests that more research and experiments are needed to investigate the impacts of vermicomposting on the abundance of ARGs and ARMs in vermicomposts. Furthermore, the processes of antimicrobial resistance acquisition during vermicomposting are unknown; hence additional research in this area is needed.
- Even though ARMs and related ARGs traverse a variety of pathways from vermicomposting substrates through vermicomposting till they reach the environment via vermicompost, the fate of vermicomposting-derived antimicrobial resistant microorganisms and genes has received little attention. As a result, it is critical to thoroughly explain the fate of ARMs and ARGs to protect the environment, ecosystems, and humans.
- It has been shown that vermicomposting dramatically reduces the presence of ARMs and ARGs in vermicomposting waste. The persistence of ARGs and ARMs following vermicomposting, on the other hand, is common. In this chapter, we recommended using bio-drying, composting at higher temperatures of about 55–70 °C, and anaerobic digestion in the presence of zerovalent iron to remove ARGs and ARMs from vermicompost. However, the application of these methodologies to vermicompost and their effects on vermicompost nutrient composition remain to be investigated. As a result, we encourage research into ways for removing ARGs and ARMs from vermicompost.
- There is a significant disconnect between vermicompost-associated ARGs and ARBs and public health. In light of this scenario, more studies into the impact of vermicompost-derived ARGs and ARMs in human health are required.

18.9 Conclusion and Outlook

ARMs and ARGs have been found in the intestines of an earthworm and the verminreactor during composting in several investigations (Domfnguez 2004; Ibrahim et al. 2016). Microorganism suppression due to intestinal enzymatic cleavage, secretion of coelomic with antimicrobial effects, preferential grazing, and indirect actions such as activation of inherent microbes, and phagocytic action are linked to the decline of ARB and ARGs. Environmental factors, however, play a critical role in removing ARB and ARGs from vermicomposting substrates and vermicomposts and physical and biological processes. Although several studies have shown a decrease in ARGs and ARBs during vermicomposting, the ARGs and ARBs are not eliminated (Cui et al. 2018; Huang et al. 2018, 2020; Tian et al. 2021; Zhao et al. 2022) and thus persist in the vermicomposts, vermicasts, and vermicompost tea. Since these persistent ARGs and ARGs/ARMs might potentially spread to human beings through various routes, including plant-derived foods, pose a severe public health concern. Tools to remove and monitor antimicrobial resistance microorganisms and ARGs are available, including composting and metagenomics. However, if vermicompost ARGs and ARMs/ARB are appropriately managed, only benefits (e.g., lowering the C:N ratio, producing homogeneous biofertilizer, improving soil physicochemical and biological parameters and improving plant growth) can be derived from vermicomposts.

References

Alberts B, Johnson A, Lewis J, Raff M, Roberts K, Walter P (2002) Introduction to pathogens. In: Molecular biology of the cell, 4th edn. Garland Science

Alekshun MN, Levy SB (2007) Molecular mechanisms of antibacterial multidrug resistance. Cell 128(6):1037–1050

Amábile-Cuevas CF (2021) Antibiotic resistance from, and to the environment. AIMS Environ Sci 8(1):18–35

Amarasiri M, Sano D, Suzuki S (2020) Understanding human health risks caused by antibiotic resistant bacteria (ARB) and antibiotic resistance genes (ARG) in water environments: current knowledge and questions to be answered. Crit Rev Environ Sci Technol 50(19):2016–2059

Anand U, Reddy B, Singh VK, Singh AK, Kesari KK, Tripathi P, Simal-Gandara J et al (2021) Potential environmental and human health risks caused by antibiotic-resistant bacteria (ARB), antibiotic resistance genes (ARGs) and emerging contaminants (ECs) from municipal solid waste (MSW) landfill. Antibiotics 10(4):374

Ansari AA, Ori L, Ramnarain YI (2020) An effective organic waste recycling through vermicompost technology for soil health restoration. In: Soil health restoration and management. Springer, Singapore, pp 83–112

Awasthi MK, Chen H, Duan Y, Liu T, Awasthi SK, Wang Q, Zhang Z et al (2019) An assessment of the persistence of pathogenic bacteria removal in chicken manure compost employing clay as additive via meta-genomic analysis. J Hazard Mater 366:184–191

Bengtsson-Palme J, Larsson DJ, Kristiansson E (2017) Using metagenomics to investigate human and environmental resistomes. J Antimicrob Chemother 72(10):2690–2703

Blaser MJ (2016) Antibiotic use and its consequences for the normal microbiome. Science 352(6285):544–545

Blouin M, Hodson ME, Delgado EA, Baker G, Brussaard L, Butt KR, Brun JJ et al (2013) A review of earthworm impact on soil function and ecosystem services. Eur J Soil Sci 64(2):161–182

Boolchandani M, Patel S, Dantas G (2017) Functional metagenomics to study antibiotic resistance. In: Antibiotics. Humana Press, New York, pp 307–329

Cardoso VL, Ramírez CE, Escalante EV (2008) Vermicomposting technology for stabilizing the sewage sludge from rural waste water treatment plants. Water Pract Technol 3(1)

Chen JY, Xia H, Huang K, Wu Y (2019a) Effects of tetracycline on microbial communities and antibiotic resistance genes of vermicompost from dewatered sludge. Huan Jing ke Xue=Huanjing Kexue 40(7):3263–3269

Chen Z, Zhang W, Yang L, Stedtfeld RD, Peng A, Gu C, Li H et al (2019b) Antibiotic resistance genes and bacterial communities in cornfield and pasture soils receiving swine and dairy manures. Environ Pollut 248:947–957

Coleman BL, Salvadori MI, McGeer AJ, Sibley KA, Neumann NF, Bondy SJ, ARO Water Study Group et al (2012) The role of drinking water in the transmission of antimicrobial-resistant *E. coli*. Epidemiol Infect 140(4):633–642

Cox G, Wright GD (2013) Intrinsic antibiotic resistance: mechanisms, origins, challenges and solutions. Int J Med Microbiol 303(6–7):287–292

Cui G, Li F, Li S, Bhat SA, Ishiguro Y, Wei Y, Yamada T, Fu X, Huang K (2018) Changes of quinolone resistance genes and their relations with microbial profiles during vermicomposting of municipal excess sludge. Sci Total Environ 644:494–502. https://doi.org/10.1016/j.scitotenv.2018.07.015

Cui G, Bhat SA, Li W, Wei Y, Kui H, Fu X, Gui H, Wei C, Li F (2019) Gut digestion of earthworms significantly attenuates cell-free and -associated antibiotic resistance genes in excess activated sludge by affecting bacterial profiles. Sci Total Environ 691:644–653. https://doi.org/10.1016/j.scitotenv.2019.07.177

Cui G, Fu X, Bhat SA, Tian W, Lei X, Wei Y, Li F (2022) Temperature impacts fate of antibiotic resistance genes during vermicomposting of domestic excess activated sludge. Environ Res 112654:112654

Dai X, Wang X, Gu J, Bao J, Wang J, Guo H, Lei L et al (2021) Responses of bacterial communities and antibiotic resistance genes to nano-cellulose addition during pig manure composting. J Environ Manag 300:113734

Deng WJ, Li N, Ying GG (2018) Antibiotic distribution, risk assessment, and microbial diversity in river water and sediment in Hong Kong. Environ Geochem Health 40(5):2191–2203

Domfnguez J (2004) 20 state-of-the-art and new perspectives on vermicomposting research. In: Earthworm ecology. CRC Press, Boca Raton, pp 401–424

Domínguez J, Aira M, Gómez-Brandón M (2010) Vermicomposting: earthworms enhance the work of microbes. In: Microbes at work. Springer, Berlin, pp 93–114

Du Plessis R (2010) Establishment of composting facilities on landfill sites (Doctoral dissertation, University of South Africa)

Dungan RS, Strausbaugh CA, Leytem AB (2019) Survey of selected antibiotic resistance genes in agricultural and non-agricultural soils in south-Central Idaho. FEMS Microbiol Ecol 95(6): fiz071

Durso LM, Miller DN, Wienhold BJ (2012) Distribution and quantification of antibiotic resistant genes and bacteria across agricultural and non-agricultural metagenomes. PLoS One 7(11): e48325

Eckstrom K (2018) Evaluating the resistome and microbial composition during food waste feeding and composting on a vermont poultry farm. The University of Vermont and State Agricultural College

Elmalı M, Can HY (2018) The prevalence, vancomycin resistance and virulence gene profiles of Enterococcus species recovered from different foods of animal origin. Veterinarski arhiv 88(1): 111–124

Ezugworie FN, Igbokwe VC, Onwosi CO (2021) Proliferation of antibiotic-resistant microorganisms and associated genes during composting: an overview of the potential impacts on public health, management and future. Sci Total Environ 784:147191

Friedman ND, Temkin E, Carmeli Y (2016) The negative impact of antibiotic resistance. Clin Microbiol Infect 22:416–422

Fu Y, Zhang A, Guo T, Zhu Y, Shao Y (2021) Biochar and hyperthermophiles as additives accelerate the removal of antibiotic resistance genes and mobile genetic elements during composting. Materials 14(18):5428

Gajalakshmi S, Ramasamy EV, Abbasi SA (2002) Vermicomposting of paper waste with the anecic earthworm *Lampito mauritii* Kinberg

Gómez-Brandón M, Domínguez J (2014a) Changes of microbial communities during the vermicomposting process and after application to the soil. Crit Rev Environ Sci Technol 44: 1289–1312

Gómez-Brandón M, Domínguez J (2014b) Recycling of solid organic wastes through vermicomposting: microbial community changes throughout the process and use of vermicompost as a soil amendment. Crit Rev Environ Sci Technol 44(12):1289–1312

Grantina-Ievina L, Rodze I (2020) Survival of pathogenic and antibiotic-resistant bacteria in vermicompost, sewage sludge, and other types of composts in temperate climate conditions. Biol Composts:107–124

Grantina-Ievina L, Andersone U, Berkolde-Pīre D, Nikolajeva V, Ievinsh G (2013) Critical tests for determination of microbiological quality and biological activity in commercial vermicompost samples of different origins. Appl Microbiol Biotechnol 97(24):10541–10554

Guo A, Gu J, Wang X, Zhang R, Yin Y, Sun W, Tuo X, Zhang L (2017a) Effects of superabsorbent polymers on the abundances of antibiotic resistance genes, mobile genetic elements, and the bacterial community during swine manure composting. Bioresour Technol 244(Pt 1):658–663

Guo J, Li J, Chen H, Bond PL, Yuan Z (2017b) Metagenomic analysis reveals wastewater treatment plants as hotspots of antibiotic resistance genes and mobile genetic elements. Water Res 123: 468–478

Haben Fesseha M (2020) Principles and applications of deoxyribonucleic acid microarray: a review

Heinonen-Tanski H, Mohaibes M, Karinen P, Koivunen J (2006) Methods to reduce pathogen microorganisms in manure. Livest Sci 102(3):248–255

Heuer H, Schmitt H, Smalla K (2011) Antibiotic resistance gene spread due to manure application on agricultural fields. Curr Opin Microbiol 14(3):236–243

Hong PY, Julian TR, Pype ML, Jiang SC, Nelson KL, Graham D, Manaia CM et al (2018) Reusing treated wastewater: consideration of the safety aspects associated with antibiotic-resistant bacteria and antibiotic resistance genes. Water 10(3):244

Huang K, Xia H, Wu Y, Chen J, Cui G, Li F, Wu N et al (2018) Effects of earthworms on the fate of tetracycline and fluoroquinolone resistance genes of sewage sludge during vermicomposting. Bioresour Technol 259:32–39

Huang K, Xia H, Zhang Y, Li J, Cui G, Li F, Wu N et al (2020) Elimination of antibiotic resistance genes and human pathogenic bacteria by earthworms during vermicomposting of dewatered sludge by metagenomic analysis. Bioresour Technol 297:122451

Ibrahim MH, Quaik S, Ismail SA (2016) General introduction to earthworms, their classifications, and biology. In: Prospects of organic waste management and the significance of earthworms. Springer, Cham, pp 69–103

Jasovský D, Littmann J, Zorzet A, Cars O (2016) Antimicrobial resistance-a threat to the world's sustainable development. Ups J Med Sci 121(3):159–164. https://doi.org/10.1080/03009734.2016.1195900

Kralik P, Ricchi M (2017) A basic guide to real time PCR in microbial diagnostics: definitions, parameters, and everything. Front Microbiol 8:108

Kui H, Jingyang C, Mengxin G, Hui X, Li L (2020) Effects of biochars on the fate of antibiotics and their resistance genes during vermicomposting of dewatered sludge. J Hazard Mater 397: 122767. https://doi.org/10.1016/j.jhazmat.2020.122767

Kumar S (2011) Composting of municipal solid waste. Crit Rev Biotechnol 31(2):112–136

Lee LH, Wu TY, Shak KPY, Lim SL, Ng KY, Nguyen MN, Teoh WH (2018) Sustainable approach to biotransform industrial sludge into organic fertilizer via vermicomposting: a mini-review. J Chem Technol Biotechnol 93(4):925–935

Leopold SJ, van Leth F, Tarekegn H, Schultsz C (2014) Antimicrobial drug resistance among clinically relevant bacterial isolates in sub-Saharan Africa: a systematic review. J Antimicrob Chemother 69(9):2337–2353

Li B, Yang Y, Ma L, Ju F, Guo F, Tiedje JM, Zhang T (2015) Metagenomic and network analysis reveal wide distribution and co-occurrence of environmental antibiotic resistance genes. ISME J 9(11):2490–2502

Li H, Duan M, Gu J, Zhang Y, Qian X, Ma J, Wang X et al (2017) Effects of bamboo charcoal on antibiotic resistance genes during chicken manure composting. Ecotoxicol Environ Saf 140:1–6

Liao H, Lu X, Rensing C, Friman VP, Geisen S, Chen Z, Zhu Y et al (2018) Hyperthermophilic composting accelerates the removal of antibiotic resistance genes and mobile genetic elements in sewage sludge. Environ Sci Technol 52(1):266–276

Liew CS, Yunus NM, Chidi BS, Lam MK, Goh PS, Mohamad M, Lam SS et al (2022) A review on recent disposal of hazardous sewage sludge via anaerobic digestion and novel composting. J Hazard Mater 423:126995

Lim SL, Wu TY, Clarke C (2014) Treatment and biotransformation of highly polluted agro-industrial wastewater from a palm oil mill into vermicompost using earthworms. J Agric Food Chem 62(3):691–698

Lim SL, Wu TY, Lim PN, Shak KPY (2015) The use of vermicompost in organic farming: overview, effects on soil and economics. J Sci Food Agric 95(6):1143–1156

Liu K, Sun M, Ye M, Chao H, Zhao Y, Xia B, Hu F et al (2019) Coexistence and association between heavy metals, tetracycline and corresponding resistance genes in vermicomposts originating from different substrates. Environ Pollut 244:28–37

Liu H, Huang Y, Duan W, Qiao C, Shen Q, Li R (2020) Microbial community composition turnover and function in the mesophilic phase predetermine chicken manure composting efficiency. Bioresour Technol 313:123658

Luby E, Ibekwe AM, Zilles J, Pruden A (2016) Molecular methods for assessment of antibiotic resistance in agricultural ecosystems: prospects and challenges. J Environ Qual 45(2):441–453

Manyi-Loh CE, Mamphweli SN, Meyer EL, Makaka G, Simon M, Okoh AI (2016) An overview of the control of bacterial pathogens in cattle manure. Int J Environ Res Public Health 13(9):843

Marti R, Scott A, Tien YC, Murray R, Sabourin L, Zhang Y, Topp E (2013) Impact of manure fertilization on the abundance of antibiotic-resistant bacteria and frequency of detection of antibiotic resistance genes in soil and on vegetables at harvest. Appl Environ Microbiol 79(18):5701–5709

McIntyre ABR, Ounit R, Afshinnekoo E, Prill RJ, Hénaff E, Alexander N, Mason CE et al (2017) Comprehensive benchmarking and ensemble approaches for metagenomic classifiers. Genome Biol 18(1):182. https://doi.org/10.1186/s13059-017-1299-7

Mena KD, Gerba CP (2009) Risk assessment of Pseudomonas aeruginosa in water. Rev Environ Contam Toxicol 201:71–115

Mullany P (2014) Functional metagenomics for the investigation of antibiotic resistance. Virulence 5(3):443–447. https://doi.org/10.4161/viru.28196

Munnoli PM, Da Silva JAT, Saroj B (2010) Dynamics of the soil-earthworm-plant relationship: a review. Dyn Soil Dyn Plant 4(1):1–21

Negreanu Y, Pasternak Z, Jurkevitch E, Cytryn E (2012) Impact of treated wastewater irrigation on antibiotic resistance in agricultural soils. Environ Sci Technol 46(9):4800–4808

Pedersen JC, Hendriksen NB (1993) Effect of passage through the intestinal tract of detritivore earthworms (Lumbricus spp.) on the number of selected gram-negative and total bacteria. Biol Fertil Soils 16(3):227–232

Peed LA, Nietch CT, Kelty CA, Meckes M, Mooney T, Sivaganesan M, Shanks OC (2011) Combining land use information and small stream sampling with PCR-based methods for better characterization of diffuse sources of human fecal pollution. Environ Sci Technol 45(13): 5652–5659

Pepper IL, Brooks JP, Gerba CP (2018) Antibiotic resistant bacteria in municipal wastes: is there reason for concern? Environ Sci Technol 52(7):3949–3959

Philip JM, Aravind UK, Aravindakumar CT (2018) Emerging contaminants in Indian environmental matrices–a review. Chemosphere 190:307–326

Pierre-Louis RC, Kader M, Desai NM, John EH (2021) Potentiality of vermicomposting in the South Pacific Island countries: a review. Agriculture 11(9):876

Poole K (2002) Mechanisms of bacterial biocide and antibiotic resistance. J Appl Microbiol 92: 55S–64S

Prestinaci F, Pezzotti P, Pantosti A (2015) Antimicrobial resistance: a global multifaceted phenomenon. Pathogens Glob Health 109(7):309–318

Qiu T, Wu D, Zhang L, Zou D, Sun Y, Gao M, Wang X (2021) A comparison of antibiotics, antibiotic resistance genes, and bacterial community in broiler and layer manure following composting. Environ Sci Pollut Res 28(12):14707–14719

Raith MR, Kelty CA, Griffith JF, Schriewer A, Wuertz S, Mieszkin S, Shanks OC et al (2013) Comparison of PCR and quantitative real-time PCR methods for the characterization of ruminant and cattle fecal pollution sources. Water Res 47(18):6921–6928

Reygaert WC (2018) An overview of the antimicrobial resistance mechanisms of bacteria. AIMS Microbiol 4(3):482

Roberts MC (2005) Update on acquired tetracycline resistance genes. FEMS Microbiol Lett 245(2): 195–203

Samreen, Ahmad I, Malak HA, Abulreesh HH (2021) Environmental antimicrobial resistance and its drivers: a potential threat to public health. J Glob Antimicrob Resist 27:101–111. https://doi. org/10.1016/j.jgar.2021.08.001

Sanganyado E, Gwenzi W (2019) Antibiotic resistance in drinking water systems: occurrence, removal, and human health risks. Sci Total Environ 669:785–797

Sharpton TJ (2014) An introduction to the analysis of shotgun metagenomic data. Front Plant Sci 5. https://doi.org/10.3389/fpls.2014.00209

Sim EYS, Wu TY (2010) The potential reuse of biodegradable municipal solid wastes (MSW) as feedstocks in vermicomposting. J Sci Food Agric 90(13):2153–2162

Singh RP, Embrandiri A, Ibrahim MH, Esa N (2011) Management of biomass residues generated from palm oil mill: vermicomposting a sustainable option. Resour Conserv Recycl 55(4): 423–434

Sinha RK, Herat S, Valani D, Chauhan K (2010) Earthworms- the environmental engineers: review of vermiculture technologies for environmental management and resource development. Int J Glob Environ Issues 10(3–4):265–292

Smith SD, Colgan P, Yang F, Rieke EL, Soupir ML, Moorman TB, Howe A et al (2019) Investigating the dispersal of antibiotic resistance associated genes from manure application to soil and drainage waters in simulated agricultural farmland systems. PLoS One 14(9): e0222470

Song T, Li H, Li B, Yang J, Sardar MF, Yan M et al (2021) Distribution of antibiotic-resistant bacteria in aerobic composting of swine manure with different antibiotics. Environ Sci Eur 33 (1):1–13

Soobhany N (2019) Insight into the recovery of nutrients from organic solid waste through biochemical conversion processes for fertilizer production: a review. J Clean Prod 241:118413

Sun W, Qian X, Gu J et al (2016) Mechanism and effect of temperature on variations in antibiotic resistance genes during anaerobic digestion of dairy manure. Sci Rep 6:30237

Suthar S (2008) Bioconversion of post harvest crop residues and cattle shed manure into valueadded products using earthworm Eudrilus eugeniae Kinberg. Ecol Eng 32(3):206–214

Swati A, Hait S (2017) Fate and bioavailability of heavy metals during vermicomposting of various organic wastes—A review. Process Saf Environ Prot 109:30–45

Swati A, Hait S (2018) A comprehensive review of the fate of pathogens during vermicomposting of organic wastes. J Environ Qual 47(1):16–29

Tadesse BT, Ashley EA, Ongarello S, Havumaki J, Wijegoonewardena M, González IJ, Dittrich S (2017) Antimicrobial resistance in Africa: a systematic review. BMC Infect Dis 17(1):1–17

Tenover FC (2006) Mechanisms of antimicrobial resistance in bacteria. Am J Med 119(6):S3–S10

Thanner S, Drissner D, Walsh F (2016) Antimicrobial resistance in agriculture. mBio 7(2):e02227–e02215. https://doi.org/10.1128/mBio.02227-15

Tian X, Han B, Liang J, Yang F, Zhang K (2021) Tracking antibiotic resistance genes (ARGs) during earthworm conversion of cow dung in northern China. Ecotoxicol Environ Saf 222: 112538. https://doi.org/10.1016/j.ecoenv.2021.112538

Tripathi YC, Hazarika P, Pandey BK (2005) Vermicomposting: an ecofriendly approach to sustainable. Verms Vermitechnol 23

Wang H, Li H, Gilbert JA, Li H, Wu L, Liu M, Zhang Z et al (2015) Housefly larva vermicomposting efficiently attenuates antibiotic resistance genes in swine manure, with concomitant bacterial population changes. Appl Environ Microbiol 81(22):7668–7679

Wang H, Sangwan N, Li HY, Su JQ, Oyang WY, Zhang ZJ, Zhang HL et al (2017) The antibiotic resistome of swine manure is significantly altered by association with the *Musca domestica* larvae gut microbiome. ISME J 11(1):100–111

Wichmann F, Udikovic-Kolic N, Andrew S, Handelsman J (2014) Diverse antibiotic resistance genes in dairy cow manure. MBio 5(2):e01017–e01013

World Health Organization (2017) Global antimicrobial resistance surveillance system (GLASS) 943 report: early implementation 2016–2017. Switzerland, Geneva

Xavier BB, Das AJ, Cochrane G, De Ganck S, Kumar-Singh S, Aarestrup FM, Malhotra-Kumar S et al (2016) Consolidating and exploring antibiotic resistance gene data resources. J Clin Microbiol 54(4):851–859. https://doi.org/10.1128/jcm.02717-15

Xia H, Chen J, Chen X, Huang K, Wu Y (2019) Effects of tetracycline residuals on humification, microbial profile and antibiotic resistance genes during vermicomposting of dewatered sludge. Environ Pollut 252:1068–1077

Xu J, Sangthong R, McNeil E, Tang R, Chongsuvivatwong V (2020) Antibiotic use in chicken farms in northwestern China. Antimicrob Resist Infect Control 9(1):1–9

Yadav A, Garg VK (2011) Industrial wastes and sludges management by vermicomposting. Rev Environ Sci Biotechnol 10(3):243–276

Zarei-Baygi A, Harb M, Wang P, Stadler LB, Smith AL (2019) Evaluating antibiotic resistance gene correlations with antibiotic exposure conditions in anaerobic membrane bioreactors. Environ Sci Technol 53(7):3599–3609

Zhang Z, Shen J, Wang H, Liu M, Wu L, Ping F, Xu X et al (2014) Attenuation of veterinary antibiotics in full-scale vermicomposting of swine manure via the housefly larvae (*Musca domestica*). Sci Rep 4(1):1–9

Zhang T, Yang Y, Pruden A (2015) Effect of temperature on removal of antibiotic resistance genes by anaerobic digestion of activated sludge revealed by metagenomic approach. Appl Microbiol Biotechnol 99(18):7771–7779. https://doi.org/10.1007/s00253-015-6688-9

Zhang J, Chen M, Sui Q, Tong J, Jiang C, Lu X, Wei Y et al (2016) Impacts of addition of natural zeolite or a nitrification inhibitor on antibiotic resistance genes during sludge composting. Water Res 91:339–349

Zhang M, He LY, Liu YS, Zhao JL, Liu WR, Zhang JN, Chen J, He LK, Zhang QQ, Ying GG (2019) Fate of veterinary antibiotics during animal manure composting. Sci Total Environ 650: 1363–1370

Zhang R, Yang S, An Y, Wang Y, Lei Y, Song L (2022) Antibiotics and antibiotic resistance genes in landfills: a review. Sci Total Environ 806:150647

Zhao X, Shen JP, Shu CL, Jin SS, Di HJ, Zhang LM, He JZ (2022) Attenuation of antibiotic resistance genes in livestock manure through vermicomposting via *Protaetia brevitarsis* and its fate in a soil-vegetable system. Sci Total Environ 807(Pt 1):150781. https://doi.org/10.1016/j.scitotenv.2021.150781

Zheng NG (2016) Effects of high temperature composting process on antibiotic resistance genes from pig manure. ACTA Sci Circumstance 37:1986–1992

Zhou X, Qiao M, Su JQ, Zhu YG (2019) High-throughput characterization of antibiotic resistome in soil amended with commercial organic fertilizers. J Soils Sediments 19(2):641–651

Zou Y, Tu W, Wang H, Fang T (2020) Anaerobic digestion reduces extracellular antibiotic resistance genes in waste activated sludge: the effects of temperature and degradation mechanisms. Environ Int 143:105980

Zrnčić SNJEŽANA (2020) European Union's action plan on antimicrobial resistance and implications for trading partners with example of National Action Plan for Croatia. Asian Fish Sci S 33: 75–82

Chapter 19
Potential Transformation of Organic Waste in African Countries by Using Vermicomposting Technology

Parveen Fatemeh Rupani, Asha Embrandiri, Hupenyu Allan Mupambwa, and Jorge Domínguez

Abstract In recent decades rising populations and rapid urbanization have changed the lifestyles of many people. This has resulted in the generation of vast quantities of different types of solid waste. In Africa, the harsh environment (hot weather and acidic soil) and the scant awareness of organic waste recycling have led to a decline in soil quality and to uncollected waste being piled up in streets, public places, and drains. Various studies have revealed that disposal of about 90% of municipal solid waste (MSW) is uncontrolled and that waste is dumped in open areas and landfill sites, creating problems for public health and the environment. This situation has increased the need for long-term green strategies in agricultural engineering and sustainable waste management. Hence, this chapter highlights vermicomposting as a sustainable, economical approach to disposing of the waste generated in African countries and discusses the value of the technology for improving agricultural practices and soil bioremediation. Current vermicomposting scenarios practiced in Africa and their future impact are also considered. The chapter concludes with a suggestion to governmental bodies including authorities and scientists to consider ways of enhancing the practice of vermicomposting in African countries.

P. F. Rupani (✉)
Department of Chemical Engineering, Process and Environmental Technology Lab,
KU Leuven, Sint-Katelijne-Waver, Belgium
e-mail: parveenfatemeh.rupani@kuleuven.be

A. Embrandiri
Department of Environmental Health, Wollo University, Dessie, Amhara, Ethiopia

H. A. Mupambwa
Sam Nujoma Marine and Coastal Resources Research Center, University of Namibia,
Henties Bay, Namibia

J. Domínguez
Grupo de Ecoloxía Animal (GEA), Universidade de Vigo, Vigo, Spain

© The Author(s), under exclusive license to Springer Nature Singapore Pte Ltd. 2023
H. A. Mupambwa et al. (eds.), *Vermicomposting for Sustainable Food Systems in Africa*, Sustainable Agriculture and Food Security,
https://doi.org/10.1007/978-981-19-8080-0_19

Keywords Vermicomposting · Africa · Sustainability · Green technology · Earthworms · Circular economy · Soil fertility

19.1 Introduction

The role of earthworms in breaking down organic matter and thus releasing nutrients has been recognized since the end of the nineteenth century (Edwards 2007). Darwin's book, "Formation of vegetable mould through the action of worms with observations on their habits," published in 1892, was the first research text to mention earthworms (Feller et al. 2003). In the preceding period, i.e., between 1870 and 1889, very few studies concerning earthworms were published. However, to date, there are more than 10,000 publications related to earthworms. Due to the proven success of vermicomposting, this method and other similar treatments using earthworms to break down organic matter to produce valuable soil-like additives and proteins for animal feed have expanded rapidly since 1970. In the 1990s, research focused on the ecology and biology of earthworms and their use to process different types of waste. The increase in researchers' interest in vermicomposting grew alongside the interests of commercial organizations throughout the world, in developed countries such as the USA, Canada, UK, Australia, Russia, and Japan, as well as in developing countries such as India, China, Chile, Brazil, Mexico, Argentina, and the Philippines (Edwards 2007).

The vermicomposting technique has specifically been practiced in countries with high levels of nutrient mining. Sub-Saharan African countries suffer from soil degradation and a significant decline in soil fertility that adversely affects crop yield and food production (Gebrehana et al. 2022). For example, soil nutrient losses from agricultural systems are very high in Ethiopia, accounting for about 30–60 kg ha^{-1} NPK per year (Nguru et al. 2020). Vermicomposting is an environmentally friendly treatment that meets sustainability goal 7 of the UN Millennium Development Goals (MDG7) and is consistent with the concept of natural resource management and sustainability to meet human food requirements. The application of vermicompost to agricultural land has been reported to improve crop production. For instance, plant growth studies on garlic (Gichaba et al. 2020) and maize (Coulibaly 2020) have demonstrated increased growth and crop yields related to the use of vermicompost. Likewise, lettuce yield and total uptake of phosphorus (P), potassium (K), calcium (Ca), and magnesium (Mg) were highest in plants grown in coir-based vermicompost (Schröder et al. (2021). In general, the vermicomposting technique is simple and inexpensive, making it feasible in low-income regions, especially in developing countries with low per capita income such as a many of the African countries classified on the basis of the human development index. This chapter reports important information based on applicability and advances in developing vermicomposting techniques in African countries towards sustainable waste management.

19.1.1 The Role of Earthworms in Breaking Down Organic Matter

The most important role of earthworms in biological processes includes breaking down solid organic matter (Atiyeh et al. 2000), thereby releasing a portion of the organic matter into earthworm biomass and respiration products (Dominguez et al. 2001) and rendering nutrients available to plants (Sun 2003). The role of earthworms in breaking down organic matter is attributed to three processes that occur in the worm gut: interaction with gut microorganisms, digestion of enzymes, and physical grinding.

Earthworm species each have a unique type of digestive system, but the gut structure is similar in all. The digestive system of earthworms consists of the buccal cavity, crop, gizzard, and intestine. Food adheres to mucus extruded by the buccal epithelium. Pressure on the buccal cavity wall is then released, which establishes a partial vacuum whereby materials are transported through the crop, gizzard, and intestine (Sun 2003). The time taken for the food to pass through the worm gut can vary between 3 and 5 h in *E. fetida* and between 12 and 20 h in *Lumbricus terrestris* (Edwards 2007).

Different earthworm species exist in almost all regions except for areas with extreme climates, such as deserts and glaciers. Earthworm species have different life cycles and behavioral and environmental requirements. They are classified into three major ecological categories on the basis of their feeding and burrowing strategies: epigeic, endogeic, and anecic (Ismail 2009; Sun 2003). Only epigeic earthworms are mostly relevant in relation to vermicomposting, though other classes of earthworms like anecic and endogeic have also been used (Edwards 2007). About 8000 species of earthworm have been described as epigeic worldwide, from 800 genera belonging to the order Oligochaeta (Edwards 2004). Among these, seven earthworm species are used in vermicomposting: *Eisenia fetida, Dendrobaena veneta, Dendrobaena rubida, Lumbricus rubellus, Perionyx excavatus, Eudrilus eugeniae,* and *Polypheretima elongata* (Edwards 2007; Sun 2003). These species show positive growth on organic waste and exhibit different life cycles and cocoon production patterns. For example, the growth and life cycle of *Eudrilus eugeniae* was found to vary from between 52 and 60 days in cocoa and cashew residues, with 88% cocoon hatchability (Coulibaly et al. (2019).

19.1.2 Vermicompost, the "Black Gold" in Agriculture

Vermicompost, the end product of vermicomposting, is a homogeneous, odorless compost with high porosity and water-holding capacity. The advantages of using vermicompost in agriculture have been widely demonstrated. Vermicompost is sometimes referred to as "black gold," because of its extremely valuable properties in relation to plant health and growth. In the last few decades, vermicomposting has been used to convert different bio-waste materials into vermicompost. The process has attracted much interest from researchers, consumers, and producers as it is an inexpensive source of a wide range of bioactive compounds (Fontana et al. 2013; Galanakis 2012). Vermicompost is an excellent plant growth promoter, with demonstrated positive effects on various aspects of agricultural and horticultural development (Gómez-Brandón et al. 2015). Vermicomposting is included among strategies used to produce value-added products for potential agricultural use; Sanchez-Hernandez et al. (2019) reported that the vermicomposting technique plays a dual role in bioremediation and environmental detoxification. For instance, vermicompost can be used for the bioremediation of pesticide-contaminated soils (Sanchez-Hernandez et al. 2019). Gómez-Brandón et al. (2020) studied the effectiveness of vermicomposting for bioconversion of by-products from the wine industry (grape marc) and studied the potential toxicity of the product. These authors concluded that the total polyphenol content of the grape marc decreased significantly to a minimum of 12.8 $\pm$ 0.4 mg in 70 days, with no adverse effects on earthworm density, with maximum growth of earthworms being reached within two months. In addition, the aforementioned authors raised safety concerns concerning the use of vermicomposted human waste for agriculture; however, other researchers noted a significant decrease in fecal coliforms and a reduction in mass of about 67% after 4 weeks of vermicomposting with *Eudrilus eugeniae* (Acquah et al. 2021; Nsiah-Gyambibi et al. 2021). Earthworms have been reported to reduce the total abundance of human pathogenic bacteria and modify their diversity, resulting in a higher quantity of Enterobacteriaceae in sludge vermicompost (Huang et al. 2020). Likewise, salmonella has been found to be less abundant than other bacteria in vermicompost, and vermicompost tea samples collected from different farms tested negative for the presence of possible pathogens (Atanda et al. 2018), indicating the safety of vermicompost when produced under hygienic conditions.

19.2 Potential of Vermicomposting in Africa

Agriculture is one of the main economic activities in Africa. Apart from providing more than 60% of food for domestic needs, agriculture is an essential source of income for most of the population, demonstrating the great importance of farming in African countries. Only 6% of all land in Africa is dedicated to agriculture. Therefore, it is vital to improve the soil fertility to enable sustainable production of crops in most African countries. While there is a need to increase food production hugely, to meet the demands of a rapidly growing population remains paramount.

Various types of food are cultivated in countries such as Egypt, Nigeria, and South Africa (Table 19.1). Increasing the volume of cultivated crops will result in massive amounts of agricultural waste generated. By applying the concept of "waste is wealth," these countries could benefit from the amount of waste produced by converting it into value-added products.

Table 19.1 Main crops cultivated in different countries in Africa (Robert K. A. Gardiner 2018)

Corn	Rice	Wheat	Legumes	Root crops	Oil palm	Coconut	Tobacco
Egypt	Egypt	Ethiopia	Egypt	Ethiopia	Congo	Côte d'Ivoire	Nigeria
Mauritius	Senegal	Kenya	South Africa	Kenya	Côte d'Ivoire	Ghana	Tanzania
Reunion	Côte d'Ivoire	Nigeria	Sudan	Madagascar	Nigeria	Nigeria	Zimbabwe
	Guinea	South Africa					
	Madagascar						
	Mali,						
	Nigeria						
	Sierra Leone						
	Tanzania						

The most beneficial approach would be to enhance the use of bioremediation technologies in these countries. Vermicomposting is a well-known cost-effective technology that does not require investment in expensive instruments. However, survival of the earthworms is necessary for success of the process. Earthworm populations depend on both physical (temperature, moisture, aeration, and texture) and chemical (pH) properties of the soil, as well as on food availability, and the ability of the organisms to reproduce and disperse. Hengl et al. (2021) mapped African soil properties and nutrients at 30 m spatial resolution and the results have shown wide variability. Although these researchers faced several problems in gathering the information, the following figure provides an overall picture of soil properties in the African continent.

According to Hengl et al. (2021), diverse soil types are found in Africa, although obtaining data from specific areas such as Congo still remains challenging. Nevertheless, African soils generally have a sandy-loam soil texture with high organic carbon content as indicated in Fig. 19.1 and can provide a niche for some earthworm

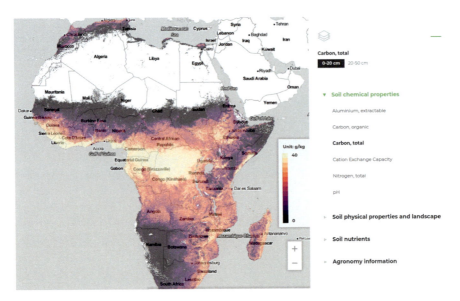

Fig. 19.1 Map showing the soil total carbon for most African countries predicted at 30 m resolution at 0–20 and 20–50 cm depths (source https://zenodo.org/record/4088064#.YqOtnZBzIV)

species. For example, *Keffia penetrabilis* n. sp. and *K. proxipora* n. sp. have been described in collections from various localities in Nigeria. Overall, the diverse soil properties and the limited access in some areas (local government security policies) make compiling a database for the whole continent challenging. In addition, the conditions in many African countries are not always favorable to earthworm growth due to the hot climate and dry soil. However, earthworms are highly tolerant and can survive well in slightly acidic environments with adequate moisture levels. Difficulties in maintaining optimal growth conditions to produce high-value fertilizers can discourage smallholders from using the vermicomposting technique to make fertilizers for agricultural purposes. Therefore, there is a need to establish awareness-raising strategies and provide farmers with an in-depth understanding of the concept of soil erosion, degradation, and sustainable farming. This could motivate farmers to develop vermicomposting plants on their farms as a simple, feasible, and economically beneficial strategy for their business.

19.2.1 Vermicomposting Practices in African Countries

Growth of the global population has led to the expansion of agricultural production and industrialization to meet current needs without further damage to the environment. In response, the demand for animal manure and food has led to the inappropriate disposal of waste (Katakula et al. 2021). Declining soil fertility is a challenge to sustainable agricultural production in many parts of the world (Fróna et al. 2019). Seeking eco-friendly ways of replenishing soils has become the main focus of research projects, with composting and vermicomposting gaining momentum due to the limited ecological footprint of these processes (Vyas et al. 2022). These techniques have also been promoted due to increasing awareness that organic nutrient sources play an important role in improving soil quality (Aulakh et al. 2022). Vermicompost can balance the entire soil environment while also providing a desirable habitat for soil microbes (Saha et al. 2022). As an amendment, vermicompost is regarded as an excellent replacement for synthetic fertilizers (Kaur 2020).

Many studies have been conducted in Africa on vermicomposting with various types of substrates, including sludge, paper waste, agricultural waste (rice straw, leaf litter, sawdust, banana peel, etc.) (Mtui 2009), winery solid waste (Masowa 2020), and poultry waste (Adetunji et al. 2021). Researchers observed that using fresh organic substrates results in elevated moisture contents in the mixtures; however, the addition of dry matter such as straw or sawdust can usually restore the moisture content to adequate levels (Al-Assiuty et al. 2021; Mahboub Khomami et al. 2021). Various studies carried out in African cities with different substrates are summarized in Table 19.2. In Kampala, Durban, and Windhoek, similar results were obtained using food waste and animal dung with relatively stable pH or homogenized compost. The findings have been corroborated by studies in other parts of the world (Torrijos et al. 2021; Zhang et al. 2022). Similarly, other research studies have revealed that African night crawlers (*Eudrilus eugeniae*) and tiger worms (*Eisenia fetida*) can digest large volumes of waste, including human excreta (Belmeskine et al. 2020; Watako et al. 2016), food waste and paper waste (Mupondi et al. 2018), producing vermicompost with a high nutrient content (Coulibaly 2020; Jjagwe et al. 2019; Katakula et al. 2021) relative to that produced by other worm species. In study carried out in South Africa, vermicompost addition to contaminated soil yielded a significant reduction in heavy metal concentrations in soils after 8 weeks (Mupondi et al. 2018). Based on the studies conducted by (Wang et al. 2022), bioavailability of Cu/Zn decreases during the vermicomposting process in substrate residue (Wang et al. 2022).

In Nigeria, native plants *Melissa officinalis L* (lemon balm) and *Sida acuta* (stubborn weed) used as additives in vermicomposting showed a high potential for phytoremediation and phytostabilization of contaminated mining soils in Madaka District (Ijah et al. 2021). This was brought about by the need to develop eco-friendly remediation technologies to restore contaminated soils in the country. X-ray fluorescence (XRF) analysis of the remediated soils revealed relatively low levels of elements including Cd, As, and Pb (Kodom et al. 2012). In addition, Lukashe et al. (2020) noted that uptake of Fe, Mn, Zn, and Cr in *Chloris gayana* (Rhodes grass) was significantly reduced by the addition of vermicompost. Similarly, *Vetiveria zizanioides* (vetiver grass) grown with vermicompost amendments displayed a good potential for phytostabilization in semiarid regions (Laxman et al. 2014).

Table 19.2 Vermicomposting scenario of various bio-waste in different African countries

Place, Country	Waste biomass	Time (days/weeks)	Earthworms	Remarks	References
Durban, South Africa	Vegetables, local topsoil, and digested sludge	Two weeks.		The successful vermicomposting results in high organic compounds.	Gårdefors and Mahmoudi (2015)
Kampala, Uganda	Cow manure and food waste	172 days.	*Eudrilus eugeniae*	45.9% reduction in material on total solid basis	Lalander et al. (2015)
Uganda	Cow manure		*Eudrilus eugeniae*	Emission factors found for the vermicompost unit were 10.8, 62.3, and 12.8 g/Mg bio-waste for methane, nitrous oxide, and ammonia, respectively	Komakech et al. (2016)
Namibia	Goat manure and vegetable food	12 weeks	*Eisenia fetida*	1—Significant difference ($P < 0.05$) in humification parameters across treatments. 2—An average 60% increase in phosphorus content with no significant differences ($P > 0.05$) in nitrate/nitrite concentrations.	Katakula et al. (2021)
Abidjan, Côted'Ivoire	Animal waste (chicken, cow, sheep and pig) waste	90 days	*Eudrilus eugenia Kinberg*	Chicken waste yielded the greatest number of earthworm hatchlings and biomass, followed by cow, sheep, and pig wastes, with the highest rate of cocoons hatching in pig waste.	Coulibaly et al. (2011)

(continued)

19 Potential Transformation of Organic Waste in African Countries by...

Table 19.2 (continued)

Place, Country	Waste biomass	Time (days/weeks)	Earthworms	Remarks	References
Liberia	Toilet waste slurry	50 days	*Eudrilus eugeniae*	Significant digestion of human excreta	Watako et al. (2016)
Uganda	Cattle manure	12 months	*Eudrilus euginea*	–	Jjagwe et al. (2019)
South Africa	Paper waste	8 weeks	*Eisenia Fetida*	Significant reduction of heavy metals to below permissible concentration of potentially toxic elements in soils after.	Mupondi et al. (2018)
Manyatta, Kenya	Goat manure	120 days	*Eisenia Fetida*	–	Gichaba et al. (2020)
South Africa	Jatropha Curcas cake	30 days	*Eisenia Fetida*	Increased NPK content in vermicompost and vermi-wash	Manyuchi et al. (2018)
Chenoua, Algeria	Sewage sludges	21 days	*Eisenia fetida*	The vermicomposting caused a decrease of fecal coliforms number	Belmeskine et al. (2020)
Accra, Ghana	Fresh human excreta and anal cleansing materials	Four weeks	*Eisenia fetida*, *Eudrilus eugeniae.*	The study showed 12.3% and 26.2% reduction in volatile solids and 60% of mass reduction	Acquah et al. (2021)
Tanta University, Egypt	Rice straw, leaf litter, sawdust, kitchen waste, and banana peel) and cow dung.		*Eisenia fetida* and *Aporrectodea caliginosa*	The quality of vermicompost was higher in vermicomposting processed by E. fetida than in case of A. caliginosa.	Al-Assiuty et al. (2021)
Kumasi, Ghana.	Fecal sludge			Low concentration of Fe, Pb, and Al in the vermicompost. Less than 16% earthworm mortality was recorded.	Nsiah-Gyambibi et al. (2021)

(continued)

Table 19.2 (continued)

Place, Country	Waste biomass	Time (days/ weeks)	*Earthworms*	Remarks	References
Holeta, Ethiopia	Manures (cattle manure and donkey manure)	90 days		–	Mnalku and Tamiru (2020)

19.3 Management and Challenges of Vermicomposting in African Countries

Although vermicomposting is promoted as a sustainable solution to soil fertility in Africa, there remain challenges regarding the successful application of this technology. A lack of general knowledge about earthworm biology is the first challenge regarding this technology. Knowledge of earthworm biology is critical for determining the type and quality of food that earthworms require and their temperature, salinity and moisture requirements. In terms of substrate quality, most earthworms used in vermicomposting do not favor organic waste rich in proteins (e.g., meat, fish waste, and dairy products), highly acidic materials (e.g., citrus fruits and onions), and materials with high salt contents (e.g., unwashed seaweed). Earthworms are also very sensitive to pesticide residues, and exposure to pesticide-contaminated soil can lead to their death. However, most smallholders are not aware of these characteristics and often feed earthworms organic materials that are rich in protein or that are acidic or salty, resulting in collapse of the vermicomposting ecosystem.

In addition, most smallholders also find it very difficult to identify the correct earthworm species suitable for vermicomposting. Therefore, using invalidated species in compost heaps has resulted in unsuccessful vermicomposting. This also leads to the production of sub-standard compost, of little value for agricultural purposes. Appropriate conditions are also essential for effective vermicomposting. For example, Dominguez and Edwards (2011)) indicated that *Eisenia fetida* has a temperature requirement of 25 °C, with a tolerance of between 0 °C and 35 °C, while *Dendrobaena veneta* has a rather low-temperature optimum and is less tolerant to extreme temperatures. The optimal temperature for *E. eugeniae* and *P. excavatus* is around 25 °C, and these earthworms die at temperatures below 9 °C. Such information is crucial in vermicompost ecosystem management, and most farmers have a limited understanding of this requirement.

The earthworm species mainly utilized in the vermicomposting process are epigeic earthworms, including *Eisenia fetida, Eisenia andrei, Eudrilus eugeniae,* and *Perionyx excavatus*. These earthworms can survive by feeding on fresh dead organic material (Gómez-Brandón et al. 2012; Mupambwa et al. 2020). However, anecic and endogeic species that live in soil or deep underground in burrows are prone to being affected by predators, such as birds, rats, ants, centipedes, ants and chickens, or by parasites, such as carabid beetles (Dominguez and Edwards 2011).

Control of these predators is challenging for many smallholders, who typically install vermireactors in open spaces in mixed farming systems where predators like birds and rats eat the earthworms. This is even a challenge in closed container vermi-bins where predators like rats may find their way into the compost.

Within the small-scale farming sector, many farmers practice windrow composting with earthworms; however, as this is done on the ground, there is a loss of vermi-leachate. The leachate contains high levels of nutrients as well as elevated levels of plant growth hormones, and it is very valuable as a liquid fertilizer. Our resource-poor farmers must understand the critical dynamics in effective vermicomposting, which can be a key driver in the adoption of this technology in preference to traditional composting.

19.4 Conclusion

Population growth and industrial expansion may lead to food shortages in Africa, and sustainable food production is therefore vital. Mind-set shifts are required in order to promote sustainable development, which can be done through educational programs. Awareness-raising programs must be directed at professionals, companies, and stakeholders, encouraging them to reduce waste and develop plans to convert different types of waste into valuable products. Vermicomposting technology is a promising approach to achieving sustainable agricultural goals. One of the main reasons for the great potential of vermicomposting is that it is an innovative biotechnology that does not call for expensive laboratories or sophisticated industrial equipment. Moreover, vermicomposting is an environmentally friendly process that meets sustainability goal 7 of the UN Millennium Development Goals (MDG7) and is consistent with the ecological sanitation concept. The simple methodology, together with the low investment required, makes vermicomposting applications possible in low-income areas, mainly tropical countries.

One of the key factors contributing to efficient vermicomposting strategies is the level of satisfaction expressed by communities. There is therefore an excellent opportunity to initiate specific initiatives involving people, private companies, and governments in order to develop solutions for sustainable waste management. Public-private partnerships (PPPs) seem to be the solution to tackling waste management issues (particularly in developing and underdeveloped countries). Other major constraints to efficient, sustainable waste management are the lack of up-to-date information and resources for planning waste treatment infrastructure and the lack of leadership to make projections and estimations for future generations. It is apparent from the existing policy-making strategies in Africa that there is still a large gap between policy-making and implementation. Hence, all stakeholders must adhere to solid regulations to ensure the establishment of efficient waste management systems.

References

Acquah MN, Essandoh HMK, Oduro-Kwarteng S, Appiah-Effah E, Owusu PA (2021) Degradation and accumulation rates of fresh human excreta during vermicomposting by Eisenia fetida and Eudrilus eugeniae. J Environ Manag 293:112817

Adetunji CO, Olaniyan OT, Bodunrinde RE, Ahamed MI (2021) Bioconversion of poultry waste into added-value products. In: Sustainable bioconversion of waste to value added products. Springer, Cham, pp 337–348

Al-Assiuty BA, Abdel-Lateif HM, Khalil MA, Khalifa AE, Ageba MF (2021) Potential utilization of various organic waste additives in vermicomposting using two different earthworm species. Egypt J Exp Biol 17(1):19–28

Atanda A, Adeleke R, Jooste P, Madoroba E (2018) Insights into the microbiological safety of vermicompost and vermicompost tea produced by South African smallholder farmers. Indian J Microbiol 58(4):479–488

Atiyeh RM, Domínguez J, Subler S, Edwards CA (2000) Changes in biochemical properties of cow manure during processing by earthworms (Eisenia andrei, Bouché) and the effects on seedling growth. Pedobiologia 44(6):709–724

Aulakh CS, Sharma S, Thakur M, Kaur P (2022) A review of the influences of organic farming on soil quality, crop productivity and produce quality. J Plant Nutr 45:1–22

Belmeskine H, Ouameur WA, Dilmi N, Aouabed A (2020) The vermicomposting for agricultural valorization of sludge from Algerian wastewater treatment plant: impact on growth of snap bean Phaseolus vulgaris L. Heliyon 6(8):e04679

Coulibaly SS (2020) Effect of the application timing of compost and vermicompost on mays. Zea Mays

Coulibaly SS, Ndegwa PM, Ayiania M, Zoro IB (2019) Growth, reproduction, and life cycle of Eudrilus eugeniae in cocoa and cashew residues. Appl Soil Ecol 143:153–160

Coulibaly SS, Kouassi KI, Tondoh EJ, Zoro Bi IA (2011) Impact of the population size of the earthworm Eudrilus eugeniae (Kinberg) on the stabilization of animal wastes during vermicomposting. Philippine Agric Sci 94:359–367

Dominguez J, Edwards CA (2011) Biology and ecology of earthworm species used for vermicomposting. In: Vermiculture technology: earthworms, organic waste and environmental management. CRC Press, Boca Raton, pp 27–40

Dominguez J, Edwards CA, Dominguez J (2001) The biology and population dynamics of Eudrilus eugeniae (Kinberg)(Oligochaeta) in cattle waste solids. Pedobiologia 45(4):341–353

Edwards CA (2007) Earthworm ecology, 2nd edn. CRC Press, Boca Raton

Edwards CA (2004) Earthworm ecology (Second edition). CRC Press, Boca Raton

Feller C, Brown GG, Blanchart E, Deleporte P, Chernyanskii SS (2003) Charles Darwin, earthworms and the natural sciences: various lessons from past to future. Agric Ecosyst Environ 99(1):29–49

Fontana AR, Antoniolli A, Bottini R (2013) Grape pomace as a sustainable source of bioactive compounds: extraction, characterization, and biotechnological applications of phenolics. J Agric Food Chem 61:8987–9003

Fróna D, Szenderák J, Harangi-Rákos M (2019) The challenge of feeding the world. Sustainability 11(20):5816

Galanakis CM (2012) Recovery of high added-value components from food wastes: conventional, emerging technologies and commercialized applications. Trends Food Sci Technol 26:68–87

Gårdefors C, Mahmoudi N (2015) Urine diverting vermicomposting toilets for Durban, South Africa

Gardiner RKA (2018) Agriculture of Africa. https://www.britannica.com/place/Africa/Agriculture

Gebrehana ZG, Gebremikael MT, Beyene S, Wesemael WM, De Neve S (2022) Assessment of trade-offs, quantity, and biochemical composition of organic materials and farmer's perception towards vermicompost production in smallholder farms of Ethiopia. J Mater Cycles Waste Manag 24:1–13

Gichaba V, Muraya M, Odilla G, Ogolla F (2020) Preparation and evaluation of goat manure-based vermicompost for organic garlic production in manyatta sub-county, Kenya, pp 5 and 51

Gómez-Brandón M, Lores M, Domínguez J (2012) Species-specific effects of epigeic earthworms on microbial community structure during first stages of decomposition of organic matter. PLoS One 7(2):e31895

Gómez-Brandón M, Lores M, Martínez-Cordeiro H, Domínguez J (2020) Effectiveness of vermicomposting for bioconversion of grape marc derived from red winemaking into a value-added product. Environ Sci Pollut Res 27(27):33438–33445

Gómez-Brandón M, Vela M, Martínez-Toledo M, Insam H, Domínguez J (2015) Effects of compost and vermicompost teas as organic fertilizers. In: Advances in fertilizer: technology synthesis, vol 1. Studium Press, New Delhi, pp 300–318

Hengl T, Miller MA, Križan J, Shepherd KD, Sila A, Kilibarda M, Antonijević O, Glušica L, Dobermann A, Haefele SM (2021) African soil properties and nutrients mapped at 30 m spatial resolution using two-scale ensemble machine learning. Sci Rep 11(1):1–18

Huang K, Xia H, Zhang Y, Li J, Cui G, Li F, Bai W, Jiang Y, Wu N (2020) Elimination of antibiotic resistance genes and human pathogenic bacteria by earthworms during vermicomposting of dewatered sludge by metagenomic analysis. Bioresour Technol 297:122451

Ijah UJJ, Abioye OP, Bala JD (2021) Vermicompost assisted phytoremediation of heavy metal contaminated soil in Madaka District, Nigeria using Melissa officinalis L (Lemon balm) and Sida acuta (Stubborn weed). Int J Environ Sci. Technol. https://doi.org/10.1007/s13762-022-04105-y

Ismail SA (2009) The earthworm book. Other India Press, Mapusa

Jjagwe J, Komakech AJ, Karungi J, Amann A, Wanyama J, Lederer J (2019) Assessment of a cattle manure vermicomposting system using material flow analysis: a case study from Uganda. Sustainability 11(19):5173

Katakula AAN, Handura B, Gawanab W, Itanna F, Mupambwa HA (2021) Optimized vermicomposting of a goat manure-vegetable food waste mixture for enhanced nutrient release. S Afr 12:e00727

Kaur T (2020) Vermicomposting: an effective option for recycling organic wastes. In: Organic agriculture. IntechOpen, Hampshire, pp 1–10

Kodom K, Preko K, Boamah D (2012) X-ray fluorescence (XRF) analysis of soil heavy metal pollution from an industrial area in Kumasi, Ghana. Soil Sediment Contam Int J 21(8):1006–1021

Komakech AJ, Zurbrügg C, Miito GJ, Wanyama J, Vinnerås B (2016) Environmental impact from vermicomposting of organic waste in Kampala, Uganda. J Environ Manag 181:395–402

Lalander CH, Komakech AJ, Vinnerås B (2015) Vermicomposting as manure management strategy for urban small-holder animal farms–Kampala case study. Waste Manag 39:96–103

Laxman N, Nair P, Kale RD (2014) Effect of vermicompost amendment to goldmine tailings on growth of Vetiveria zizanioides. Int J Adv Pharm Biol Chem 3(2):341–351

Lukashe NS, Mnkeni PNS, Mupambwa HA (2020) Growth and elemental uptake of Rhodes grass (Chloris gayana) grown in a mine waste-contaminated soil amended with fly ash-enriched vermicompost. Environ Sci Pollut Res 27(16):19461–19472

Mahboub Khomami A, Haddad A, Alipoor R, Hojati SI (2021) Cow manure and sawdust vermicompost effect on nutrition and growth of ornamental foliage plants. Cent Asian J Environ Sci Technol Innov 2(2):68–78

Manyuchi MM, Mbohwa C, Muzenda E (2018) Bio stabilization of Jatropha Curcas Cake to bio fertilizers through vermicomposting. In: Paper presented at the proceedings of the international conference on industrial engineering and operations management, Paris, France

Masowa MM (2020) Assessment of maize productivity and soil health indicators following combined application of winery solid waste compost and inorganic fertilizers. North-West University (South Africa)

Mnalku A, Tamiru G (2020) Interplay of bedding materials and harvesting time on vermicompost yield and quality. Ethiop J Agric Sci 30(2):99–108

Mtui GY (2009) Recent advances in pretreatment of lignocellulosic wastes and production of value added products. Afr J Biotechnol 8(8):1398

Mupambwa HA, Ravindran B, Dube E, Lukashe NS, Katakula AA, Mnkeni PN (2020) Some perspectives on Vermicompost utilization in organic agriculture. In: Earthworm assisted remediation of effluents and wastes. Springer, Singapore, pp 299–331

Mupondi LT, Mnkeni PNS, Muchaonyerwa P, Mupambwa HA (2018) Vermicomposting manure-paper mixture with igneous rock phosphate enhances biodegradation, phosphorus bioavailability and reduces heavy metal concentrations. Heliyon 4(8):e00749

Nguru WM, Ng'ang'a SK, Gelaw F, Kanyenji GM, Girvetz EH (2020) Survey data on factors that constrain the adoption of soil carbon enhancing technologies in Ethiopia. Sci Data 7(1):1–6

Nsiah-Gyambibi R, Essandoh HMK, Asiedu NY, Fei-Baffoe B (2021) Valorization of fecal sludge stabilization via vermicomposting in microcosm enriched substrates using organic soils for vermicompost production. Heliyon 7(3):e06422

Saha P, Barman A, Bera A (2022) Vermicomposting: a step towards sustainability. IntechOpen, London

Sanchez-Hernandez JC, Cares XA, Domínguez J (2019) Exploring the potential enzymatic bioremediation of vermicompost through pesticide-detoxifying carboxylesterases. Ecotoxicol Environ Saf 183:109586

Schröder C, Häfner F, Larsen OC, Krause A (2021) Urban organic waste for urban farming: growing lettuce using vermicompost and thermophilic compost. Agronomy 11(6):1175

Sun Z (2003) Vermiculture & vermiprotein. China Agricultural University Press, Beijing

Torrijos V, Dopico DC, Soto M (2021) Integration of food waste composting and vegetable gardens in a university campus. J Clean Prod 315:128175

Vyas S, Prajapati P, Shah AV, Varjani S (2022) Municipal solid waste management: dynamics, risk assessment, ecological influence, advancements, constraints and perspectives. Sci Total Environ 814:152802

Wang F, Yao W, Zhang W, Miao L, Wang Y, Zhang H, Ding Y, Zhu W (2022) Humic acid characterization and heavy metal behaviour during vermicomposting of pig manure amended with 13C-labelled rice straw. Waste Manag Res 40(6):736–744

Watako D, Mougabe K, HEATH T (2016) Tiger worm toilets: lessons learned from constructing household vermicomposting toilets in Liberia. Innovation in Sanitation 35:136–147

Zhang W, Kong T, Xing W, Li R, Yang T, Yao N, Lv D (2022) Links between carbon/nitrogen ratio, synergy and microbial characteristics of long-term semi-continuous anaerobic co-digestion of food waste, cattle manure and corn straw. Bioresour Technol 343:126094

Chapter 20
Earthworms in Bioremediation of Soils Contaminated with Petroleum Hydrocarbons

Abdullah Ansari, Jonathan Wrights, and Sirpaul Jaikishun

Abstract Engine oil contaminated soils present a significant environmental hazard, due to the complex constituents of hydrocarbons and the toxic effect it places on the natural ecosystem and human health. It is essential to develop environmentally safe approaches to remove these contaminants from local soil. The effect of *Eisenia fetida* on the physicochemical properties of the soil, bacteria population, and earthworm survival is proven by bioaccumulative role of *Eisenia fetida* in the remediation process and has been recorded by numerous researchers. *Eisenia fetida* generally enhanced the physicochemical properties of contaminated soil. pH and electrical conductivity were significantly improved by earthworm activity in bioremdiation process. The presence of *Eisenia fetida* significantly improved nitrogen, phosphorous, and potassium concentration in soil. The experiments conducted in Guyana assessed the potential of earthworm species, *Eisenia fetida* in the removal of petroleum hydrocarbons from local soils contaminated with used engine oil in a vermiwash system. The study found that earthworm species "*Eisenia fetida*" can bioremediate local soils contaminated with used engine oil and enhance nutrient availability and microbial population.

Keywords Ecosystem · Earthworms · Vermiwash · Petroleum hydrocarbon · Vermiremediation

20.1 Introduction

The environment faces severe threats as a result of anthropogenic activities. The natural equilibrium of our ecosystem is incessantly disturbed encouraging many adverse consequences, often in the name of development. A pervasive impact that suppresses our biological and ecological systems stems from pollution caused by the

---

A. Ansari (✉) · J. Wrights · S. Jaikishun
University of Guyana, Georgetown, Guyana
e-mail: abdullah.ansari@uog.edu.gy

© The Author(s), under exclusive license to Springer Nature Singapore Pte Ltd. 2023
H. A. Mupambwa et al. (eds.), *Vermicomposting for Sustainable Food Systems in Africa*, Sustainable Agriculture and Food Security,
https://doi.org/10.1007/978-981-19-8080-0_20

usage, extraction, and production of petroleum and petroleum products. Petroleum constitutes diverse compounds made up mainly of hydrogen and carbons; hence referred to as petroleum hydrocarbons. Many countries have experienced the devastating effects of an oil spill and the extravagant costs financially, socially, and environmentally. Some of the worst oil spills recorded today include Torrey Canyon oil spill (1967) in Scilly Isles U.K which resulted in 180 miles of coastland being directly affected, killing more than 15,000 seabirds and numerous aquatic animals. A chemical (toxics solvent based cleaning agents) was used by the navy to quell the spill but instead accelerated environmental damages which resulted in the use of bombs to burn out the oil (Moss 2010). Thus, it distorted the air, land, and aquatic environment as a result of severe pollution by chemicals in various forms. A more recent occurrence is that of the Gulf of Mexico deepwater horizon oil rig (2010). Gulf of Mexico deepwater horizon oil spill was the most massive in history. It resulted in an approximate 206 million gallons of oil being spilled over 572 miles of Gulf shoreline, killing 11 workers on the rig and devastating the Gulf coast up to present day (Moss 2010).

The damaging effect on the environment is evident, as even on a smaller scale lays an ever-growing scourge that impedes the health of local environs and humans. The daily usage of petroleum and petroleum-based products and their improper disposal often result in pollution of soil and groundwater (Adewoyin et al. 2013). This practice directly impacts the productive capacity of the land and the viability for consumption of goods produced. However, this is often overlooked by many in society, specifically within the local context of Guyana where there is a wide range of mechanic shops that fail to implement proper disposal of spent oil from motor vehicles. The effects may be subtle yet pose severe negative impacts over a prolonged period. According to literature, prolonged exposure of petroleum hydrocarbon may have an acute to a severe adverse effect on human health (Gay et al. 2010). Additionally, the hydrocarbons present in used engine oil was found to harm the male reproductive systems in rats (Akintunde et al. 2015) and as such infer same on humans. Hence, it is essential to ensure proper disposal and where necessary effective remediation measures to prevent its presence in the soil and groundwater, which potentially compromises human health.

Anthropogenic activities that involve petroleum-based products pose a threat of severe adverse effects on the environment. It was found associated with the poor handling of used oil and ignorance of the disposers. In a study carried out the majority of automobile mechanics admitted to the disposal of used oil to the land (Zitte et al. 2015). Hence, contributing to petroleum hydrocarbon pollution of soil occurs daily and there is need for more sustainable means of rehabilitating these polluted soils, an area where limted research has focused on.

20.2 Petroleum Remediation Techniques

The common practice once employed for the remediation of sites contaminated with crude oil (petroleum hydrocarbons) was geared mostly towards physical and chemical methods which were often quite costly and not very environmentally friendly, as seen in the Torrey Canyon oil spill (Moss 2010). However, as time progresses, numerous techniques were adopted for remediation of petroleum hydrocarbon and recently attention towards different bioremediation strategies as an option to remove petroleum hydrocarbon from soil (Das and Das 2015).

Vermiremdiation is a socially acceptable, environmentally friendly and economically viable method for bioconversion of organic wastes into nutrient-rich compounds (Sinha et al. 2010a). Vermeremediation is based primarily on the unique ecological, biological, and behavioral characteristics of earthworms in soil (Ma et al. 1995; Chaoui et al. 2003). Thus, earthworms play a significant role in the field of waste remediation, and they were also an essential component in this study since they would provide a cost-effective and safer remediation techniques for contaminated soil.

Guyana is now on the verge of exploiting its recently discovered petroleum hydrocarbon source. It would be amiss to overlook the potential impacts that could arise from this resource and its effect on the natural environment. Therefore, it is vital to explore preventative and remediation measures to preserve the environment. A consequence of this resource is a likely increase in the availability and use of petroleum-based products in Guyana. The improper disposal of petroleum-based products which occur daily by many small businesses and mechanic workshops is already of concern to environmental preservation. The production of petroleum products by Guyana is likely to accelerate this practice and also present the possibility of disasters faced by other countries. Thus, it is on the ideology of "*failing to prepare is preparing to fail*" that the researcher sought to investigate an environmentally friendly approach for the remediation of petroleum hydrocarbon in native soil. More so, the use of readily available species of earthworms to facilitate the rehabilitation of petroleum hydrocarbon from the soil will provide a cost-effective, ecologically safe and socially acceptable method of protecting the environment. In this study, the potential of earthworms to facilitate the bioremediation of engine oil contaminated soils, and their effect on soil nutrients and natural bacteria community was investigated. The research involved the acquisition of soil, the addition of petroleum hydrocarbon source (used motor vehicle engine oil) to soil at 0.5% (w/w), followed by the addition of earthworm species, *Eisenia fetida* to treatment soils. The use of a structured and controlled vermiwash system in the process had significant impact on contaminated soils. An assessment of the physicochemical characterization of soil and concentration of TPH in soil at twenty days intervals allowed for the evaluation of earthworms in the removal of TPH.

Petroleum products are used during daily activities. They range from household chemicals to the more common motor vehicle lubricants and fuels. The masses often overlook these products as an environmental threat, and this drives their improper usage and disposal (Zitte et al. 2015). In Guyana, the most visible consequence is the eyesore caused by improper disposal of petroleum-based products from a motor vehicle by mechanic shops, commonly referred to as waste oil. The physical outlook is a small insight into what occurs under the surface. According to the Agency for Toxic Substance and Disease Registry, petroleum hydrocarbons enter the environment via accidental occurrences from industries or as byproducts from commercial or private uses, both causing adverse consequences (Agency for Toxic Substances and Disease Registry ATSDR 1999). Contamination may occur via direct contact with water or soil. When released directly to water, some total petroleum hydrocarbon (TPH) fractions float and form thin surface films, while other denser portions accumulate in the sediment at the bottom of the water directly affecting bottom-feeding fishes and other organisms. TPH released directly to soil may progress downward to the groundwater, with individual compounds separated from the mixture based on chemical properties. Some of the compounds evaporate into the air, others dissolve in the groundwater dispersing to different areas, some are broken down by soil organisms, and others remain in the soil for a prolonged period (ATSDR 1999).

From a broader perspective, petroleum hydrocarbon is a severe environmental threat. More so, The United States Environmental Protection Agency (USEPA) identifies the most serious hazardous waste sites in their nation. These sites make up their National Priorities List (NPL) targeted for long-term federal cleanup activities. USEPA reported TPH at 34 of the 1519 current or former NPL sites (ATSDR 1999). The constituent of used engine (mineral-based) oil is a mixture of metals and PAHs, that increases in its carcinogenic and mutagenic properties as motor oil undergoes thermal decomposition with resultant formation of gasoline combustion products (Bingham 1988; Ingram et al. 1994). Additionally, petroleum hydrocarbons categorized as polycyclic aromatic hydrocarbons (PAH) are known contaminants formed via combustion. These compounds tend to accumulate in oil by a factor up to one thousand; they are characteristically very toxic environmental compounds with carcinogenic and mutagenic properties. Consequently, they pose a significant threat to human health. These compounds enter the environment through various sources; these include particulates, oil leaks, and poorly controlled oil changes. There are one hundred and forty different PAHs in used oil of crankcase lubricated engines and lower quantities in new or fresh oil (Van Donkelaar 1990). Thus, it emphasizes the threat to human health and the environment as a consequence of improper disposal of motor vehicle oil and petroleum products by mechanic shops in absentia of appropriate remediation techniques.

20.2.1 Earthworms

Earthworms are soil fauna recognized in association with the modification and enhancement of soil productivity (Buckerfield 1998). The mechanical, biological, and chemical effects exerted on the soil during their normal activities aid their ability

20 Earthworms in Bioremediation of Soils Contaminated with Petroleum Hydrocarbons 353

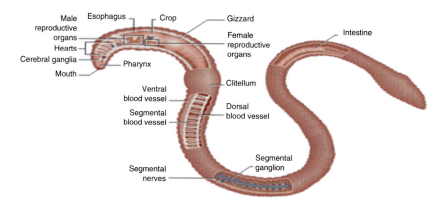

Fig. 20.1 Anatomy (internal structure-function) and morphology (external appearance) of an earthworm; source (Biernbaum 2014)

Fig. 20.2 Reproduction and life cycle of *Eisenia fetida*. Source (Biernbaum 2014)

to enhance soil properties (Edwards and Bohlen 1996). The various structure and life cycles of earthworm are schematically shown in Figs. 20.1, 20.2, and 20.3.

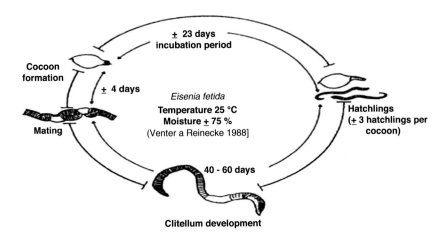

Fig. 20.3 Schematic representation of *Eisenia fetida* life cycle. Source (Venter and Reinecke 1988)

20.2.2 Ecology of Earthworms

Earthworm researchers have adopted a lexicon-based classification system that primarily focuses on the behavioral characteristics and niche of earthworms in particular ecology rather than phylogenetic traits even though they often coincide (Edwards and Bohlen 1996). The three main paradigms adopted within this classification system are anecic, endogeic, and epigeic. Anecic earthworms are known for their vertical burrowing into the soil, which results in permanent burrows which contrasts that of endogeic earthworms. They are known to migrate between the different strata of the soil to acquire nutrients. They are physically known to have dark colored heads ranging from red to brown and pale tail endings (Edwards and Bohlen 1996). Endogeic earthworms live within the soil and depend on its nutrients for survival. These earthworms are often characterized by their behavioral horizontal burrowing into the earth to attain a suitable stratum for their residence. They generally possess vivid color difference from the epigeic species, usually having a pale complexion of gray, pink, green, or blue (Edwards and Bohlen 1996). Epigeic earthworms are characteristically known for their presence in surface and near-surface soil, commonly among plant debris. They are anatomically small with a high level of pigmentation and are generally ubiquitous. *Eisenia fetida* is a known prolific species of epigeic earthworm that is a located within different climatic conditions globally. Epigeic earthworms display essential adaptive ability and play a critical role in the degradation of debris to topsoil in many ecosystems (Edwards and Bohlen 1996). The extent of changes induced by earthworms on the soil's physical, chemical, and biological properties are highly dependent on their ecological and scientific classification (Lavelle 1988).

20.3 Petroleum Hydrocarbon

Petroleum is liquid oil commonly found in deposits within the earth's sub-layers. It is formed from rock minerals and the decomposition of living tissues of organisms that died ages ago. Hence, it is organic and is one-third of the three primary fossil fuels, the others being coal and natural gas (Oil Spill Intelligence Report 1997). The chemical constituent of petroleum is primarily a combination of hydrogen and carbons referred to as hydrocarbons. Petroleum also consists other compounds and trace elements in negligible amounts. (Hassanshahian and Capello 2012). Some of the most common byproducts of crude oil or petroleum are gasoline, diesel, and kerosene, all of which consist many diverse petroleum hydrocarbon compounds (Fig. 20.4).

The high demand for petroleum makes its availability in the environment very prevalent and the potential for petroleum hydrocarbon pollution the same. An estimated quantity in excess of a million ton of oil enters the sea annually, some of which are naturally degraded by aquatic microorganisms, other cycles into the sediment and around the ecosystem (Hassanshahian and Capello 2012). Pollution of this scale negatively affects the natural equilibrium of food webs. It is therefore essential to investigate the measures taken to curb pollution caused by petroleum and petroleum-based products. More so, to develop means to propel remediation measures for this pollutant in the environment further.

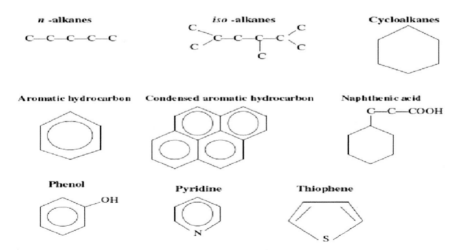

Fig. 20.4 Structural classification of some major petroleum hydrocarbon components. Source (Hassanshahian and Capello 2012)

20.4 Petroleum Hydrocarbon Pollution in the Environment

From the commencement of the petroleum industry in 1859, the environment faced the task of absorbing the adverse effects of this resource production. Oil spills and underground breaches in wells and pipelines have resulted in significant disturbance of the natural environment and even human health (U.S. Department of Health and Human Services 1999). Petroleum pollution occurs on site and away from the source of production, an indication of the modes of entry for damaging effects on the natural ecosystem. The two primary ways for the introduction of petroleum pollution are point source pollution (includes spills from tankers, pipelines, and oil wells) where the origin of contamination is from a single identifiable point, and non-point source pollution where small quantities of oil originating from various sources build up over time to have wide spanning effects. Nonpoint source accounts for the majority of petroleum pollution (Oil Spill Intelligence Report 1997).

20.5 Strategies for the Removal of Petroleum Hydrocarbon from Soil

Over the years the effects of petroleum pollution (Figs. 20.5 and 20.6) have become very pervasive and have caused an upsurge of research in the field of remediation techniques. As the world became more modernized and technology advanced, various methods were developed. These methods involved chemical, physical, biological and in some case a combination of more than one approach. The conventional method of treating the issue of hydrocarbon pollution involved burning and removal of contaminated soil, which was found to be more of a hazard to the environment since it merely transformed these toxic chemicals from one phase (liquid to gas) to another without changing their chemical structure and caused relocation to new sites (Das and Das 2015). More so, this method required excessive financial input and provided little to no environmental improvement. In recognition of such research was geared towards more environmentally friendly approaches.

Fig. 20.5 Effect of Nigeria's Agony Dwarfs Gulf oil spill on land vegetation (Vidal, 2010)

During

After

Fig. 20.6 Effect of the Deepwater Horizon oil spill on seabirds (Stanton 2015)

Consequently, emphasis on chemical and biological treatments for petroleum pollutant became evident.

Chemical techniques applied for the degradation of hydrocarbons in soil were photocatalytic, photo-oxidation, and ozone treatment. Photocatalytic degradation, which involved the use of artificial UV light rays in the presence of titanium dioxide to degrade Phenanthrene (PHE), Pyrene (PYRE), and Benzo[*a*]pyrene (BaP) on soil surface (Zhang et al. 2008), the study showed positive output as UV rays in synergy with titanium oxide reduced the half lives from 533.15 to 130.77 h, 630.09 to 192.53 h, and 363.22 to 103.26 h, respectively. Photo-oxidation was done to assess the biodegradation of four PAHs [anthracene (ANT), acenaphthene (ACE), phenanthrene (PHE), and benz[a]anthracene (BaA)] using a GaN:ZnO solution with platinum as a catalyst which showed activity that favored the reduction of PAHs via complete removal of all four PAHs after 8 hours (Koua et al. 2010). Ozone treatment which utilized molecular ozone successfully degraded 95% of phenantrene in soil in 2.3 hours at ozone flux of 250 mg/h and 91% of pyrene after 4 hours at an ozone fluz of 600 mg/h (Masten and Davies 1997). Chemical treatments of petroleum hydrocarbon in soil indicated positive outcomes in a very short time for most approaches but emphasis was not placed on the potential adverse effects on the soil organism and natural ecosystem that may result from the use of unstable chemicals. Additionally, chemical methods were observed to be cost intensive as most approaches require the use of specialized equipment and highly sensitive chemicals to carry out investigation.

Biological approaches applied to hydrocarbon remediation from soil that attracted research attention over the years included; land farming, bioaugmentation, vermiremediation, and phytoremediation. Land farming entails the use of soil microorganisms and sustainable agricultural practices under aerobic conditions to treat contaminated soil through transformation, immobilization, and detoxification processes. Land farming is categorized on the principles of intensive land farming and passive land farming (Harmsen 2004). Both of which requires a lengthy period of time for a successful output, with limitations of oxygen supply, availability of appropriate microorganisms, and bioavailability of pollutants to microorganisms (Harmsen et al. 2007). Bioaugmentation involves multile approaches inclusive of: the use of cells encapsulated in a carrier (alginate); gene bioaugmentation which focused on transferral of remediaition genes to indigenous microoganisms by added innoculant; rhizosphere bioaugmentation, a microbial inoculant is added to a site along with a plant that serves as a niche for the inoculant's growth and phytoaugmentation, that utilize the engineered insertion of remediation genes directly into a plant for use in remediation without a microbial inoculant. The survival of exogenous microorganisms, competition with native organisms, and the potential adverse effect on the natural ecosystems are uncontrolled constraints to the use of bioaumentation (Gentry et al. 2004). Vermeremediation involves the use of earthworms in the reclamation of contaminated sites, such as polluted land and sewage sludge (Sinha et al. 2009) to an enviromrentally safer form for human interaction and as organic fertilizers for crop production, via earthworm physical and biological actions (Sinha et al. 2002). Phytoremediation requires the use of plant in the remediation of soil pollutant. Phytoremediation was successfully exploited againt petroleum hydrocarbon in soil with aid of *Mirabilis Jalapa L.* The plant removed 41.61%–63.20% of petroleum hydrocarbons from soil over a period of 127 days (Peng et al. 2009). Biological applications for hydrocarbon removal from soil appear to be more environmentally friendly and from a broader viewpoint cost-effective. However, the observed drawbacks observed with biological applications were time inefficiency when compared to some chemical applications. Albeit such, this paper was based on the investigation of petroleum hydrocarbon removal using a biological approach, since the focus of the study was initiated towards an environmentally safe approach.

20.6 Bioremediation

The process of bioremediation for soil contamination has been in practice since the 1940s, but the field gained global recognition preceding the Exxon Valdez oil spill in 1989 (Margesin and Schinner 1997). Exxon Valdez oil spill propelled widespread studies that proved bioremediation to be useful in the treatment of sites contaminated by hydrocarbons. Further, earthworm speices, *Eisenia fetida* were found to double the reduction rate of some of the most abundant and persisitement intermediate products of hydrocarbons present in soil within a period of 28 days (Coutiño-González et al. 2010).

Bioremediation functions primarily on biodegradation, defined as the total mineralization of organic contaminants into environmentally safe compounds or transformation of complex organic pollutants to simpler organic compounds by biological agents such as bacteria in the genus *Acinobacter, Aeromicrobium, Brevibacterium, Burkholderia, Dietzia, Gordonia,* and *Mycobacterium* (Das and Chandran 2010). Many indigenous bacteria and fungi in the aquatic and soil environs facilitate the degradation of hydrocarbon contaminants based on the type of hydrocarbon chain (Das and Chandran 2010). Within the domain of bioremediation, bioaugmentation and bio-stimulation are the two important sub-categories. Bio-stimulation focuses on organisms present within an area of contamination; it refers to the introduction of organic or inorganic compounds to stimulate indigenous microorganisms to break down pollutants in the environment. Bioaugmentation involves the isolation and transferral of foreign organisms to the contaminated site to achieve restoration (Gentry et al. 2004). Bio-stimulation and bioaugmentation may be used to enhance bioremediation based on the type of enviroment and the availability of hydrocarbon degrading bacteria. Researchers purport that bioremediation strategies provide more benefits in comparison to physical and chemical approaches since its implementation can be in situ, its straightforward, less intrusive and often more cost-effective (Das and Das 2015).

20.7 Earthworms as a Beneficial Organism

Earthworms are organisms that hold significant benefits towards a sustainable society. They are thought to have the potential of offering cost-effective solutions to many social, economic, and environmental issues being imposed on the human environs. More specifically, they possess the ability to aid in the natural management of municipal and industrial organic waste away from landfill sites and reduce greenhouse gas emission (Sinha et al. 2010b). The anatomy and physiology of earthworms aids in their capacity to bioremediate toxic chemicals from soil and aquatic environment. Their bodies act as biofilters that can purify, disinfect and detoxify municipal and industrial wastewaters. Thus, facilitates significant reductions in the biological oxygen demand, chemical oxygen demand and total dissolved solids of wastewater. They can eradicate endocrine disrupting chemicals from sewage, bio-accumulate and bio-transform chemical contaminants from the soil and rejuvenate contaminated lands for development. Their excreta and secretions are utilized for the restoration and improvement of soil fertility. Furthermore, earthworms are being explored even within the medical fields for the development of cures for major ailments affecting humans (Sinha et al. 2010a).

Earthworms possess the ability to hasten the removal of contaminants from the soil through the regulation of soil properties via their mechanical actions increasing contact between contaminants and soil microorganisms (Roriguez-Campos et al. 2014); and internal biological processes inclusive of desorption of contaminants as it passes through their gut and enzyme action to facilitate breakdown (Azadeh and Zarabi 2015). Earthworms ability to exist in a highly toxic environment renders them beneficial in the rejuvenation of diverse ecosystems. Sinha et al. (2009) articulated that earthworms have the innate ability to absorb chemicals through their moist body wall into their interstitial water and also via their mouth during the passage of soil through their gut. The chemical contaminants passed through their stomach are either bio-transformed or biodegraded rendering them harmless in their bodies. Earthworms simultaneously enhance soil quality through their burrowing actions and the excretion of nutrient-rich materials constituting Nitrogen, Phosphorous, Potassium (NPK) and micronutrient in the form of vermicasts. It was promulgated that earthworms utilize their digestive system in the biodegradation of heavy metals by the detachment of complex aggregates between ions and humic substances in waste as it decays. Enzyme driven processes facilitate the assimilation of metal ions by earthworms trapping them within their tissues preventing their release into the environment. Thus, preventing the entry of heavy metal to organic waste (Pattnaik and Reddy 2012).

20.8 Earthworms in Bioremediation of Petroleum Hydrocarbons

In the field of environmental sciences, earthworms were investigated for their use in the bioremediation of petroleum hydrocarbons. The documented researches reviewed was noted to be done primarily within the continent of Africa and Asia. Earthworms are cosmopolitan and thus gives scope for more global research. To date, research has label various earthworm species as beneficial organisms in bioremediation, notable species includes *Lumbricus rubellus*, *Pheretima hawayana*, *Perionyx excavates*, *Hyperiodrilus africanus*, *Eudrilus eugeniae*, and *Eisenis fetida*.

20.8.1 *Lumbricus Rubellus*

Ma et al. (1995) investigated the influence of earthworm species *Lumbricus rubellus* on the disappearance rate of polycyclic aromatic hydrocarbons in soil and the bioaccumulation of these compounds in the body of earthworms. The study found that *Lumbricus rubellus* accelerated the disappearance of specific polycyclic aromatic hydrocarbons and also bioaccumulated fluoranthene and phenanthrene in their body tissues. However, bioaccumulation reduced by changes in bioavailability associated with aging of the PAH-amended soil. Hence, natural attenuation may also be associated with the reduction. Additionally, *Lumbricus rubellus* enhanced bioaccumulation of PAHs in conditions of food limitations.

20.8.2 *Pheretima Hawayana*

In a study conducted for the bioaugmentation of Polychlorinated Biphenyls (PCB) degrading microorganisms (*Ralstonia eutrophus* H850 and *Rhodococcus spp.* strain ACS), *Pheretima hawayana* was found to be instrumental in the successful biodegration of PCB in soil. Further, it improved soil aeration which aided mineralization of PCB and also enhanced the dispersal of microorganisms which attributed to a 55% reduction in soil PCB as compared to a 39% reduction in soil without earthworms. Further, soil containg earthworms showed a 65% PCB degradation at subsurface depths as compared to a 44% in soil without earthworms (Singer et al. 2001) The findings of this study depicted *Pheretima hawayana* role in the bioaugmentation of microbial organisms. Moreover, it illustrated the multifaceted benefits of earthworms as a biological agent of soil pollutants, through its ability to distribute microbes throughout contaminated soil and alter soil structure to enhace the availabity of contaminants for microbial action.

20.8.3 *Perionyx Excavates*

Subash and Sasikumar (2014) investigated the potential of earthworm species *Perionyx excavates* for the removal of polyaromatic hydrocarbons from contaminated soils and restoration of soil fertility. The finding from the study showed favorable results, via the reduction of total petroleum hydrocarbon in soil and improved soil quality indicated by increased total nitrogen, total potassium, total phosphorous, and total magension. The improvement of soil quality may be attributed to the burrowing characteristics of earthworms and the nutrient-rich excreta (known as vermicasts) produced as a result of the internal bioconversion of toxic to beneficial and mineralized compounds.

20.8.4 Hyperiodrilus Africanus

Ekperusi and Aigbodion (2015) investigated the potential of earthworm species *Hyperiodrilus africanus* for the bioremediation of petroleum hydrocarbons in soil. The result showed a considerable decrease in the total petroleum hydrocarbons, benzene, toluene, ethylbenzene, and xylene. Analysis of the earthworm tissues showed 57.35/27.64% of TPH bioaccumulation, respectively. Hence, it suggests that *H. africanus* displays the capability for the bioremediation of petroleum hydrocarbons within the Niger Delta region of Nigeria. *H. africanus* further supports the notion of earthworms ability to biaccumulate petroleum hydrocarbons in the process of soil remediation.

20.8.5 Eudrilus Eugeniae

Vermi-assisted bioremediation of petroleum hydrocarbons in soil found *Eudrilus eugeniae* to be successful as a biocatalyst in the bioremediation process. *Eudrilus eugeniae* in contaminated mechanic workshop soil showed a maximum reduction of 44.18% and 39.10% in two sites after 35 days (Ameh et al. 2013). Similarly, research conducted under laboratory conditions assessed *Eudrilus eugeniae* ability for the bioremediation of diesel-contaminated soil, in which an 84.99% reduction in total petroleum hydrocarbon was observed in soil containing this species (Abrahami and Iruobe 2015). The study also showed a reduction in a variety of other harmful chemicals in the soil, such as mercury, chromium, arsenic, and lead. Thus, validated *Eudrilus eugeniae* use in petroleum hydrocarbon and other pollutant remediation in soil. Contrary to these results, a study that investigated the potential of *Eudrilus eugeniae* for the bioremediation of soil contaminated by used engine oil, amended with poultry manure showed that *Eudrilus eugeniae* did not accelerate the breakdown of TPH in the soil, since samples with worms showed slower remediation rates than those without (Alewo et al. 2012). The variation in the results yielded maybe attributed to the amendment with poultry manure by Alewo et al. (2012) which was not in the other studies. Additionally, environmental parameter such as soil type or nutrient availability to earthworms along with the number of earrhwomrs used in the study may have limited the earthworms performance in the bioremediation process. Since, the study only used an earthworm range of 5–20 worms per kg and saw a reduction in TPH as the number of earthworms increased in the soil.

20.8.6 Eisenis Fetida

Coutiño-González et al. (2010) conducted research that investigated the ability of *Eisenia fetida* to accelerate the degradation of 9, 10-anthraquinone, a persistent intermediate product formed from anthracene in soil. The study found soils that contained *Eisenis fetida* saw a 93% reduction of anthracene as compared to a 41% reduction in soil without *Eisenis fetida* after 70 days. Further, an accelerated reduction of 9, 10-anthraquinone within 28 days was observed for treatments with *Eisenis fetida*. The reduction of 9, 10 anthracene was linked to oxygenation of anthracene by *Eisenis fetida* autochthonous microoganisms. Further, investigation on the bioaccumulation and biodegradation of phenanthrene from polluted soil with a bacterial consortium intandem with *Eisenia fetida,* proved successful as bioaccumulation and bioaugmentation alone had a removal efficiency of 60.24% and 50.3%, respectively and in combination, a total phenanthrene removal efficiency was 63.81% (Asgharnia et al. 2014). The result of this study was in agreement with that of Njoku et al. (2016) in the assessment of *Eisenia fetida* in the vermiremediation of soils contaminated with diesel, gasoline, and spent engine oil, which showed a consistent decline of petroleum hydrocarbon in soils containing *Eisenia fetida*. Hence, *Eisenia fetida* use in the remediation of soils contaminated with mixtures of petroleum products was proven beneficial. The capacity of *Eisenia fetida* in the remediation of soil pollutants was proven to be extensive, as they were also successful against toxic chemicals such as chromium and cadmium at specific levels in the soil (Aseman et al. 2015).

20.9 Factors Affecting Bioremediation

Hamby (1997) postulated that earthworms are diverse and very cooperative organisms given that they function both independently and also aid the function of bacteria (bioaugmentation) and plants (phytoremediation) in the removal of harmful chemicals from the environment. Some parameters may influence the bioaccumulation rate and effect of earthworms in the environs. Soil type and soil compounds were found to affect the bioavailability of contaminants such as PAHs for microbial degradation and degrading capabilities of microbial communities. However, studies indicate that indigenous soil microbial activity is critical for changes in hydrocarbon availability to invertebrates. Since, it was observed that in non-sterile soils, microorganisms degraded readily available contaminants very quickly and after several days only the non-available fraction is left for uptake by earthworms or other invertebrates (Šmídová 2013). Hence, natural, chemically unaltered soils would provide better bioaccumulation of hydrocarbon compounds by earthworms. Further, the effect of soil type on bioaccumulation suggests an understanding that the soil textural class is essential for bioremediation of petroleum hydrocarbons.

Hickman and Reid (2008), found earthworm assisted bioremediation to be a viable approach for reclamation of soil contaminated by hazardous chemicals. They emphasized the significance of earthworm characteristics in the process, stating niche type (epigeic, endogeic and anecic), food availability and soil type as significant environmental parameters that influence their response and behavior in soils.

The most significant areas noted in the review contributes to the mechanical development of the abiotic systems and biotic systems (i.e., microorganism promotion). The former noticeably augments soil conditions rendering earthworms applicable to soil improvement practices.

Hongjian (2009) investigated the impacts of parameters; contact time on extractability, availability of petroleum hydrocarbon (hexachlorobenzene [HCB]) in different soils (paddy soil, red soil, and fluvo-aquic soil), and bioaccumulation in *Eisenia fetida*. The result showed that the aging rate of HCB varied depending on soil type. More so, most extractable HCB declined in the first 60 days after being spiked into soils. Paddy soil showed a higher aging rate than in fluvoic-aquic soil or the red soil. Hence, soil type was observed to affect earthworm bioremediation. Therefore, bioremediation studies require specific empahsis on areas of varying soil type, geography, and climatic conditions.

Dendoovena et al. (2011) found that earthworms ability to enhance the removal of PAHs and the degradation products from soil may be limited by the number of earthworms, feed material, and moisture content of the soil. These factors are all essential for the successful bioremediation of polluted soil. They are all inextricably linked since the availability of feed material would influence the livelihood of earthworms and thus affect their survival which directly affects the quantity available for bioremediation. Further, the mechanical activity of earthworms and their casts affect the soil moisture.

Krishna et al. (2011) assessed the potential of three earthworm species: *Eisenia fetida, Eudrillus eugeniae,* and *Anantapur species* for the bioremediation of phenol in the soil. *Eisenia fetida* and *Eudrillus eugeniae* were the most tolerant species, while the *Anantapur species* was found to be least tolerant of the initial concentration of phenols raging from 20 ppm to 100 pm during experiment. As the concentration of phenol increased from 20 ppm tp 100 ppm across treatments, the bioremediation potential increased for *Eisenia fetida,* but the opposite occured in *Eudrillus eugenia and Anantapur species.* Further, *Eisenia fetida* was capable of uptaking 100 ppm of phenol from soil within 72 hours. Hence, species type and contaminant level were limiting factors in the study. Futher, Dada et al. (2016) opined that earthworms were diverse in the environment and depicted variations in the type of pollutants under which they are capable of remediating. Species specificity was observed as a significant factor in bioremediation and pollutant type may also affect bioremediation based on species variation.

Al-Haleem and Khalid (2016) conducted a study that revealed the trend of degradation of total petroleum hydrocarbon in soil (when compounded with punch waste and nutrients) to increase significantly with an increase in earthworms number of five, ten, and twenty earthworms respectively per treatment and with an increase in a number of tested days of 0, 7, 14, and 28 days. The total petroleum hydrocarbon (TPH) concentration (20,000 mg/kg) was reduced to 13,200 mg/kg, 9800 mg/kg, and 6300 mg/kg in treatments with five, ten, and twenty earthworms respectively. Additionally, TPH concentration (40,000 mg/kg) was reduced to 22,000 mg/kg, 10,100 mg/kg, and 4200 mg/kg in treatments with the above number of earthworms, respectively. Hence, the factors of time and quantity of earthworms available were essential in the bioremediation process.

There are numerous treatments available for soils polluted with petroleum hydrocarbons. However, among the options bioremediation seems to be highly favored as an effective and environmentally friendly technology for the removal of petroleum hydrocarbons from soil. The benefit of earthworms in the remediation of pollutants from soil has been highly recognized over the years and continues to grow in the field of environmental research. Bioremediation of petroleum hydrocarbons from soils utilizing earthworms faces many limitations, inclusive of but not limited to earthworm species, food availability, soil type, and time. Over the past three decades, many experimental studies have been undertaken to acquire knowledge on the potential of earthworm species in the bioremediation of petroleum hydrocarbons.

20.10 Conclusion

The bioremediation of soil contaminated with petroleum hydrocarbons by indigenous earthworm species (Eisenia fetida) is an effective method. The earthworms effectively reduced total petroleum hydrocarbon levels in the soil. Further, earthworms coexist with soil bacteria within the soil ecosystem, thereby carrying out the essential degradation processes. They work synergistically to facilitate the bioremediation of petroleum hydrocarbons. Earthworms also reduced the adverse effect of used engine oil on soil parameters such as pH and EC. Moreoever, they significantly improved three main macronutrients (Nitrogen, Phosphorus, and Potassium) in soil essential for plant growth and development and increased microbial population over time. The increase in CFU/g for bacteria in soil that contained earthworms support the cohesive relationship among these organisms in the bioremediation of petroleum hydrocarbons. The ability of earthworms to thrive under adverse conditions was indicated by their high survival percentage in soils contaminated by used engine oil. This is a significant demonstration for the use of earthworm in environmentally safe technologies towards the cleanup of soils polluted with petroleum-based contaminants.

References

Abrahami E, Iruobe AF (2015) Bioremediation of heavy metals and petroleum hydrocarbons in diesel contaminated soil with the earthworm: *Eudrilus eugeniae*. Springerplus 4:540

Adewoyin OA, Hassan AT, Aladesida AA (2013) The impacts of auto-mechanic workshops on soil and groundwater in Ibadan metropolis. Afr J Environ Sci Technol 7(9):891–898

Agency for Toxic Substances and Disease Registry (1999) Toxicological profile for total petroleum hydrocarbons. Department of Health and Human Services, Public Health Service, Atlanta

Akintunde WO, Olugbenga OA, Olufemi OO (2015) Some adverse effects of used engine oil (Common Waste Pollutant) on reproduction of male sprague dawley rats. Maced J Med Sci 3(1): 46–51

Alewo A, Mohammed-Dabo IA, Ibrahim S, Yahuza T, Baba AJ, Bello TK (2012) Effect of earthworm inoculation on the bioremediation of used engine oil contaminated soil. Int J Biol Chem Sci 6(1):493–503

Al-Haleem AA, Khalid AM (2016) Treating of oil-based drill cuttings by earthworms. Res J Pharm Biol Chem Sci 7:2088–2094

Ameh AO, Mohammed-Dabo IA, Ibrahim S, Ameh JB (2013) Earthworm-assisted bioremediation of petroleum hydrocarbon contaminated soil from mechanic workshop. Afr J Sci Technol 7(6): 531–539

Aseman E, Mostafaii GR, Sayyaf H, Asgharnia H, Akbari H, Iranshahi L (2015) Bioremediation of the soils contaminated with cadmium and chromium, by the earthworm *Eisenia fetida*. Iran J Health Environ 8(3). https://doi.org/10.11137/2014_2_216_222

Asgharnia H, Jafari AJ, Kalantary RR, Nasseri S, Mahvi A, Yaghmaeian K, Shahamat YD (2014) Influence of bioaugmentation on biodegradation of phenanthrene-contaminated soil by earthworm in lab scale. J Environ Health Sci Technol 12:150

Azadeh F, Zarabi M (2015) Combining vermiremediation with different approaches for effective bioremediation of crude oil and its derivatives. In: Global issues in multidisciplinary academic research (GIMAR), vol 1, pp 1–12

Biernbaum AJ (2014) Vermicomposting biological, environmental and quality parameters of importance. Environment, Quality, Michigan. http://www.hrt.msu.edu/uploads/535/78622/Vermicomposting-Bio-Enviro-Quality-13-pgs.pdf

Bingham E (1988) Carcinogenicity of mineral oils. Ann N Y Acad Sci 534:452–458

Buckerfield JC (1998) Earthworm are indicators of sustainable production. In: Sixth international symposium, vol vol 95. Earthworm Ecology, Vigo

Chaoui HI, Zibilske LM, Ohno T (2003) Effects of earthworms cast and compost on soil microbial activity and plant nutrient availability. Soil Biol Biochem 35(2):295–302

Coutiño-González E, Hernández-Carlos B, Gutiérrez-Ortiz R, Dendooven L (2010) The earthworm *Eisenia fetida* accelerates the removal of anthracene and 9, 10-anthraquinone, the most abundant degradation product in soil. Int Biodeterior Biodegradation:525–529

Dada EO, Njoku KL, Osuntoki AA, Akinola MO (2016) Heavy metal remediation potential of tropical wetland earthworm, *Libyodrilus violaceus* (Beddard). Iran J Energy Environ 7(3): 247–254

Das N, Chandran P (2010) Microbial degration of petroleum hydrocarbon: an overview. Biotechnol Res Int 2011:941810

Das N, Das D (2015) Strategies for remediation of polycyclic aromatc hydrocarbons from contaminated soil: an overview. J Crit Revi 2(1):20–25

Dendoovena L, Alvarez-Bernalb D, Contreras-Ramosc SM (2011) Earthworms, a means to accelerate removal of hydrocarbons (PAHs) from soil? A mini-review. Pedobiologia - Int J Soil Bio 54:S187–S192

Edwards C, Bohlen P (1996) Biology and ecology of earthworms, 3rd edn. Chapman & Hall, London. http://www.allaboutworms.com/epigeic-endogeic-and-anecic-earthworms-a-guide

Ekperusi OA, Aigbodion FI (2015) Bioremediation of petroleum hydrocarbons from crude oil contaminated soil with the earthworm: hyperiodrilus africanus. Biotech 5:957–965

Gay J, Shepherd O, Thyden M, Whitman M (2010) The health effects of oil contamination: a compilation of research. Worcester Polytechnic Institute, Washington, pp 1–211

Gentry TJ, Rensing C, Pepper IL (2004) New approaches for bioaugmentation as a remediation technology. Environ Sci Technol 34:447–494

Hamby DM (1997) Site remediation techniques supporting environmental restoration activities: a review. Sci Total Environ 191(3):203–224

Harmsen J (2004) Landfarming of polycyclic aromatic hydrocarbons and mineral oil contaminated sediments. Alterra:1–344

Harmsen J, Rulkens WH, Sims RC, Rijtema PE, Zweers AJ (2007) Theory and application of landfarming to remediate polycyclic aromatic hydrocarbons and mineral oil contaminated sediments: beneficial reuse. J Environ Qual 36:1112–1122

Hassanshahian M, Capello S (2012) Crude oil biodegradation in the marine environments. In: Chamy R (ed) Bidegradation - enginerring and technology, vol 5. IntechOpen, London, pp 1–28

Hickman AZ, Reid JB (2008) Earthworm assisted bioremediation of organic contaminants. Environ Int 34:1072–1081

Hongjian G (2009) Bioaccumulation of hexachlorobenzene in *Eisenia foetida* at different aging stages. J Environ Sci 21:948–953

Ingram AJ, Scammells DV, May K (1994) An investigation of the main mutagenic components of a carcinogenic oil by fractionation and testing in the modified Ames assay. J Appl Toxicol 14(3): 173–179

Koua J, Li Z, Guo Y, Gao J, Yang M, Zoua Z (2010) Photocatalytic degradation of polycyclic aromatic hydrocarbons in GaN: ZnO solid solution-assisted process: Direct hole oxidation mechanism. J Mol Catal A Chem 325:48–54

Krishna MV, Hrushikesh N, Sreehari K, Aravind KT, Vidyavathi N, Pallavi A (2011) Study on bioremediation of phenol by earthworm. Int J Environ Sci 1(6):1268–1273

Lavelle P (1988) Earthworm activities and the soil system. Biol Fertil Soils 6:237–251

Ma W, Imerzeel J, Bodt J (1995) Earthworm and food interactions on bioaccumulation and disappearance of PAHs: studies on phenanthrene and flouranthene. Ecotoxicol Environ Saf 32:226–232

Margesin R, Schinner F (1997) Laboratory bioremediation experiments with soil from a diesel-oil contaminated site significant role of cold-adapted microorganisms and fertilizers. J Chem Biotechnol 70:92–98

Masten SJ, Davies SH (1997) Efficacy of in-situ ozonation for the remediation of PAH contaminated soils. J Contam Hydrol 28:327–335

Moss L (2010) 13 largest oil spills in history. Mother Nature Work: http://www.mnn.com/earth-matters/wilderness-resources/stories/the-13-largest-oil-spills-in-history

Njoku LK, Akinola OM, Angibogu CC (2016) Vermiremediation of soils contaminated with mixture of petroleum products using eisenia fetida. J Appl Sci Environ Manage 20(3):771–779

Oil Spill Intelligence Report (1997) Oil spills from vessels (1960–1995): an international historical perspective. Aspen Publishers, New York

Pattnaik S, Reddy M (2012) Remediation of heavy metals from urban waste by vermicomposting using earthworms: *Eudrillus eugeniae*, *Eisenia fetida* and *Perionyx excavatus*. Int J Environ Waste Manag:284–296. ScienceDaily: www.sciencedaily.com/releases/2012/08/12081 6133420.htm

Peng S, Zhou Q, Cai Z, Zhang Z (2009) Phytoremediation of petroleum contaminated soils by Mirabilis Jalapa L. in a greenhouse plot experiment. J Hazard Mater 168:1490–1496

Roriguez-Campos J, Dendooven L, Alvarez-Bernal D, Contreras-Ramos S (2014) Potential of earthworm to accelerate removal of organic contaminants from soil: a review. Appl Soil Ecol 79:10–25

Singer A, Jury W, Leupromchai E, Yahng C-S, Crowley DE (2001) Contribution of earthworms to PCB bioremediation. Soil Biol Biochem 33:765–775

Sinha R, Herat S, Asadi R, Carretero E (2002) Vermiculture technology for environmental management: Study of action of earthworms *Eisenia fetida*, E*udrilus euginae* and *Perionyx excavatus* on biodegradation of some community wastes in India and Australia. Environmentalist 22:261–268

Sinha R, Herat S, Bharambe G, Brahambhatt A (2009) Vermistabilization of sewage sludge (biosolids) by earthworms: converting a potential biohazard destined for landfill disposal into a pathogen free, nutritive & safe bio-fertilizer for farms. Waste Manag Res 28:872

Sinha RK, Agarwal S, Chauhan K, Chandran V, Soni BK (2010a) Vermiculture technology: reviving the dreams of Sir Charles Darwin for scientific use of earthworms in sustainable development programs. Technol Invest 1:155–172

Sinha RK, Herat S, Valani D, Singh K, Chauhan K (2010b) Vermitechnology for sustainable solid waste management: a comparative study of vermicomposting of food & green wastes with conventional composting systems to evaluate the efficiency of earthworms in sustainable waste management with reduction in GHG emission. NOVA Science Publications, Hauppauge

Šmídová K (2013) Bioavailability of persistent organic pollutants in soils with regard to soil properties and contamination aging. Res Centre Toxic Compounds Environ:1–93

Stanton D (2015) Five years after the deepwater horizon oil spill. Awesome Ocean: http://awesomeocean.com/2015/09/23/awesome-research-five-years-deepwater-horizon-oil-spill/

Subash N, Sasikumar C (2014) Bioremediation of PAHS contaminated soil by utilizing an indigenous earthworm species, *Perionyx excavatus*. Int J Pharm Bio Sci 5(3(B)):449–455

U.S. Department of Health and Human Services (1999) Toxicological profile for total petroleum hydrocarbons (TPH). Agency for Toxic Substances and Disease Registry, Atlanta

Van Donkelaar P (1990) Environmental effects of crankcase and mixed lubrication. Sci Total Environ 92:165–179

Venter JM, Reinecke A (1988) The life-cycle of the compost worm *Eisenia fetida*. S Afr J Zool 23(3):161–165

Zhang L, Li P, Gong Z, Li X (2008) Photocatalytic degradation of polycyclic aromatic hydrocarbons on soil surfaces using TiO2 under UV light. J Hazard Mater 158:478–484

Zitte L, Awaadu GD, Okorodike CG (2015) Used-oil generation and its disposal along east-west road. J Waste Res 6:195

Printed in the United States
by Baker & Taylor Publisher Services